D0102800

Edited by JUDITH A. McGAW

Early American Technology

Making and Doing Things from

the Colonial Era to 1850

Published for the Omohundro

Institute of Early American History and

Culture at Williamsburg, Virginia, by the

University of North Carolina Press,

Chapel Hill and London

The Omohundro

Institute of Early

American History

sponsored jointly

by the College of

William and Mary

and the Colonial

Williamsburg

Foundation.

© 1994 The University of North Carolina Press

All rights reserved

Manufactured in the United States of America

Library of Congress Cataloging-in-Publication Data

Early American technology : making and doing things from
the colonial era to 1850 / edited by Judith A. McGaw.

p. cm. — (Institute of early American history and
culture)

Includes bibliographical references and index.

ISBN-13: 978-0-8078-2173-2 (cloth: alk. paper)

ISBN-10: 0-8078-2173-X (cloth: alk. paper)

ISBN-13: 978-0-8078-4484-7 (pbk.: alk. paper)

ISBN-10: 0-8078-4484-5 (pbk.: alk. paper)

1. Technology—United States—History—18th century.

2. Technology—United States—History—19th century.

I. McGaw, Judith A., 1946– . II. Series.

T21.E24   1994

303.48′3′0973—dc20   94-4913

CIP

*This book received indirect support from an
unrestricted book publication grant awarded to the
Institute by the L. J. Skaggs and Mary C. Skaggs
Foundation of Oakland, California.*

09 08 07 06 05   7 6 5 4 3

For BROOKE HINDLE

with thanks for allowing us

to stand on your shoulders

Preface

When I initially proposed editing a collection of original essays in early America's technological history, Merritt Roe Smith kindly warned me how much work and how many headaches it would entail. Perhaps it is fortunate for scholarship that one can appreciate such warnings only belatedly. In any case, I thank him for his counsel, all of which proved sage.

Fortunately, during the darkest hours of my work I could count on Fredrika J. Teute, Editor of Publications at the Institute of Early American History and Culture, for enthusiasm when mine flagged, prodding when I needed it, patience when mine was in short supply, and, in general, an unfailing sense of the important issues in early American history, in this collection of essays, and in editing. Her close readings and judicious comments make each of these a better essay than it would otherwise have been. Gil Kelly's fine copyediting also helped.

Likewise, I have been inspired to persevere by the example of Brooke Hindle, whose numerous editorial efforts helped keep early American technological history a vital and viable scholarly enterprise. All of us are indebted to him for his pioneering efforts in bibliography and in defining the shape of the field. Our dedication of this volume to him is but a small token of our great esteem.

Some of the time spent editing this collection was made available through a National Science Foundation grant. More generally, a year-long grant funded jointly by the National Science Foundation and the National Endowment for the Humanities enabled several of us to pursue research in early American technological history. I am especially indebted to Ron Overman of the NSF and Dan Jones of the NEH for their assistance in formulating that proposal. Several other scholars whose work is included in this volume were able to pursue early American historical research through the NEH-funded Transformation of Philadelphia Project, which focused year-long attention on technological history in 1989–1990. The Project also brought additional scholars to Philadelphia for brief presentations of their work. I thank Michael Zuckerman for making this collaborative effort possible and for bringing his enormous enthusiasm and invigorating curiosity to the enterprise. Likewise, Richard Dunn, Richard Beeman, and other members of the Philadelphia Center for Early American Studies contributed their wealth of expertise to our ongoing discussions. And Edward C. Carter II and the staff of the American Philosophical Society hosted and supported a conference that offered a preliminary rendition of several of the essays presented here.

Finally, Stephanie G. Wolf provided me with so much knowledge of early American history and so much support for this enterprise that it is impossible for me fully to acknowledge my debt to her. In addition, Jeanne Boydston, Mary Kelley, Lynne Lamstein, Richard K. Lieberman, Pat Malone, Patricia E. Mikols, Billy Smith, Catherine Stouch, Tom Waldman, and Joan White each did much to help make this volume a reality. I also thank the several historians who served as anonymous preliminary referees for the essays included herein.

Contents

Early American Technology

JUDITH A. MCGAW

Introduction

The Experience of Early American Technology

More than a quarter of a century has elapsed since Brooke Hindle penned the first systematic survey and evaluation of the history of early American technology. At that time he offered a convincing case that technology "belongs very close to the center" of early American history. Nonetheless, his assessment that "the central role of technology in early American history has only . . . begun to receive some of the attention it requires" remains largely true today. And, as documented by Nina Lerman's survey of books in early American technological history, which follows this volume's concluding essay, many of Hindle's more specific characterizations of the field remain equally applicable. With a few notable exceptions the categories he identified several decades ago encompass most of the books published since. Likewise, his observations concerning the relative contributions to the field by scholars versed in the history of science, engineers, economic historians, practitioners of local history, and those approaching technology through literary, archaeological, and aesthetic methods remain sufficiently timely that his essay, "The Exhilaration of Early American Technology," still rewards close reading. Indeed, as Robert Post's essay substantiates, Hindle's classic statement continues to offer a provocative

For comments on earlier versions of this essay, the author especially wishes to thank Mary Kelley, Gary Kulick, Pat Malone, and Fredrika Teute.

blueprint for early American technological history. The volume in which it appeared having gone out of print, it has been reprinted here.[1]

The limited growth and development of early American technological history stands in especially stark relief against the rich proliferation of scholarship on America's post-1850 technological development. Although in 1966 Hindle expected new interest in early American technology to be stimulated by the recently organized Society for the History of Technology and its journal, *Technology and Culture,* the overwhelming subsequent predominance of interest in later technological history is evident both in the chronological distribution of published articles and in the chronologically concentrated array of session topics constituting the society's annual meeting programs. If anything, books nominated for the Society's Dexter Prize during the past several years reflect an even greater preoccupation with recent technological history.[2]

The principal goal of this volume is to stimulate and abet increased attention to America's early technological history. To that end it offers an array of essays intended to exemplify the diverse and provocative assortment of potential approaches and topics in the field. And it includes a bibliographical essay designed to make extant resources in the field accessible to early Americanists unacquainted with technological history and to students of recent technology unfamiliar with early American history.

Before discussing the scholarly possibilities suggested by the essays that follow, it is worth musing for a moment on the larger question: Why has the history of early American technology attracted relatively little attention? Some answers seem obvious. The years since 1966 have formed an era during which "relevance," often defined quite narrowly, became a prominent criterion in selecting worthwhile historical projects. The history of technology is sufficiently young that most of its practitioners came of age intellectually in the wake of the scholarly revolution of the sixties. To many of us, industrial and postindustrial technology obviously posed great social, environmental, and cultural challenges. It made sense to examine that technology and its consequences directly.

1. Brooke Hindle, *Technology in Early America: Needs and Opportunities for Study* (Chapel Hill, N.C., 1966), quotes on 3, 28 (or below, 40, 67).

2. Observations concerning the Dexter Prize book submissions are based on my service on the Prize Committee from 1990 through 1992. John S. Staudenmaier, *Technology's Storytellers: Reweaving the Human Fabric* (Cambridge, Mass., 1985), offers an extensive discussion of the topical distribution of articles published in *Technology and Culture.*

Equally important, I suspect, is the *exhilaration* of *modern* American technology, to borrow a notion from Hindle. If, as Hindle maintains, the best studies of technology come from those with the capacity "to stand at the center of the technology—on the inside looking out" and if, as he also speculates, "technology was not so much a tool or a means as it was an experience—a satisfying emotional experience," then it is hardly surprising to find many outstanding scholars producing works about the technologies that provided them with seminal emotional experiences: space vehicles, ballistic missiles, nuclear power plants, television, airplanes, automobiles, computers, automated facilities, or modern household technology, to name but a few.[3]

Set against these obvious attractions of studying recent technology are the considerable difficulties of examining technology's early history. Hindle alluded to many of them. He singled out especially the inescapable division between craft and early industrial technology. And he noted that the historical challenge of comprehending these two very diverse technological experiences is often compounded by institutional specialization in one or the other by museums and archives. He was equally astute in noting that answering many of the most important questions in America's early technological history must involve bridging the divide; defining, understanding, and assessing the American Industrial Revolution requires intimate knowledge of both craft and machine technology.

Coinciding chronologically with the advent of salient early industrial technologies is the other great divide scholars of early technological history must bridge: the Revolutionary transition between the colonial era and the era of the early Republic. On a purely technical level this means that the historian of America's technological transition must master the very different records produced by various political entities, including the different currencies, legal and business terminologies, and rhetorical conventions that followed political change. More generally, studying early

---

3. Hindle, *Technology in Early America*, 24 (below, 62–63). Certainly other, more mundane factors such as the relatively more abundant funding opportunities to study recent technological development; the greater number of jobs in recent American technological history; and the more ready popular and student interest in modern high technology have all played a part in shaping the distribution of work. My emphasis here on the scholar's ability and willingness to *experience* technology is not meant to deny their influence, but it is meant to suggest that in the history of technology the need to cultivate a special personal relationship with the technology one studies may exert quite a profound influence.

technology means entering a field in which all records are fragmentary, including, as Hindle underscores, the important material remains that "retain a primacy [in the history of technology] that does not adhere to the physical objects associated with science, religion, politics, or any intellectual or social pursuit."[4]

Several of the essays that follow well illustrate another significant barrier restricting entrance to the study of early technological history: it generally entails learning a new language. Susan Klepp's careful reading of the very different terms in which early women spoke of issues we would reduce simply to contraception offers a model both of the difficulties entailed in comprehending alternative views of apparently similar technological activities and of the rewards that accrue from such attentive listening. Likewise, Michal McMahon wrestles with the evidence that both the American landscape and the language through which its disposition has been debated have undergone substantial modification virtually from the moment of settlement. Similar intellectual and linguistic challenges are evident in Robert Gordon's and Carolyn Cooper's studies of the early industrial era. Gordon's work reminds us of a general need for students of early extractive industries to appreciate the geological understandings through which early promoters viewed their technological choices. Similarly, Cooper's scholarship makes evident that we will continue to misunderstand such an apparently transparent notion as "invention" until we reckon with the manner in which patent arbitration defined novelty and shaped its marketability. In a more prosaic vein, Patrick O'Bannon indicates what a different linguistic world early Americans inhabited, depicting the very different meaning of such a simple term as "beer."

These various obstacles are sufficiently formidable to account for the dearth of practitioners of early American technological history, but I surmise that a subtler, yet more imposing, barrier deters many as well. Whereas the exhilaration of our personal emotional experience with recent technology engages many of us in its study, the profoundly alien quality of colonial technology, craft manufacture, frontier agriculture, or early mechanical processes leaves many of us looking at such technology from the outside. And an outsider's perspective remains inherently limited.

At the same time, becoming an insider—centering oneself in the experience of early technology—is extraordinarily difficult. So, as Hindle notes,

4. Hindle, *Technology in Early America*, 10 (below, 47–48).

we have mostly looked "through the eyes of science, economics, political reflections, social results, or literary antagonisms. There is no telling what related factors and forces will appear once the historian develops . . . insights from the center; it has hardly been tried."[5]

At bottom, then, the difficulty is not so much an intellectual as an experiential one. It is rather like the dilemma D. H. Lawrence identified as accounting for our failure to hear what "the old-fashioned American classics," the literary products of the era of early American technology, have to say to us. "It is hard to hear a new voice," he observes, "as hard as it is to listen to an unknown language. We just don't listen. . . . Why? — Out of fear. The world fears a new experience more than it fears anything. Because a new experience displaces so many old experiences. And it is like trying to use muscles that have perhaps never been used, or that have been going stiff for ages. It hurts horribly." By contrast: "The world doesn't fear a new idea. It can pigeon-hole any idea. But it can't pigeon-hole a real new experience."[6]

Not coincidentally, Lawrence notes our penchant for dismissing the early American classics as children's books. Much the same attitude characterizes the abbreviated accounts of early technology that serve to introduce studies of later, industrial technologies. Characteristically, early technology appears as simple, safe, and sanitized; it seems like child's play—like the reassuring technological images in such children's tales as *Ox-Cart Man, Little House on the Prairie,* or *The Little Engine That Could.* So simple does it appear that intellectually there seems little reason to dig deeper.

The *experience* of early American technology is an entirely different matter. It is redolent with the aromas of sexuality and resounds with cries of violence. It exposes the things we try to hide from the children—things we seek through our modern technology to obscure or control. Early American technology offers all of the uncomfortable sensuous and visceral experiences we have worked for decades to eradicate: the pervasive smell of human and animal excrement, the constant irritation of wool and vermin close to the skin, the alternate need to breathe clouds of dust or wade through mud along commonly traveled paths or roadways, the feeling of hands raw from exposure to harsh chemicals or from working in water in the dead of winter, the cries of animals being slaughtered and the death

---

5. Hindle, *Technology in Early America,* 24–25 (below, 63).
6. D. H. Lawrence, *Studies in Classic American Literature* (New York, 1977), 7.

rattle in the throat of a neighbor, and, generally, the physical and emotional pain inherent in continuous, relatively unmediated interaction with nature—a nature that only our subsequent technological insulation has enabled us to romanticize. It is these and other fundamental aspects of the human condition that experiencing early American technology threatens to remind us of. Understandably, we would rather experience ourselves as the Promethean creators of rockets to free us from gravity and of automatons to free us from the curse of Adam than as sapient animals closely bound to the cycles of heat and cold, deluge and drought, feast and famine, birth and death.

WHATEVER has kept us from experiencing early American technology, there are good reasons why even those bent only on understanding our industrial and postindustrial society must begin looking at early America's technological experience. The essays that follow reveal numerous ways in which we will fail to understand contemporary experience until we examine that of our remote ancestors. Among other things, they suggest that at least some of our "modern technological problems" cannot be laid at the door of recent, sophisticated technology, but derive from a far older and more deeply rooted array of social and legal decisions; they offer evidence that many of the new possibilities "opened by technology" actually represent a narrowing of technological choice; and they document that we miss much of what goes on technologically when we focus exclusively on the public sphere and on hardware, ignoring the private, knowledge-based technologies often associated with women.

In other words, these essays force us to define technology more broadly than we are commonly accustomed to. This redefinition is summarized by Hindle's phrase, incorporated in this volume's title: "making and doing things," a phrase that reminds us that products are only one outcome of technological processes. Technologies also serve to move things from place to place, kill enemies, or limit fertility, to name but a few obvious alternative uses of tools.

But the study of early American technology also reminds us forcefully that technology is more than tools. My own, working definition of early American technology describes it as the tools, skills, and knowledge needed to make and do things. As the essays that follow illustrate, such a definition is essential simply because early American tools could not operate by themselves. Considered in isolation, the things of early American technology are relatively meaningless. The route of an early road, the tools listed in a farm's or a brewery's inventory, or the existence of a root cellar informs us only minimally about how people laid out transportation routes, grew crops, brewed beverages, or preserved food. Experiencing early American technology is often literally essential for the scholar to recover some sense of the skills or knowledge that no longer adhere to the artifacts or persist in written form. By underscoring the *-ology* of technology, this redefinition carries important implications for our understanding of contemporary technological dilemmas as well. It alerts us that the invisible aspects of technology may make such tasks as solving environmental problems, fostering flexible work schedules, or stimulating technological creativity far more complex than engineering logic might suggest.

APPRECIATING what these essays have to tell us necessarily begins with an appreciation of their limitations. As I have noted, the needs and opportunities for study far exceed the accomplishments in the history of early American technology. Thus, it seemed to me appropriate to solicit essays that indicate, through the power of example, the substantial unrealized possibilities for scholarship in the field. More than anything else, then, what these essays have in common is that they treat subjects or issues on which, as the bibliography indicates, there is currently little or no serious scholarship. In consequence, these essays share both the excitement and the limitations of the pathbreaker: much of what they show us is novel, but they cannot survey the terrain broadly, since there are no giants on whose shoulders to stand.

Nonetheless, this is not a random survey of the terrain. Most of the essays focus on the mid-Atlantic region, arguably both the least studied early American locale and the area best suited to explore central aspects of early American technology. Unlike New England, often treated as the prototypic American region in earlier work, the mid-Atlantic was generally well endowed for the practice of agriculture, the overwhelmingly predominant early American technological activity, and for the various extractive industries that colonial policy favored. And, unlike New England, the mid-Atlantic colonies and states featured an extensive frontier, making them generally representative of eighteenth- and nineteenth-century America in such quintessentially American technological activities as their land use patterns and transportation developments. Moreover, by the late eighteenth and early nineteenth centuries the mid-Atlantic offered a higher level of affluence and urbanization than any other region, making possible a highly developed craft manufacturing system as well as an extremely diverse assortment of early industries. Finally, the mid-Atlantic's extraordinary ethnic, religious, and cultural diversity made it a comparatively hospitable home for the immigrants who transferred many new technologies to America, and make it a good place to examine the processes of technological diffusion and cross-fertilization that are often taken to be distinctively American.[7]

Chronologically, this volume maintains Hindle's definition of early American technological history as extending from earliest settlement to about 1850. Scholarship since 1966 has sustained his observation that "many lines of technological development . . . seem to attain measures of fulfillment in the 1840's and . . . 1850's"; the initial phase of industrialization was generally complete by that time. More problematic, ultimately, is the notion that American technological activity began with European settlement. Although comparatively little scholarship before or since 1966 has examined technological activity antedating European settlement, a few notable exceptions challenge the notion that we can assess even the nature and meaning of European technological activity without appreciating the prior and contemporaneous technological activities of Native Americans.[8]

---

7. Indeed, except in the much-studied textile industry and in the development of interchangeable parts technology, it is hard to argue for the centrality of New England's experience. Rather, the literature at hand suggests that, even in these areas, our impression of New England as the model for American technological development derives principally from our lack of knowledge of developments in other regions.

In part the geographic focus of this volume reflects the fact that many of its contributors offered early versions of their work at various seminars, conferences, and informal talks held in 1989–1990, the year in which the Transformation of Philadelphia Project of the Philadelphia Center for Early American Studies focused on technological change in the mid-Atlantic. Gordon, Jackson, O'Bannon, McGaw, Cooper, and Lerman each received funding through grants from the National Endowment for the Humanities or the National Science Foundation or both, enabling them to participate in most of the Project's activities. Most of the other authors of essays in this collection presented their work at Project seminars or informal gatherings, as did a number of other scholars.

8. Hindle, *Technology in Early America*, 16 (below, 54). William Cronon, *Changes in the Land: Indians, Colonists, and the Ecology of New England* (New York, 1983); and Patrick M. Malone, *The Skulking Way of War: Technology and Tactics among the New England Indians* (Lanham, Md., 1991) suggest that understanding Native American technology can prove crucial to comprehending early Euro-American technological choices.

Chronologically as well, the emphasis of this volume is upon periods most neglected in previous work. In practice, that has produced relatively even coverage of the era, although less attention has been accorded initial settlement and the process of industrialization, the two phases of early American technological activity to receive most prior attention. And those studies that focus on aspects of industrial development—the essays by Cooper and Gordon—emphasize its neglected aspects. Cooper focuses, not on the initial stages of mechanical development and innovation, but on machinery's interregional diffusion, and Gordon emphasizes how mining's economically insignificant preindustrial history shaped fundamentally its later development. Likewise, although Donald Jackson's work treats an aspect of the relatively familiar transportation revolution, his topic—turnpikes—has been all but ignored by previous scholars.

Topically, insofar as possible, this collection favors the quotidian, whereas previous scholarship has often featured the exceptional: the leading sector, the large enterprise, the heroic inventor, the science-based development. This is very much in accord with Hindle's perception that "once . . . the intensity of concentration on invention is relaxed, a number of pervasive themes within technology come to mind, some of them not even progressive in nature."[9] In particular, the essays by Judith McGaw and Sarah McMahon focus on the technological activities of agricultural households, the most commonplace of all early American activities. Likewise, the essays by Michal McMahon and O'Bannon devote long-overdue attention to the small-scale industrial processing of agricultural products, the craft manufacturing activities that were arguably most familiar, if only through their olfactory manifestation, to early Americans and are, by contrast, least familiar to modern scholars.

To readers who come to this volume expecting technological history to be a familiar tale of "men, metal, and machines," this topical selection may appear to miss the main action. One response is to direct such readers to Lerman's and Hindle's bibliographies as well as to *Technology and Culture*'s annual bibliography of articles. These afford access to a number

---

The dearth of work in progress on this topic precluded inclusion of work on Native American technology in this volume. Similarly, Afro-American technology and the transfer of technology from Africa have received negligible attention from historians of technology working on any period.

9. Hindle, *Technology in Early America*, 25 (below, 63).

of extant works on mechanization, military technology, ironmaking and metalworking, interchangeable parts development, and famous inventors. But I would also urge such readers to set aside their preconceptions and recognize through the coverage of these essays that early Americans, like citizens of most underdeveloped countries today, engaged especially in extractive industries, including agriculture, and in those technological activities designed to supply daily essentials. Until we are acquainted with the habits of hand and mind that Americans acquired through such routine activities, we will have no sense of how much they influenced the main lines of later technological development. In consequence, my own wish list of topics this volume might have incorporated would include small-scale grain milling (the nation's leading manufacturing activity throughout the era), fishing and fish preservation, ferry construction and operation, and occult techniques for manipulating the future. This is by no means an exhaustive list, and it is my hope, at least, that, in emphasizing relatively neglected topics, this volume will commend other such areas to scholars' attention.[10]

Having set aside the assumption that we can best trace the history of modern technology by studying its most obvious ancestors—the large enterprise, the complex mechanism, the famous "first"—readers committed to relevance may be surprised to discern much that is "advanced" and modern within the commonplace activities of early Americans. For example, the now familiar battle between vigorous private-profit interests and relatively weak defenders of the public good manifests itself repeatedly in early American land and water use patterns (Michal McMahon, Gordon) and in decisions about public works (Jackson, Michal McMahon). Likewise, many of these essays document how early and pervasive was the role of the mere investor, as opposed to the astute practitioner, in shaping technological choice (Gordon, Jackson, O'Bannon, Cooper). More generally, readers may find their view of contemporary technological issues enhanced by these essays' ability to make us see around us the physical and social survivals of eighteenth- and nineteenth-century technological practice, ranging from the chemical content early mining waste contributes to our water (Gordon), to the distribution of

10. The phrase "men, metal, and machines" derives from discussions with Professor Deborah Fitzgerald, a student of 20th-century agricultural technology. Her field well illustrates that narrow, men-and-hardware definitions restrict scholarship on recent technological topics as well.

modern settlement along routes chosen by men long dead (Jackson), to the paucity of wetlands (Michal McMahon), to the presence within many homes of spaces laid out for obsolete technological purposes (Sarah McMahon, McGaw), to the early legal precedents that continue to shape technological practice (Cooper), to the tortured public language in which we discuss the linked issues of women's health, contraception, and abortion (Klepp).

Among the mundane aspects of early American technology, none had received less attention at the time Hindle wrote than those associated with women. The present volume especially points the way toward a greatly enriched history of early American technology in its concern with women's technological activities. Moreover, the emphasis in the essays by Klepp, McGaw, and Sarah McMahon on women's reproductive and domestic food-processing activities departs from the main lines of research to date, as chronicled in the bibliography. Although the years since 1966 have seen more attention to women and technological change, much of that attention has focused on women's work for the market, especially women's work outside the home in the early industrial period. For example, as Lerman notes, despite the relative abundance of works on textile factories and women's work therein, the far more common household and craft production of yarn and cloth remains a scholarly lacuna. In sum, scholars have told us mostly about those aspects of women's work that resembled men's.[11]

Through their attentiveness to women's private technological activities, the essays in this volume suggest how central is the understanding of women's technologies to answering the most important questions in early American technological history. For example, as Hindle noted, debates

---

11. The few notable exceptions are dissertations, including, especially, Gail Barbara Fowler, "Rhode Island Handloom Weavers and the Effects of Technological Change, 1780–1840" (Ph.D. diss., University of Pennsylvania, 1984); and Adrienne Dora Hood, "Organization and Extent of Textile Manufacture in Eighteenth-Century, Rural Pennsylvania: A Case Study of Chester County" (Ph.D. diss., University of California–San Diego, 1988).

For a fuller discussion of these issues, see Judith A. McGaw, "No Passive Victims, No Separate Spheres: A Feminist Perspective on Technology's History," in Stephen H. Cutcliffe and Robert C. Post, eds., *In Context: History and the History of Technology: Essays in Honor of Melvin Kranzberg* (Bethlehem, Pa., 1989), 172–191; and McGaw, "Women and the History of American Technology," in Sandra Harding and Jean F. O'Barr, eds., *Sex and Scientific Inquiry* (Chicago, 1987), 47–77.

about the reality, nature, and influence of "labor scarcity" have figured prominently in discussions of the distinctively American shape of various technologies. Klepp's essay, by suggesting the real possibilities of effective contraceptive practice, also suggests the need to refigure discussions of early American population so as to take full account of all the costs of childbearing and -rearing. Similarly, discussions of industrialization have often dwelt on its consequences for "women's work," arguing that technology either opened men's work to women or that it removed women's work from household to factory. By focusing on the ambiguous area of food preservation, Sarah McMahon alerts us that the gendered division of labor was not always so clear. Perhaps its anomalous nature accounts in part for our relative neglect of such a central aspect of early American manufacturing as food processing. Likewise, the essays by McGaw and Sarah McMahon make evident that almost any change in agriculture entailed a change in the household organization of labor. This added element of complexity may help to explain why early agricultural innovation, America's most common technological activity, has received such limited and superficial scrutiny.

These essays also remind us that, whereas studies of industrialization and of recent technological innovation often focus on changes in hardware, understanding early American technology, especially women's technology, generally entails giving fuller attention to the invisible aspects of technology—to its software, if you will. In this respect, as well, the study of women's technology illuminates a central feature of early American technology generally: it appears simple only if we limit our attention to the hand tools early Americans generally wielded. Yet none of these tools worked effectively without a considerable body of skill and knowledge in the hand, eye, and mind of its manipulator. Until we have chronicled fully these aspects of technology, we will continue to compare apples and oranges in our discussions of the transition from craft to machine production. We will especially miss the deleterious and monotonous features that were often central to the prescribed uses craftsmen and women made of their tools, features generally absent from the romanticized portraits of craftsmanship that have grown up since industrialization and that have exerted substantial influence on studies of the transition to industrialization. Ultimately, the sensitivity that serious attention to early American technology forces us to cultivate toward the software side of technology should help us to recapture the continuing influence of skill and knowledge in making later, mechanized production possible.

The essays in this volume are also well chosen to redress the scholarly balance in other important respects. For example, most attention to the "dark side," or negative consequences, of technological change has focused on how industrialization affected labor. Yet the thrust of early American technology was far more often directed toward substituting abundant resources for both scarce capital and scarce labor than it was toward substituting capital for labor. Thus, the principal and most deeply rooted negative heritages of early American technological development have been environmental. Both Gordon and Michal McMahon address these issues, and they do so by focusing on mining and water pollution, whereas most of the limited previous work in early American environmental history has devoted attention to lumbering and agriculture. On those topics, as well, this collection makes important contributions. Cooper's work presents the story of woodworking's technological development as far more complex than that told in previous scholarship wherein wood's abundance simply fostered a resource-intensive technology. Similarly, whereas previous scholarship suggested that abundant land allowed farmers to wait generations before soil exhaustion forced them to adopt new and better methods, McGaw's essay indicates that the use of land-conserving or restorative agricultural techniques was not the mere offspring of necessity. It also derived from non-English ethnic traditions and varied as ethnic settlement patterns either fostered or discouraged inter–ethnic group borrowing.

Likewise, these essays serve to complement and correct the overemphasis on the scientific roots of technology that has characterized historical study of technology generally. Gordon, especially, makes evident that scientific knowledge was and often is ill equipped to breed better technology. By contrast, Cooper calls into question the entire progressive vision of technology that supports the notion of science-based technology as inherently better. At the same time, she reveals a more germane system of knowledge shaping technology: the complex legal structure built around patent law. In like manner, Jackson moves away from conventional notions of engineering knowledge to uncover the sophisticated design considerations that help account for the fact that modern roads generally follow in the footsteps of their ancestors. More generally, virtually all of these essays shift emphasis away from the drama of discovery and invention and toward an appreciation of technological innovation as selective change, slow adaptation.

Nonetheless, for early Americanists unacquainted with the history of technology and for historians of technology unfamiliar with pre-twentieth-century technology, perhaps the most important theme of the present volume is the sophistication, the complexity of early American technological decision making. Even so simple a process as brewing represented an array of products, equipment, and technological choices (O'Bannon); "birth control" comprised a diversity of techniques, many of which we have too quickly dismissed as primitive (Klepp); food preservation entailed choices about crops, marketing, and the construction of storage spaces as well as about the practices women performed after harvest (Sarah McMahon); agricultural and domestic tools were far more numerous and varied than the standardized inventory we customarily envision (McGaw); and decisions about the organization of space—the siting of industry and roads as well as the social construction of landownership systems—shaped both the technologies of production and their environmental consequences (Gordon, Jackson, and Michal McMahon). Indeed, by taking knowledge and social choice into account, most of these essays suggest that a decline in technological complexity may have accompanied industrialization and specialization.

Broadening the narrow tool- or machine-oriented view of technology and, simultaneously, revealing the sophistication of early American technology should enable this collection to serve two larger purposes. It can encourage new attention to technology by early Americanists, scholars especially well prepared to comprehend early technology as knowledge. It may also stimulate new concern with early America among historians of recent technology, who may have assumed early technology to be too simple to be intellectually engaging. Of course, such outside scholars may find it hard "to stand at the center of the technology"—to capture the "satisfying emotional experience" it entailed. Fortunately, as was true in 1966, the history of early American technology is too rich and diverse to be encompassed by any single approach, including Hindle's. His words still ring true: "Since we are not yet entirely sure what are the 'right' questions or the 'right' approaches, it is doubly necessary that no doors be closed and that the insights of all who have looked in this direction be sought." [12]

Most of all, I hope it is the rich potential, opening to a diversity of approaches, that this collection of essays captures. For it may well be that the most important thing the study of early American technology can tell

12. Hindle, *Early American Technology*, 28 (below, 67).

us is that we live in a world culturally impoverished by a technology that favors "one best way" and that generally serves to restrict both intellectual and economic competition. This is not to argue that early American technology was somehow better. It is to underscore that it was different, and in experiencing more fully that difference lies the genuine possibility of envisioning the real alternative technologies we need.

Robert C. Post

Technology in Early America

A View from the 1990s

In 1966 the University of North Carolina Press published a small book for the Institute of Early American History and Culture, *Technology in Early America*. Along with the interpretive essay by Brooke Hindle that is reprinted below, it comprised a sixty-five-page critical bibliography, also by Hindle, and a directory of artifact collections by Lucius F. Ellsworth of the Eleutherian Mills–Hagley Foundation. *Technology in Early America* was part of a series, Needs and Opportunities for Study, sponsored by the Institute, earlier volumes having been written by Whitfield J. Bell, Jr. (on science), William N. Fenton (on Indian-white relations), Bernard Bailyn (on education), and Walter Muir Whitehill (on the arts). Each book had been "the outgrowth of a conference to explore a special historical field which scholars have neglected or indifferently exploited or in which re-newed interest has developed in our own times." The aim was to promote research "by providing a wide-ranging view that correlates what has been done with what needs to be done."[1]

Jointly sponsored by the Institute of Early American History and Cul-

---

1. "The Institute Conferences," in Brooke Hindle, *Technology in Early America: Needs and Opportunities for Study* (Chapel Hill, N.C., 1966), vii; for further background on the conference, see Brooke Hindle, "A Retrospective View of Science, Technology, and Material Culture in Early American History," *William and Mary Quarterly*, 3d Ser., XLI (1984), esp. 427.

ture and the Eleutherian Mills–Hagley Foundation, the conference on Technology in Early American History was convened on October 15, 1965, at the Eleutherian Mills Historical Library in Wilmington, Delaware. Some three dozen scholars participated, all of whom shared Hindle's interest in technology, whether their field was history, historical archaeology, museology, or material culture. Presiding was Melvin Kranzberg, the spark plug for a new graduate program in the history of technology at the Case Institute of Technology in Cleveland. Hindle's essay, "The Exhilaration of Early American Technology," had been circulated; Whitfield Bell of the American Philosophical Society summarized, and there was commentary from Eugene S. Ferguson of Iowa State University and Daniel J. Boorstin of the University of Chicago. The group also discussed Hindle's critical bibliography, and after the conference he set about preparing his material for publication.

In a foreword to *Technology in Early America,* Lester J. Cappon, director of the Institute of Early American History and Culture, noted the growth of a body of scholarship that conceptualized "the development of technology as an integral part of cultural history." The immediate backdrop was Oxford University Press's publication of five volumes of the collaborative work, *A History of Technology,* between 1954 and 1958, the founding of the Society for the History of Technology (SHOT) under Mel Kranzberg's leadership in 1958, and the debut in 1959 of SHOT's quarterly journal, *Technology and Culture.* Cappon also suggested that the history of technology's newfound strength owed much to "the revitalized role of the museum as a creative educational institution." And now there were Hindle's provocative essays, together "cutting a benchmark for future measurements and assessments, a new starting point for scholarly research, a stimulus to historical synthesis." [2]

2. In Hindle, *Technology in Early America,* x, xi. The Oxford series edited by Charles Singer, E. J. Holmyard, A. Rupert Hall, and Trevor I. Williams defined technology at the outset as "how things are commonly done or made" and "what things are done and made." On the founding of SHOT, see John M. Staudenmaier, *Technology's Storytellers: Reweaving the Human Fabric* (Cambridge, Mass., 1985), chap. 1; and Robert C. Post, "Missionary: An Interview with Melvin Kranzberg," *American Heritage of Invention and Technology,* IV (Winter 1989), 34–39. Participants in the 1965 conference at the Hagley included Carl Condit, Cyril Stanley Smith, Robert S. Woodbury, W. David Lewis, and Thomas P. Hughes, among others of Kranzberg's early academic SHOT cohort, as well as a significant complement of museum professionals, including Robert P. Multhauf of the newly opened National

As well as the first of those two essays reprinted here, readers are urged to scrutinize the listings from Hindle's "Bibliography of Early American Technology" in the Appendix. However real the neglect of this field, I suspect many will share my own sense of surprise at how rich this literature was, nearly three decades ago. In addition to the citations of contemporary sources such as travel narratives, the general surveys, and antiquarian treatises on the order of the Williamsburg Craft Series on colonial trades, Hindle also took note of what would easily amount to a five-foot shelf of books by critical scholars whose work remains valuable to this day.[3] Several of these works were broadly synthetic—Leo Marx's *The Machine in the Garden: Technology and the Pastoral Ideal in America* (1964) and Boorstin's *The Americans: The National Experience* (1965), to cite just two examples. Others such as Carl Bridenbaugh's *The Colonial Craftsman* (1950) covered a lot of ground but without the overt ambitions of Marx or Boorstin.

None of those scholars would have defined his specialty as technological history—nor would Hindle, who then as always saw himself as a generalist but was convinced of the need for integrating the historical analysis of technology and culture. This was a conviction shared by Boorstin, who was Hindle's contemporary, by Marx, a little younger, and by his revered mentor Bridenbaugh. All were of a generation influenced by Charles Beard's mode of progressive synthesis, a dualism suggestive of connections between seemingly disparate historical events. All were influenced by (or, perhaps, helped shape) the subsequent school of interpretation that derived from Tocqueville's sense "of the ineluctable singularity of American development, his stress on the preformed character of our democratic institutions, the importance of the democratic revolution that never had to happen."[4] Muting class conflict and the role of ideology, this school

---

Museum of History and Technology (NMHT). Between the late 1960s and the late 1970s, Multhauf and Hindle would each serve as director of NMHT, their terms bracketing Boorstin's.

3. I would put the list of such books at no fewer than 60. Hindle also took modest account of his own *David Rittenhouse* (Princeton, N.J., 1964) and *The Pursuit of Science in Revolutionary America: 1735–1789* (Chapel Hill, N.C., 1956) as well as the work of two SHOT stalwarts of a slightly younger generation: Carroll W. Pursell's Berkeley dissertation on stationary steam power was not yet published, and Bruce Sinclair's on the Franklin Institute was still in the works at Case Institute (see n. 23, below).

4. "Except for him," Hindle said of Bridenbaugh in the acknowledgments to *The Pur-*

encouraged the formulation of bold generalizations about "the American experience."

When *Technology in Early America* appeared in print, however, the pendulum was swinging yet again. A new generation of historians was coming of age, a generation with revised priorities regarding need and opportunity, many with different perceptions of history's utility, some who took the historian's role to be a matter of addressing, not progress, but paradox, conflict, even failure. Although it was a truism that every generation writes its own history, certain older historians did not understand at all. In his presidential address to the American Historical Association in 1962, for example, Bridenbaugh had expressed dismay at the incursion of historians "whose emotions not infrequently get in the way of historical reconstructions." Within a few years the American historical profession would be in the throes of an upheaval that was chronicled even in the popular press and whose effects still reverberate to this day, an upheaval Peter Novick has termed "the collapse of comity."[5]

By 1966, synthetic touchstones that had enjoyed broad scholarly accord were being, as Hindle later lamented, "challenged and often cancelled." Among these was a generality that, in America, technology was something fundamentally beneficent, or at least that Americans had proved adept at "creating an endless series of satisfying community experiences that kept pace with advancing technology."[6] Technology might have precipitated

---

suit of *Science in Revolutionary America,* vii, "this book would not have been written." Quote about Tocqueville in Richard Hofstadter, *The Progressive Historians: Turner, Beard, Parrington* (New York, 1968), 445.

5. Peter Novick, *That Noble Dream: The "Objectivity Question" and the American Historical Profession* (New York, 1988), chap. 13. For suggestive interpretations of Bridenbaugh's presidential address, see Novick, 339–340, and Lawrence W. Levine, "Reflections on Recent American Historiography," in Levine, ed., *The Unpredictable Past: Explorations in American Cultural History* (New York, 1993), 3–13. For an example of how this generational upheaval was depicted in the popular press, see Jack McCurdy, "Historians Clash over How to Write History," *Los Angeles Times,* Feb. 7, 1971, an account of a confrontation involving Oscar Handlin, Christopher Lasch, and David Hackett Fischer at the 85th Annual Meeting of the AHA in Boston.

6. Brooke Hindle, "Historians of Technology and the Context of History," in Stephen H. Cutcliffe and Robert C. Post, eds., *In Context: History and the History of Technology: Essays in Honor of Melvin Kranzberg* (Bethlehem, Pa., 1989), 232; William H. Goetzmann, "Time's American Adventures: American Historians and Their Writing since 1776," *American Studies International,* XIX, no. 2 (Winter 1981), 5–47, quote on 32. For more

things dark and satanic elsewhere, but in the United States such effects were mitigated by a Lockean liberalism that was so widely shared as to constitute a consensus. The consensus school's prophets included Boorstin, but, as one of his prize pupils would later observe, Brooke Hindle too was "quintessentially a consensus historian." Among the incidental casualties of the Vietnam War, which was escalating so fearfully even as Hindle was putting the finishing touches on *Technology in Early America,* was this school of interpretation and the synthetic mode of exposition that both consensus historians and the progressive school had idealized. The consequences would long endure. As a reviewer of Novick's *That Noble Dream* put it in 1989, "There is no American history now, only American histories."[7]

So it was that *Technology in Early America* appeared at the very moment when the validity of key needs Hindle posited was being questioned and often denied by scholars who were attracted to a different set of opportunities, opportunities for demonstrating the failure of the American dream, or spotlighting "negatives" anyway. Certainly the timing was ironic, and the irony was enhanced by a perception that the consensus school had often seemed inattentive to irony in history.[8] To be sure, "negativism" never came anywhere close to dominating early American historiography; indeed, there was some tendency among new left historians to *idealize* the period before 1850. During the 1970s, the most significant new mode of address to technology and culture in early America was a body of community studies exemplified by the work of scholars like Thomas Dublin, Susan Hirsch, Diane Lindstrom, Howard Rock, Carl Siracusa, Merritt Roe Smith, and Anthony Wallace—studies that often emphasized

---

detailed surveys of American historiography, see: Novick, *That Noble Dream;* Hofstadter, *The Progressive Historians;* John Higham, with Leonard Krieger and Felix Gilbert, *History* (Englewood Cliffs, N.J., 1965); Harvey Wish, *The American Historian: A Social-Intellectual History of the Writing of the American Past* (New York, 1960); and Marcus Cunliffe and Robin W. Winks, eds., *Pastmasters: Some Essays on American Historians* (New York, 1969).

7. Judith A. McGaw to the author, Apr. 30, 1991, *Technology and Culture* correspondence, National Museum of American History; David W. Noble, "Perhaps the Rise and Fall of Scientific History in the American Historical Profession," *Reviews in American History,* XVII (1989), 522.

8. See John P. Diggins, "The Perils of Naturalism: Some Reflections on Daniel J. Boorstin's Approach to American History," *American Quarterly,* XXIII (1971), 153–180.

conflict but were usually characterized less by harsh preachments than by meticulous efforts to reconstruct the interactions of technology in everyday life.

Nor did the collapse of comity seriously impede the pursuit of scholarship in most of the established specialties categorized by Hindle in 1965. Following a temporary loss of momentum that coincided with the darkest period of America's involvement in Southeast Asia and campus revolt, the production of monographic studies in early American technology regained speed and has seldom flagged since. If Nina Lerman is correct in stating—in the concluding essay of this book—that "early American historians have neglected technology, and historians of technology have neglected early America," her bibliography is scarcely indicative that they have neglected it entirely. In addition to the book-length works she enumerates, one can add a wealth of articles and essays by (among others) Carolyn Cooper, Jadviga M. da Costa Nunes, Robert Gordon, David Jeremy, Gary Kulik, Steven Lubar, Judith McGaw, Sarah McMahon, Gail Mohanty, Alex Roland, Bruce Seely, Todd Shallat, Merritt Roe Smith, and Paul Uselding.[9]

9. To limit citations to just one article on early American technology by each of the scholars named (some have published several of them): Carolyn C. Cooper, "Social Construction of Invention through Patent Management: Thomas Blanchard's Woodworking Machinery," *Technology and Culture* (hereinafter cited as *T&C*), XXXII (1991), 960–998; Jadviga M. da Costa Nunes, "The Industrial Landscape in America, 1800–1840: Ideology into Art," *IA: The Journal of the Society for Industrial Archeology*, XII, no. 2 (1986), 19–38; Robert B. Gordon, "Materials for Manufacturing: The Response of the Connecticut Iron Industry to Technological Change and Limited Resources," *T&C*, XXIV (1983), 602–634; David J. Jeremy, "Innovations in American Textile Technology during the Early Nineteenth Century," *T&C*, XIV (1973), 40–76; Gary Kulik, "A Factory System of Wood: Culture and Technological Change in the Building of the First Cotton Mills," in Brooke Hindle, ed., *Material Culture of the Wooden Age* (Tarrytown, N.Y., 1981), 300–335; Steven Lubar, "Culture and Technological Design in the Nineteenth-Century Pin Industry: John Howe and the Howe Manufacturing Company," *T&C*, XXVIII (1987), 253–282; Judith A. McGaw, "Accounting for Innovation: Technological Change and Business Practice in the Berkshire County Paper Industry," *T&C*, XXVI (1985), 703–725; Sarah F. McMahon, "A Comfortable Subsistence: The Changing Composition of Diet in Rural New England, 1620–1840," *WMQ*, 3d Ser., XLII (1985), 26–65; Gail Fowler Mohanty, "Experimentation in Textile Technology, 1788–1790, and Its Impact on Handloom Weaving and Weavers in Rhode Island," *T&C*, XXIX (1988), 1–31; Alex Roland, "Bushnell's Submarine: American Original or European Import?" *T&C*, XVIII (1977), 157–174; Bruce Seely, "Blast Furnace

And when Hindle and Lubar prepared a list of suggested readings to conclude their survey of early American technology titled *Engines of Change* (a book written in conjunction with their permanent exhibition of the same name at the Smithsonian Institution), much of the literature they cited had been published since the early 1970s. Indeed, of thirteen works they noted that were "based upon specific studies with broadening insights," eleven dated no further back than 1972.[10]

Which is all by way of saying that scholarship in early American technology has a persistent vitality. Yet the synthetic expeditions that Hindle (and Lester Cappon) regarded as so essential to enhancing historical understanding were few and far between, and, given the career exigencies of most academics in the 1990s, there was little reason to anticipate anything soon again appearing on the order of Boorstin's *The Americans,* let alone the Beards' *Rise of American Civilization.* Other matters of crucial import to Hindle in 1965 were likewise problematic, an enhanced focus on "the physical things of technology," for one. Even his definition of technology as the means "for making and doing things" (echoing the editors of the Oxford series in the 1950s) soon came under fire. " 'How things are done or made' is not, strictly speaking, a historian's question," argued a promising scholar at Northwestern, George Daniels, in an influential 1970 article. The definition must also include "*why* they are done or made the way they are rather than in a number of other possible ways and what difference it makes to the society in which the doing or making is done."[11]

Daniels was presenting a clear formulation of the concept of choice in technological design, a concept that posits that technology never evolves naturally, or inevitably, or autonomously, and indeed that "evolution" is

Technology in the Mid-Nineteenth Century: A Case Study of the Adirondack Iron and Steel Company," *IA: The Journal of the Society for Industrial Archeology,* VII, no. 1 (1981), 27–54; Todd Shallat, "Building Waterways, 1802–1861: Science and the United States Army in Early Public Works," *T&C,* XXXI (1990), 18–50; Merritt Roe Smith, "Military Entrepreneurship," in Otto Mayr and Robert C. Post, eds., *Yankee Enterprise: The Rise of the American System of Manufactures* (Washington, D.C., 1981), 63–102; Paul Uselding, "Elisha K. Root, Forging, and the 'American System,' " *T&C,* XV (1974), 543–568.

10. Brooke Hindle and Steven Lubar, *Engines of Change: The American Industrial Revolution, 1790–1860* (Washington, D.C., 1986), 285. Of the 70-odd books listed in Lubar's abbreviated exhibition catalog, two-thirds had been published since 1972.

11. George H. Daniels, "The Big Questions in the History of American Technology," *T&C,* XI (1970), 1–21, quotes on 2.

an inapt metaphor. Yet Hindle's "Exhilaration" essay was far from inno-
cent of a sense that there is rarely "one best way" or of a recognition that
certain significant themes in technological history are "not even progressive in nature." He took special note, for example, of "the spread and
conflict of different cultural traditions within the same technology," citing
specifically iron and glass production, the design of plows and fire engines, and the divergence between the Rhode Island and Waltham systems
of cotton goods manufacture.[12]

Nor was Hindle anything less than fully attuned to the ramifications
and nuances of his assertion that technology was a "question of process,
always expressible in terms of three-dimensional 'things.' " Nowadays it is
essential for scholars to understand that some of the most vital intersec-
tions of technology have to do with gender, with relationships in the workplace, with what Philip Scranton calls "the cultural imaging and political
networks involving technology as alien incursions." Nevertheless, nobody
has shown that any significant facet of technology exists entirely apart
from process, apart from three-dimensional things. And I think it is fair to
add that Hindle understood something else as well. Hindle was centrally
concerned with imparting a perception that analyzing technology could
provide a unique means of historical insight. Analytical power could be
sacrificed if the focus were to be diverted too far into the realm of institutional relationships, into the realm of values—into the realm of context, to
cite the term that John Staudenmaier would bequeath to historiographical
discourse.[13]

In 1981, one of the best of the younger historians of technology, David
Hounshell, suggested that, absent the old "internalist" precept of technology as "how things are done or made," there might not be anything to
hold the history of technology together as a "discipline." "If we grant the

12. Hindle, "The Exhilaration of Early American Technology," in *Technology in Early
America*, 25 (or below, 63–64).

13. Philip Scranton, "Theory and Narrative in the History of Technology: Comment,"
T&C, XXXII (1991), 385–393, quote on 388; the Hindle quote is from "The Exhilaration
of Early American Technology," 4–5 (below, 42). "Contextualism" has of course become
an established ideal in many historical specialties in the past quarter-century. Staudenmaier
himself was partial to the term "ambience," which he regarded as "suitably ambiguous."
See " 'The Frailties and Beauties of Technological Creativity': An Interview [with John
Staudenmaier] by Robert C. Post," *American Heritage of Invention and Technology*, VIII,
no. 4 (Spring 1993), 16–24.

claims of the contextualists," Leo Marx has asked more recently, "how can we justify segregating the history of technology . . . from the history of the societies and cultures that shape it?"[14] What if Heidegger was right when he said that "the essence of technology is by no means anything technological"?

While such an assertion may be arguable, the idea harbors seeds of absurdity, as in the observation that "the steadily accelerating pace of technological change . . . seems, on the whole, to be willed by fate." "Only a philosopher unconcerned with the messiness of historical processes could maintain such a claim," deadpans Thomas Misa.[15] Yet Hindle would of course find himself emphatically at odds with Heidegger's perception of "essence." Near the top of his agenda was the affirmation of two interrelated concepts—that the study of technological history could provide insights not otherwise attainable, and that the essence of technology lay in its artifacts, nowhere else:

> The "things" of technology retain a primacy that does not adhere to the physical objects associated with science, religion, politics, or any intellectual or social pursuit. The means of technology are physical; the objectives of technology are also physical or material. Three-dimensional physical objects are the expression of technology—in the same way that paintings and sculpture are the expression of the visual arts.[16]

Among academic specialists in early American history, Hindle's enthusiasm for "three-dimensional physical objects" made him appear to be something of a maverick, and there were even self-defined historians of technology who rarely thought in such terms. But Hindle enjoyed affirmation from an influential cohort, among the most persuasive of whom was Eugene Ferguson. At the Hagley conference Ferguson stated that "the historian should consider looking at artifacts to be . . . part of his trade."

---

14. David A. Hounshell, "Commentary: On the Discipline of the History of American Technology," *Journal of American History*, LXVII (1980–1981), 854–865; Leo Marx, review of Cutcliffe and Post, eds., *In Context, T&C*, XXXII (1991), 394–396, quote on 395.

15. Friedrich Rapp, *Analytical Philosophy of Technology*, Boston Studies in the Philosophy of Science, LXIII (Dordrecht, 1981), 184; Thomas J. Misa, "How Machines Make History, or How Historians (and Others) Help Them to Do So," *Science, Technology, and Human Values*, XIII (1988), 308–331, quote on 310.

16. "The Exhilaration of Early American Technology," 10 (below, 48).

Before joining the faculty at Iowa State, Ferguson had spent three years as a Smithsonian curator; after leaving Iowa for the University of Delaware in 1969, he trained his graduate students in the ways of museology as well as history, and certain of those students would achieve pivotal insights from the analysis of historical artifacts, Hounshell for one. During a term as a predoctoral fellow at the National Museum of History and Technology (and during Hindle's term as director, 1974–1977) Hounshell ascertained that domestic sewing machines were largely manufactured using hand-fitted rather than interchangeable parts, even when produced for a mass market. Later, Robert B. Gordon of the Kline Geological Lab at Yale, trained in metallurgy but emerging as a gifted and resourceful historian as well, put the "artificer" at center stage and interpreted surficial markings on the components of small arms as indicative of four key elements of mechanical skill: "dexterity, judgment, planning, and resourcefulness." [17] While labor history has flourished in a community studies context that was only on the horizon when Hindle wrote his essays for *Technology in Early America,* one can argue that artifactual analysis provides the single most valuable means of understanding what early American workers actually did. There *is* information "inherent" in technological artifacts. Indeed, this concept now sounds thoroughly prosaic.

Yet, in his declarations about the primacy of "things," Hindle may have been striving for something deliberately extravagant, and one cannot be certain of how much faith he himself had in the technique. On the one hand, he could observe that "the physical things of technology in many ways remain the ultimate source for the history of technology," and on the other hand he could issue a warning that no artifact could "be relied on not to 'lie.'" In his 1966 bibliographical essay he noted that "only the smallest portion of the objects of technology have been preserved and, even

17. *Ibid.,* 15 (below, 53); David A. Hounshell, *From the American System to Mass Production, 1800–1932: The Development of Manufacturing Technology in the United States* (Baltimore, 1984), appendix 2, "Singer Sewing Machine Artifactual Analysis," 337–344; Robert B. Gordon, "Who Turned the Mechanical Ideal into Mechanical Reality?" *T&C*, XXIX (1988), 744–778. Hindle's personal intercession was crucial to the award of Hounshell's fellowship. Gordon's ingenuity at integrating technology and culture was later instrumental in his garnering the Smithsonian's highest honor for a visiting scholar, a Regents Fellowship, a fellowship also held at one time by Roe Smith, a specialist in the history of small arms manufacture.

if they were more extensive, the information they present would remain limited." [18]

If even Hindle felt ambivalent about "the 'things' of technology," surely most other scholars would too. In the 1990s, just about any historian of technology would agree that artifactual analysis could yield valuable information. But the majority would be skeptical about assertions that any machine could "speak for itself" (once a cliché among museologists) and dubious of three-dimensional evidence that lacked corroboration from what Hindle called "the conventional reliances of the historian," that is, documentary sources. Hindle's call for an artifact-centered historiography was a need only imperfectly fulfilled, just as was his call for synthesizing early American technology and culture.

But there was an even higher item on his agenda, one rooted in his certitude about the importance of what he called "the internal life of technology." Hindle believed passionately that one could never understand early American technology (or any technology) in the absence of an effort "to see through the eyes and feel through the hands of the craftsman and mechanic." "The social relations of technology must not be neglected," he acknowledged; "they represent the historian's highest goal, but they are attainable only after the historian has developed a direct understanding of the men and their works." It was through this "direct understanding" that one might be able to perceive technology as something exhilarating, hence the title of his essay:

> Perhaps technology was not so much a tool or a means as it was an experience—a satisfying emotional experience. When it became possible to make thread and cloth by machine, they were so made; when boats, trains, and mills could be driven by steam, they were so driven. These things could be accomplished only to the extent that economic needs and social attitudes permitted; but is it not possible that the more elemental force was within the technology itself? It is an idea often denied. [19]

18. Brooke Hindle, "Technology through the 3-D Time Warp," *T&C*, XXIV (1983), 451 (Hindle made the latter comment with direct reference to Henry Ford's assertion that "relics of days gone past cannot lie"); Hindle, "A Bibliography of Early Technology," *Technology in Early America*, 31.

19. "The Exhilaration of Early American Technology," 24 (below, 62–63).

Is it not possible that the more elemental force was within the technology itself? More than anything else, the emergence of the history of technology in academic cloaks had been marked by an effort to banish technological determinism in any of its various guises, from Marxism ("The handmill gives you a society with the feudal lord; the steam-mill, society with the industrial capitalist") to the sociology of William F. Ogburn, who posited that technology wrought changes to which society then had to adapt, and that the resulting "lag" typically had significant historical consequences.[20] In such schemes, technology was assigned an active—a deterministic— role. The triumph of contextualism was essentially an affirmation that "technology" is a mere abstraction with no capacity for making any "impact" whatsoever. And what was determinism, then? More than just a wrongheaded theory, it was a doctrine of singular utility to reactionary politics; to Judith McGaw it was a "crippling social malaise."[21]

To speak of there being anything "within" technology smacked of determinism. But when Hindle referred to an "elemental force," he had in mind something more subtle and more profound as well. Certain technological practitioners could readily confirm what he said about "a satisfying emotional experience," and so could some historians, but how many scholars had any "*direct* understanding" of technology? After the 1960s, more and more historians would apprehend its *results* in mostly negative terms, but they ran a serious risk if they were tempted to universalize such perceptions. They risked losing sight of the reality that technological practitioners may find their pursuits every bit as exhilarating as scholars find theirs—and that, because of this, they may pursue technological ends

20. Karl Marx, *The Poverty of Philosophy* (New York, 1971), 109. For a nuanced analysis of Marx's technological determinism, see Donald MacKenzie, "Marx and the Machine," *T&C*, XXV (1984), 437–502. For expressions of the social-lag theory, see William Fielding Ogburn, *Social Change with Respect to Culture and Original Nature* (New York, 1923); *The Social Effects of Aviation* (Boston, 1946); and, with M. F. Nimkoff, *Technology and the Changing Family* (Cambridge, Mass., 1955). Ogburn strongly influenced Roger Burlingame, whose *March of the Iron Men: A Social History of Union through Invention* (New York, 1938) and *Engines of Democracy: Inventions and Society in Mature America* (New York, 1940) have been termed the first efforts by an American historian to deal with "big questions against the broad expanse of American history" (Daniels, "The Big Questions," *T&C*, XI [1970], 2).

21. Judith A. McGaw, *Most Wonderful Machine: Mechanization and Social Change in Berkshire Paper Making, 1801–1885* (Princeton, N.J., 1987), 6.

beyond all bounds of rational behavior. They may even characterize these quests in terms of being driven, obsessed, addicted.

Is there any such thing as an "elemental force" within technology, truly, rationally? This cannot be demonstrated, at least not nearly as readily as it can be shown to be a matter of cultural conditioning. Yet Hindle understood that people whose lives are concerned primarily with spoken words or words on paper are critically deficient in their range of perceptions if they fail to see how compelling the pursuit of technological progress can be—so compelling that it may *in effect* endow technology with a force of its own. If it is true, as Ferguson once suggested, that "each of us has only one message to convey," then I suggest that this was Brooke Hindle's message. Call this elemental force exhilarating, or call it something negative, either way the concept is so compelling that it may well be capable of sustaining a major synthesis of technology and culture. Hindle perhaps knew this, but in the post-Vietnam climate of opinion he was not to have the satisfaction of seeing any such synthesis attempted.[22]

Hindle published two retrospectives on *Technology in Early America* during the early 1980s. The first stemmed from a paper titled "Early American Technology, Today," in which he celebrated exhilaration on the part of technological historians:

> A week seldom passes without a book, article, or preprint that communicates enthusiasm about another range or defile in the history of technology. Books now in the pipeline from established scholars, but more especially from younger scholars, promise a continuation of pub-

22. Ferguson quoted by Melvin Kranzberg in "Technology and History: 'Kranzberg's Laws,'" *T&C*, XXVII (1986), 544. There had long been a literature calling into question the extent to which invention was impelled by economic motives or even by rationality, a literature anchored in sociological studies such as Joseph Rossman's *Industrial Creativity: The Psychology of the Inventor* (1932; rpt., New York, 1964), which indicated that "love of inventing" was a more common motivation among inventors than either financial gain, perceived necessity, altruism, or fame (152). I was influenced by this literature when I suggested that C. G. Page was driven forward primarily by his "romantic imagination" ("The Page Locomotive: Federal Sponsorship of Invention in Mid-Nineteenth-Century America," *T&C*, XIII [1972], 140–169), and I assume that Edward W. Constant must have had it in mind when he posited in *The Origins of the Turbojet Revolution* (Baltimore, 1980) that a revolutionary inventive concept was "motivated by a combination of delight in the esthetic appeal of the new idea and the fervor of a provacateur" (this apt thumbnail description is Staudenmaier's in *Technology's Storytellers*, 52).

lications which display control of the hardware technology within its larger social contexts and marked by a command of approaches from other disciplines and sub-disciplines of history. . . . The "exhilaration" expressed by the mechanics and projectors of technological development in early America appears to have been transferred to the most successful historians studying them.[23]

Here, Hindle concluded on a note of affirmation. But the other retrospective had quite a different tone. It was presented under the auspices of the Hagley Museum, where the Conference on Early American Technology had convened nearly two decades before, and appeared in a booklet titled *The History of American Technology: Exhilaration or Discontent?* Again, Hindle took note of the scholarship published since 1965—with his wife, Helen, he had tallied twenty-two articles on early American technology in *Technology and Culture* along with eighteen books in which technology was central and twice that many in which it was secondary—and he was pleased that there had been "a sharp rise in studies involving a good command of technology from the inside as well as a good understanding of the social and economic relationships." Again he sounded a clarion call for synthesis. And he reasserted the necessity for getting "inside" technology in order to gain a proper understanding of "creativity or the intellectual

23. Brooke Hindle, "Early American Technology, Today," 5 (a revised version of a presentation to the Maryland Historical Society on the occasion of the publication of *The Engineering Drawings of Benjamin Henry Latrobe,* ed. Darwin H. Stapleton, May 29, 1980). Hindle outlined several books that he identified as "possessing a breadth of approach which makes them important to general history, coupled with a true perception of the inner ways of technology": Carroll W. Pursell, Jr., *Early Stationary Steam Engines in America: A Study in the Migration of a Technology* (Washington, D.C., 1969); Bruce Sinclair, *Philadelphia's Philosopher Mechanics: A History of the Franklin Institute, 1824–1865* (Baltimore, 1974); Robert C. Post, *Physics, Patents, and Politics: A Biography of Charles Grafton Page* (New York, 1976); Merritt Roe Smith, *Harpers Ferry Armory and the New Technology: The Challenge of Change* (Ithaca, N.Y., 1977); Carl W. Condit, *The Port of New York: A History of the Rail and Terminal System, from the Beginnings to Pennsylvania Station* (Chicago, 1979); Louis C. Hunter, *Water Power in the Age of the Steam Engine,* vol. 1 of *A History of Industrial Power in the United States, 1780–1930* (Charlottesville, Va., 1979); and Anthony F. C. Wallace, *Rockdale: The Growth of an American Village in the Early Industrial Revolution* (New York, 1978). A much-attenuated version of this paper appeared as part of Hindle's essay, "A Retrospective View of Science, Technology, and Material Culture in Early American History," *WMQ,* 3d Ser., XLI (1984), 422–435.

course toward invention and innovation."[24] But his remarks bespoke a general disillusion.

"To ask what has happened in early American technology since 1965 is to evoke a memory of great change, of activity, development, and advance," he began his talk at the Hagley. The change, the activity, did not all please him, far from it, and here it was that Hindle made his own terminological bequest to historiographic discourse. "A bitterness and antagonism is shattering the consensus of most historians at work eighteen years ago," he said, and part of the reason was that certain individuals were seeking "to apply the historian's integrity to their politically oriented reading of history," a reading that stressed the "dark side."[25]

The perception that the historical discipline had fragmented, that a sense of common purpose had been lost, lost forever perhaps, was shared by many scholars of Hindle's generation, but dark-side interpretations of technology were not solely the province of unreconstructed campus radicals. Indeed, they could be found in the writings of Daniel Boorstin, the prophet of consensus. More so than any other work of mainstream history, Boorstin's *The Americans* had "given the role of technology a place of central importance in a major reassessment of modern American social and cultural development"—so said SHOT when it awarded Boorstin its premier book prize in 1974. Yet the third and final volume of his trilogy underscored "various ways in which the uses of technology have intensified the rootlessness endemic in American life, transformed our language, vitiated time-honored customs, and encouraged the resort to sweepingly quantitative criteria in the assessment of social problems and issues."[26]

24. Brooke Hindle, " 'The Exhilaration of Early American Technology': A New Look," in David A. Hounshell, ed., *The History of American Technology: Exhilaration or Discontent?* (Wilmington, Del., 1984), 11, 23. This was a subject that Hindle explored in his provocative *Emulation and Invention* (New York, 1981), a book that added insight to the work of Cyril Stanley Smith on the artistic component of technology (see, for example, Smith, "Art, Technology, and Science: Notes on Their Historical Interaction," *T&C,* XI [1970], 493–549) and Eugene S. Ferguson on spatial thinking (see Ferguson, "The Mind's Eye: Nonverbal Thought in Technology," *Science,* CXCVII [Aug. 26, 1977], 827–836).

25. " 'The Exhilaration of Early American Technology': A New Look," in Hounshell, ed., *History of American Technology,* 7, 11, 12, 13.

26. "The Dexter Prize," *T&C,* XVI (1975), 422. *The Americans* also won a Pulitzer in History in 1974. As editor of the Chicago Series in American Civilization, Boorstin had fostered publication of monographs such as John F. Stover's *American Railroads* (Chicago, 1961).

Boorstin was an anomaly, to be sure (among other things he had been a sometime speechwriter for Vice-President Spiro Agnew, who *truly* knew how to trivialize negativism). The emergence of a dark-side literature could be traced more directly to the Frankfurt School and to Herbert Marcuse's *One-Dimensional Man* (1964). It was rooted in John McDermott's 1969 essay, "Technology: The Opiate of the Intellectuals," which transformed the Vietnam War into an authoritarian technological metaphor, and in the publication a year later of Lewis Mumford's *Myth of the Machine: The Pentagon of Power*, a book that articulated a fear that "man's vital organs will all be cannibalized in order to prolong the megamachine's meaningless existence." [27]

Through such writings, and through Charles A. Reich's *Greening of America* (1970) and Theodore Roszak's *Where the Wasteland Ends* (1972), dark-side themes penetrated the popular literature of technology, and, soon, the professional literature as well, particularly in Langdon Winner's *Autonomous Technology: Technics Out-of-Control as a Theme in Political Thought* (1977) and the more widely read works of David F. Noble, *America by Design: Science, Technology, and the Rise of Corporate Capitalism* (1977) and *Forces of Production: A Social History of Industrial Automation* (1984). Noble was something of a bête noire for Hindle; when Hindle was planning "Engines of Change" at the National Museum of American History in the early 1980s, Noble was employed there as a curator, and the two clashed repeatedly. (Hindle's successor as director, Roger G. Kennedy, had both courted Noble and seen to it that the very word "technology" was dropped from the museum's name.)

Hindle was unfailingly courteous to Noble in print; his books, he wrote, "demonstrate good research and well-organized interpretations." [28] Yet he

27. John McDermott, "Technology: The Opiate of the Intellectuals," *New York Review of Books,* July 31, 1969, 25–35; Lewis Mumford, *The Myth of the Machine,* [II], *The Pentagon of Power* (New York, 1970), quote on 435. Mumford's much earlier *Technics and Civilization* (New York, 1934) remained in the 1970s, and perhaps remains even today, the most stimulating synthesis of its kind ever written, but even this book expresses fears of mankind's subservience to the machine.

28. Hindle, "Historians of Technology and the Context of History," in Cutcliffe and Post, eds., *In Context,* 239. In conversation and correspondence, Hindle was much more critical of what he called Noble's "advocacy history," and it seems certain he would have agreed with the general observation by Hofstadter that "the activist historian who thinks he is deriving his policy from his history may in fact be deriving his history from his policy, and

found such work terribly disquieting. Noble's avowed aim was to destroy America's faith in technology by depicting it "in a context of class conflict and informed by the irrational compulsions of an all-embracing ideology of progress."[29] Hindle was not denying the reality that all technologies exact costs. He would not even have denied that technology could be driven by "irrational compulsions"—how, after all, was saying something like this so different from speaking of "a great human urge to do everything that developing means permit man to do"? The damage wrought by the dark-side reading of history was, not least, to common ideals. Noble was contemptuous of the very idea of "the history of technology," but many of his most divisive views seemed to be shared by SHOT stalwarts. Carroll W. Pursell, for years the society's secretary, referred to the American Industrial Revolution's "ever-elaborating authoritarian technics." An exchange of opinion between Staudenmaier (the society's semi-official historian) and two members of the founding generation, including Mel Kranzberg himself, confirmed Hindle's apprehension that a crisis of comity "deeply infects the history of technology."[30] He was profoundly pessimistic.

Too pessimistic, it seems to me. Hindle understood that his was a generational perception; he realized that he reflected the outlook of someone "positively conditioned by the Depression and World War II," and he was conscious of "more negative conditioning, in the next generation of historians."[31] But not every member of that generation was conditioned in the same way. The commentator at the Hagley session was Stuart Leslie, a

---

may be driven to commit the cardinal sin of the historical writer: he may lose his respect for the integrity, the independence, the pastness, of the past" (*The Progressive Historians*, 465).

29. David F. Noble, *Forces of Production: A Social History of Industrial Automation* (New York, 1984), xiv. Cf. two reviews of this book: by David S. Landes in *The New Republic*, Nov. 19, 1984, 37–41, and by Wickham Skinner in the *New York Times Book Review*, Sept. 2, 1984, 10–11 (the latter a joint review with Hounshell's *From the American System to Mass Production*).

30. Carroll W. Pursell, Jr., "The Problematic Nature of Late American Technology," in Hounshell, ed., *History of American Technology*, 26; Hindle, " 'The Exhilaration of Early American Technology': A New Look," *ibid.*, 2.

31. Hindle, "Brooke Hindle: An Intellectual Autobiography," *T&C*, XXVI (1985), 586–589, quotes on 589. It is perhaps worth mentioning that Hindle had elected to culminate his career as a public historian, and, conceivably, too, he was reflecting a sensitivity to doubts among academics about public history's "compatibility with traditional norms of the profession" (Novick, *That Noble Dream*, 517).

scholar on the other side of the divide that Hindle saw as crucial to what Leslie called "The 'Lost Exhilaration' of American Technology." Leslie confirmed that the 1960s had "affected scholars on each side of the 'generation gap' somewhat differently." No longer were historians "simply boasting of the fulfillment of America's technological dreams, but instead paying more careful attention to how those dreams were fulfilled, what they cost, and who paid the price."[32]

Nevertheless, Leslie sought to reassure Hindle that the theme to which he attributed such power had not been lost at all. Take David McCullough's books on the Panama Canal and the Brooklyn Bridge, Tracy Kidder's *The Soul of a New Machine,* Tom Wolfe's *The Right Stuff*—each was brimming with evidence of "the exhilaration of those who created and those who used technology." True, professional historians had not been so attentive to this. Yet the work of Thomas Hughes was certainly relevant. His concept of "technological momentum," first enunciated in the late 1960s, would take full wing two decades later in *American Genesis,* an epic confirming Hughes's emergence as the premier American historian of technology. Hughes's focus was recent history—the period 1870–1970, which he denominated "A Century of Invention and Technological Enthusiasm"—but he began with a quote from Perry Miller about how Americans "flung themselves into the technological torrent, how they shouted with glee in the midst of the cataract, and cried to each other as they went headlong down the chute that here was their destiny."[33]

Like Hindle's "elemental force," the use of the term "destiny" hinted at technological determinism, as did Hughes's own heavy reliance on momentum as a conceptual device. And, as Leo Marx observed, Hughes was one historian of technology who seemed willing to confront this specter head-on—not in any "absurdly reifying manner," but in terms of

32. Stuart W. Leslie, "Commentary: The 'Lost Exhilaration' of American Technology," in Hounshell, ed., *History of American Technology,* 28, 29. Steve Lubar reminds me that Hindle was a vocal opponent of America's Vietnam involvement. And Hindle's assessment of what the 1960s had done to the climate of scholarship was far from the catastrophic extreme as exemplified in Lewis S. Feuer, *The Conflict of Generations: The Character and Significance of Student Movements* (New York, 1969), esp. chap. 9.

33. Leslie, "Commentary," in Hounshell, ed., *History of American Technology,* 32; Perry Miller, "The Responsibility of Mind in a Civilization of Machines," *American Scholar,* XXXI (1961–1962), 51–69, quoted in Thomas P. Hughes, *American Genesis: A Century of Invention and Technological Enthusiasm, 1870–1970* (New York, 1979), 1.

"a culturally nurtured propensity to mechanize as many aspects of life as possible." Thomas Misa, a Hughes student, indicated that Hughes's concept transformed "technologically deterministic forces into components of technological systems."[34] That is, he viewed technological momentum as a concomitant of a given society's cultural commitments. And a key aspect of momentum—a major determinant (so to speak)—is a culturally nurtured enthusiasm.

This is an important insight for several reasons. As I have suggested already, it may be central to the question of whether the history of technology is anything more than a loosely congruent set of research agendas, a shorthand convenience, within the larger historical enterprise—whether it has genuine intellectual coherence or exists simply for "careerist" aims, an accusation levied by David F. Noble.[35] If indeed it does have cohesion, this may be provided by the nature of technological enthusiasm. Indeed, enthusiasm may be taken, as Ferguson has suggested, as a means of synthesizing the relationship of technology and culture:

> Enthusiastic technologists not only have built the world we live in but . . . by and large, they themselves have hustled the support required to be able to do so. To build nearly anything, and particularly something that is difficult or hazardous, is an intensely interesting process for those who have to solve the many problems that are inevitably encountered. If we fail to note the importance of enthusiasm that is evoked by technology, we will have missed a central motivating influence in technological development.[36]

Old guard internalists, who believed that the history of technology was fundamentally about "how things were done or made," used to claim that one could not properly address a technological subject without fully

---

34. Leo Marx, review of Cutcliffe and Post, eds., *In Context, T&C*, XXXII (1991), 396; Misa, "How Machines Make History," *Science, Technology, and Human Values*, XIII (1988), 318–319.

35. David F. Noble quoted in Svante Lindqvist, *The Teaching of History of Technology in USA* (Stockholm, 1981), 13.

36. Eugene S. Ferguson, "Toward a Discipline of the History of Technology," *T&C*, XV (1974), 13–30, quote on 21. Leslie indicates that Ferguson argued his point even more directly in "Enthusiasm and Objectivity in Technological Development," a paper delivered to the 1970 meeting of the American Association for the Advancement of Science but never published.

understanding the esoteric body of pertinent technical knowledge. But that was never the important necessity, and, even before Ferguson, Brooke Hindle realized what was—articulating it in *Technology in Early America*. The necessity is to comprehend the power of enthusiasm, particularly in imparting momentum, and, to that end, appreciating how (take your pick) "exhilarating" or "intensely interesting" technology can be to its practitioners.

To say this implies nothing at all about abandoning a critical stance toward "progress talk." Technological pursuits may be socially useless at best. At worst, they can be wasteful, exploitative, dangerous, evil; in one way or another they may cost too much. Even so—particularly so—we need to understand what compels the search for the "elegant solution" to technical problems. And what about the deterministic implications of referring to forces "within technology itself," or to "technological momentum," or to a technological "destiny"? Ferguson did not shy from them; Hughes did not. Nor did Hindle: "One of the great questions that might yield to the central study of technology is the extent to which its internal character is deterministic," he wrote. "It is clear that at each stage of development, new possibilities are opened."[37]

Probably the determinism would turn out to be "soft"—in the sense of Lynn White, jr., of a door that might be open but without any compulsion to go through—although Hindle did not rule out the possibility that it could be something more than that. In certain cultural contexts, technical exigencies might appear to be imperatives. One need not approve, but one must comprehend. Leslie, a Ferguson student, perhaps put it as well as anyone:

> For those who created (and create) our technology, and those who used (and use) it, were (and are) enthusiasts of the first order. Whether or not historians of technology themselves feel this exhilaration, we must still understand and appreciate it if we are to understand and appreciate the unfolding of America's technological history.[38]

Applying tenets of the Old History–New History debate to the historiography of technology, Staudenmaier has suggested that we are not yet as critical as we must be toward "the whig reading of Western technological

37. "The Exhilaration of Early American Technology," 24 (below, 62, 63).
38. Leslie, "Commentary: The 'Lost Exhilaration' of American Technology," in Hounshell, ed., *History of American Technology*, 33.

evolution," the reading that suggests that there is some inevitability about technological change.[39] That may be so, although it is likewise arguable that we may not be sufficiently attentive to the "internal life of technology." Perhaps an appreciation for this is best fostered by firsthand experience. That can be a matter of training and background: Ferguson and Hughes were schooled as engineers, Hindle studied naval architecture and marine engineering at Massachusetts Institute of Technology and was a radar maintenance officer during World War II.[40] Or it can be imparted through the museum world's concern with artifact and process—Ferguson and Hindle often moved in museological circles. Younger scholars such as Leslie and Hounshell have been attracted to subjects whose interpretation entails considerable technical proficiency. Practitioners of historical archaeology, industrial archaeology, material culture studies in its various permutations, and other fields predicated on the assumption that "the utilization of objects as part of a research program offers an additional avenue of understanding" are likewise more inclined to appreciate the meaning and significance of exhilaration and enthusiasm.[41]

Engineers, curators, industrial archaeologists, and the like may also be more inclined to harbor a continuing affection for the wonders of technology, that is, to be technological enthusiasts themselves. Hence it is important to bear in mind that the one does not inevitably follow from the other. We can take exhilaration into account without necessarily applauding the ends toward which it is directed. Without forgetting that it may be an emotional elixir available only to people whose hands are, so to speak, on the throttle. Without glossing over complexity—the "hesitant and equivocal" response to industrialization at Harpers Ferry, for instance, compared to the enthusiasm of both masters and mechanics at

39. John M. Staudenmaier, "Recent Trends in the History of Technology," *American Historical Review*, XCV (1990), 725.

40. See Brooke Hindle, *Lucky Lady and the Navy Mystique: The "Chenango" in WW II* (New York, 1991). In a marvelously perceptive review of this book, A. Hunter Dupree (a participant in the 1965 conference at the Hagley) suggests that "in his classic works on early American science and technology and in his biography of David Rittenhouse, Hindle has built the insight gained as an actor on *Chenango* into the historical literature of his generation" (*T&C*, XXXIII [1992], 636).

41. Laurence F. Gross, "Wool Carding: A Study of Skills and Technology," *T&C*, XXVIII (1987), 804.

Springfield. Without losing sight of social costs, without abandoning our critical responsibilities, without regressing to the naive attachment to the past that is characteristic of the antiquarian. But we do need to understand—as Brooke Hindle so clearly understood when he penned "The Exhilaration of Early American Technology"—the technologist's "emotional satisfaction in reaching forward with each improvement." [42]

The explanatory power of economic incentives, by themselves, is clearly insufficient. Thinking in terms of quests for mastery and control may get us further, but Hindle suggests that men like Oliver Evans and William Norris were motivated by a need to fulfill new potentials simply because the potentials existed. He also poses the most apposite critical query: "Were they so inner-directed that they failed to exercise needful restraint in moving as fast as they did toward dependence upon machines?" [43]

The aim of the 1965 conference at the Hagley was to promote research "by providing a wide-ranging view that correlates what has been done with what needs to be done." The hope that it would encourage historical synthesis may have gone largely unfulfilled, though it needs be noted that historians are perpetually inclined to fret over a historiography "mired in details without illuminating larger issues." [44] The hope that changes in the character of American technology would be perceived, not as a story within the history of technology alone, but rather as "very close to the center as an expression and fulfillment of the American experience" may likewise have gone largely unfulfilled, so far. Since the point about being "very close to the center" was unquestionably true, however, there was no reason to abandon faith that it would eventually be broadcast. Indeed, Hindle himself had done much to establish its validity through "Engines of Change," an exhibition seen by millions of people and characterized even

42. Merritt Roe Smith, *Harpers Ferry Armory and the New Technology: The Challenge of Change*, 23, 323; "The Exhilaration of Early American Technology," 24 (below, 63).

43. "The Exhilaration of Early American Technology," 24 (below, 62).

44. The second quote is from David Thelen, "The Profession and the *Journal of American History*," *JAH*, LXXIII (1986–1987), 9. Expressions of yearning for "a redirection of scholarship away from specialization and toward greater significance of issues" (*ibid.*, 10) seem timeless—cf. "Most of our scholarly history, including the best, is not by specialists for specialists, but is by specialists for a small fraction of the specialists" (W. Stull Holt, "Who Reads the Best Histories?" *Mississippi Valley Historical Review*, XL [1953–1954], 619), and I would wager that similar laments could be found much earlier.

by an unsympathetic critic as "brave in its conception, in places sophisticated in its balancing of issues." [45] So, too, Hindle's hopes for encouraging
the study of early American history by means of historical artifacts: "Engines of Change," with its effective contextualization of the locomotive
John Bull, the Blanchard lathe, and the Howe pin machine, stood as a
sterling example of what *could* be accomplished in this realm.

In some regards, Hindle had lost all headway: Certainly the climate of
opinion in the 1990s was such that other historians of technology would
be, at best, ambivalent about his assertion in the 1960s that artifacts call
for "celebration" just like works of art. The dark-siders continued to command attention, indeed they advanced interpretations that were increasingly compelling, and Hindle was increasingly troubled. "The greatest role
of history is as our social memory," he wrote toward the end of the 1980s.
"Remembering the past as nothing but defeats and failures is a route to
disaster." [46]

But the past was scarcely being remembered as nothing but defeats and
failures. Specialized studies were still appearing whose scope and stance
were similar to those Hindle had cited in his 1966 bibliographical essay;
scholars were still cultivating the same topics he delineated, from agriculture and food processing to patents and invention. The attention to
contingency was more sensitive, however, and the context much richer.
Indeed, when one underscores the error of positing "a fixed sequence to
technological development," or the truth that "technical designs cannot
be meaningfully interpreted in abstraction from the human fabric of their
contexts," one now has a sense of voicing platitudes. Perhaps, then, the
time is at hand to turn to something more problematic—perhaps to the
question of what, exactly, there might be "within the technology itself."
Leo Marx had posed a compelling challenge, a challenge to demonstrate
that technology has "specifiable, presumably inherent attributes." [47]

---

45. "The Exhilaration of Early American Technology," 28 (below, 67); Michael Strat-
ton, "Integrating Technological and Social History: 'Engines of Change' at the National
Museum of American History," *T&C*, XXXI (1990), 271–277, quote on 277.

46. "The Exhilaration of Early American Technology," 10 (below, 48); Hindle, "Historians of Technology and the Context of History," in Cutcliffe and Post, eds., *In Context*,
230, 231.

47. Robert L. Heilbroner, "Do Machines Make History?" *T&C*, VIII (1967), 336;
John M. Staudenmaier, "What SHOT Hath Wrought and What SHOT Hath Not: Reflections on Twenty-five Years of the History of Technology," *T&C*, XXV (1984), 707–730,

A younger historian who had been quite persuasive in underscoring the power of exhilaration recently characterized *Technology and Culture* as being in a "rut" because of its single-minded devotion to a contextual style which concentrates on "institutions, social movements, intellectual currents, economic conditions, and the like." Perhaps it is, although there is evidence in the following essays that discerning historians, whether younger or older, are still coming to early American technology with a remarkable diversity of approaches. The possibility of there being something elemental within the technology itself has a differing explanatory potential with each of these essays, but I do not think it is altogether irrelevant to any of them. In concluding "The Exhilaration of Early American Technology," Hindle wrote: "Since we are not yet entirely sure what are the 'right' questions or the 'right' approaches, it is doubly necessary that no doors be closed and that the insights of all who have looked in this direction be sought."[48] Judy McGaw could not resist quoting this admirable thought, just as I cannot, and it may well be Brooke Hindle's most enduring admonition. But I would like to close by suggesting again that his primary message for historians in the 1990s may be something different. *Could* there be an elemental force within technology itself? Can this be demonstrated? *That* door, more so than any other, should definitely be held wide open.

---

quote on 711; Leo Marx, review of Cutcliffe and Post, eds., *In Context*, *T&C*, XXXII (1991), 396. Marx questioned whether anything with such "obscure boundaries" as technology could make a claim to being "the focus of a discrete field of specialized historical scholarship" (395). The point was arguable, to be sure, but one should not forget where Marx was coming from (so to speak). American studies, or more specifically the symbol-and-myth school of which he was an exemplar, had been accused of treating part of reality as if it were all of reality. See Bruce Kuklick, "Myth and Symbol in American Studies," *American Quarterly*, XXIV (1972), 435–450.

48. Robert Friedel, review of *Technology and Culture, History of Technology*, and *History and Technology: An International Journal*, in *Isis*, LXXXI (1990), 295; "The Exhilaration of Early American Technology," 28 (below, 67).

BROOKE HINDLE

The Exhilaration of Early American Technology

An Essay

The central role of technology in early American history has only recently begun to receive some of the attention it requires. Strikingly, the craftsmen, mechanics, engineers, and entrepreneurs who built that technology were enthusiastically—even ebulliently—aware of the pervasive significance of their work. Historians have not been unresponsive, but they have often been uninformed and they have usually been too preoccupied with other investigations to give it serious study. Thus, the history of technology, invigorating and stimulating as it is, has not yet reached an academic status parallel to other fields of history. But increasingly specialists who have cultivated corners of the field and general historians who have admitted an oblique interest are being joined in Europe and in America by historians competent in the field of technology.

The European interest in the history of technology opened first and probably most effectively—certainly most relevantly to the American story—in England. There it was moved forward by a train of enthusiasts, such as Samuel Smiles, and by the continuing effectiveness of such institutions as the Science Museum and the Newcomen Society. The new *Journal of Industrial Archaeology,* along with the surrounding publication and preservation projects in Great Britain, are promising steps in the

This essay is reprinted from Brooke Hindle, *Technology in Early America: Needs and Opportunities for Study* (Chapel Hill, N.C., 1966), by permission of the author.

direction of informed specialization. Many of the recent British books on the history of technology, notably Singer's five-volume *History of Technology*, offer increasingly significant contributions by academic historians.[1] Continental interest has also risen, but it remains closely associated with technological museums—as evidenced in Daumas' continuing *Histoire* and Klemm's lecture survey.[2] In the Iron Curtain countries, political dogma has encouraged an interest in the history of technology, an emphasis reflected in publications and in the 1965 International Congress of the History of Science, which met in Poland.

In the United States, the growth of a new interest and a new approach has several roots and supports. Most apparent are the formation of the Society for the History of Technology in 1958, the issuance of its journal, *Technology and Culture*, and the development of a program in the history of technology at Case Institute of Technology—all under the stimulus of Melvin Kranzberg. Similarly, at other educational institutions academic work is arising; probably the largest concentration of scholars interested in the history of technology is to be found at the Massachusetts Institute of Technology. At the same time the newly invigorated Museum of History and Technology of the Smithsonian Institution has become a center for much important activity in the field, and smaller museums are beginning to acquire academically oriented personnel. The Eleutherian Mills-Hagley Foundation is the best example of new institutions with a scholarly approach to the history of technology.

Of the numerous approaches that have been opened to the history of technology, that of the history of science seems the most directly pertinent. The History of Science Society and its organs have always professed to include technology within their scope; under its present editor, Robert P. Multhauf, *Isis* is markedly interested in the subject. George Sarton, who did so much to define the field, included technology within his great work, and it remains a component of the annual "Critical Bibliography" which he founded. Formally, then, the history of technology has a home within the history of science, yet it has long been clear that technology does not occupy a central place in the efforts of this group and that some members regard it as a distinctly alien element.

Basically, the tensions surrounding the relationship between science

1. Charles Singer *et al.*, eds., *History of Technology*, 5 vols. (London, 1954–58).

2. Maurice Daumas, *Histoire Générale des Techniques* (Paris, 1962—); Friedrich Klemm, *A History of Western Technology*, trans. Dorothy Waley Singer (New York, 1959).

and technology point to fundamental differences between the two and, even more pointedly, to differences in the historical course which each followed. Science and technology have different objectives. Science seeks basic understanding—ideas and concepts usually expressed in linguistic or mathematical terms. Technology seeks means for making and doing things. It is a question of process, always expressible in terms of three-dimensional "things." In the early American period, tools and means, as well as products, were all usually three-dimensional; but even when the product was not, as in the electric telegraph, the means were.

The relationship between science and technology has varied in time and place. During the Middle Ages the great technological advances owed little to science or to those who pursued science. On the other hand, those engaged in the space technology of the present are sometimes at a loss to draw a line between the science and the technology they use. Schofield and Gillispie have helped to illuminate the far from obvious relationships between the two in eighteenth-century Europe.[3]

In colonial America the role of science in technology was certainly very limited, but the Americans entered on independence amid many Baconian intentions of applying science. Mechanics who called for "principles" had in mind the tabular test results of modern engineering handbooks rather than science. In the nineteenth century, however, the literature reveals well-conceived efforts to use science. The results have not been seriously studied, but appearances suggest that the prevailing British generalization was correct: American technology depended but little upon the "sober reasoning of science."[4]

Richard H. Shryock has suggested that the Americans were most successful with those elements of "technology applied to specific and single objectives" and less concerned with "the more general or abstract technology." As examples, he cited the cotton gin, a machine of limited applicability, and the dental techniques in which the Americans pioneered. Technology regarded in this light might then be labeled either "applied" or "basic," following a familiar division of science. Accepting this distinction, Robert S. Woodbury pointed out that the idea of a highly specialized

3. Robert E. Schofield, "The Industrial Orientation of Science in the Lunar Society of Birmingham," *Isis*, 48 (1957), 408–15; Charles C. Gillispie, "The Discovery of the Leblanc Process," *ibid.*, 152–67.

4. Oliver Evans, *Abortion of the Young Steam Engineer's Guide* (Philadelphia, 1805), 22; Thomas Tredgold, *The Steam Engine*, 2 vols. (London, 1838), I, 43.

machine tool was distinctly American but, on the other hand, that some of the basic machine tools were also American—notably the milling machine. Carl W. Condit suggested that the concept of basic technology might be related to the emerging scientific character of technology and that the Americans were generally indifferent to this sort of theory.[5]

Yet science inescapably looms in the background of all technology and offers the historian one of his best tools for evaluating past technologies. He must certainly be cognizant of the state of science in the period whose technology he studies.

Further, the historians of science have met problems in their study which must similarly be faced by those who study the history of technology. Specifically, the old dictum that one must be a competent scientist before he can approach the history of science has generally given way to an insistence that he possess both the historian's viewpoint and a good working knowledge of the science under study. The parallel answer for technology would ask not that the student be an engineer but rather that he be a historian with a good knowledge of the technology—and the related science—he studies.

Yet a different answer is implicit in another approach to the history of technology—that of the practicing engineer with a deep interest in the history of his profession. Some of the most useful work on early American technology has been done by just such men; for example, by Roe, Bathe, and Steinman.[6] Their insights depend, inextricably, upon their experiences, and our understanding will be poorer if it cannot continue to draw from this reservoir. At the same time, as other fields of history have advanced in professionalism—the history of medicine and military history, for example—they have depended less extensively upon the practitioners of the professions under study.

Of the older approaches to the history of technology, economic history has probably been the most productive. In the absence of any field of the history of technology until very recently, economic historians have sup-

5. All of these comments were made at the Conference on Early American Technology held at the Hagley Museum, October 15–16, 1965, cited hereafter as Early American Technology Conference.

6. Among their many works are: Joseph W. Roe, *English and American Tool Builders* (N.Y., 1916); Greville and Dorothy Bathe, *Oliver Evans: A Chronicle of Early American Engineering* (Phila., 1935); and David B. Steinman, *The Builders of the Bridge: the Story of John Roebling and his Son* (N.Y., 1945).

plied many of the available surveys and monographs. Thus, the surveys of manufacturing in the United States by Clark and by Bishop are presented in terms of economic history.[7] So is the fine access to early iron technology offered by Bining, to woolen manufacturing technology by Cole, and to textile and machine tool production by Gibb and by Navin.[8] Yet technology was never more than a part of their story—one of the factors of production.

From a similar background of economic history, a few recent writers do give primary attention to technology. This is conspicuously true of Hunter in his *Steamboats on the Western Rivers,* and we may anticipate that it will be even more so in his forthcoming book on power.[9] Of the studies which start from a consideration of economic growth, two place technology at the center of their inquiry: Strassman and Habakkuk.[10] Since it is almost never possible to separate technology entirely from the economic process, economic history and even economic theory offer rich insights which even those whose central concern is physical technology should seek.

One of the less obvious but often relevant approaches is that of the local historian. Even the most general local surveys tend to show concern for social history and for physical survivals, often preserving information that might otherwise have been lost as a result of the prevailing neglect by academic historians during the late nineteenth and early twentieth centuries. A good example of this sort of repository is Scharf and Westcott's history of Philadelphia.[11] Regions, states, and towns have been delineated in this manner, while other local historians have focused their efforts topically. In a different category are more specialized local studies which can also

7. Victor S. Clark, *History of Manufactures in the United States, 1607–1870* (Washington, 1916); J. Leander Bishop, *A History of American Manufactures from 1608 to 1860,* 3 vols. (Phila., 1861–68).

8. Arthur C. Bining, *Pennsylvania Iron Manufacture in the Eighteenth Century* (Harrisburg, 1938); Arthur H. Cole, *The American Wool Manufacture,* 2 vols. (Cambridge, Mass., 1926); George S. Gibb, *The Saco-Lowell Shops: Textile Machinery Building in New England, 1813–1949* (Cambridge, Mass., 1950); Thomas R. Navin, *The Whitin Machine Works since 1831* (Cambridge, Mass., 1950).

9. Louis C. Hunter, *Steamboats on the Western Rivers* (Cambridge, Mass., 1949).

10. W. Paul Strassman, *Risk and Technological Innovation: American Manufacturing Methods during the Nineteenth Century* (Ithaca, 1959); H. J. Habakkuk, *American and British Technology in the Nineteenth Century* (Cambridge, Eng., 1962).

11. John Thomas Scharf and Thompson Westcott, *History of Philadelphia, 1609–1884,* 3 vols. (Phila., 1884).

be described as economic studies or, in the case of Green and Shlakman, almost as technological studies.[12] There have been many local studies of technology, from the early industrial surveys of Freedley to the recent small specialized studies of Harry B. Weiss.[13]

Literary and philosophical approaches to early American technology have been much neglected, although there is a considerable body of contemporary and late nineteenth-century writings to be studied and evaluated. The recent work by Leo Marx is the best examination of the literary impact of technology, a product of the American Studies approach.[14] Critiques of philosophical character tend to be recent in emphasis, as is Boorstin's, and aesthetic in direction, as is Kouwenhoven's.[15]

Very different is the anthropological approach to technology. Regarding history in terms of identifiable cultures, it is responsive to technological elements and often emphasizes them. This is distinctly true of Anthony N. B. Garvan's Index of American Cultures, being developed at the University of Pennsylvania, which is based upon selected culture regions studied intensively within brief time spans. Here, known artifacts, literary as well as three-dimensional, are photographed and described on cards. They are then keyed into an elaborate index. This may offer both specific details and suggestions on method to the historian of technology.[16]

Somewhat related is the archaeological manner of studying the past, most familiar in terms of digging for the remains of ancient civilizations. On a local level, and with relevance to more recent society, archaeology has always been better supported in England than in the United States. The English activity in industrial archaeology builds upon a rich heritage. Several restorations have supported archaeological efforts in this country; their technological relevance is great in all cases but highest when industrial plants rather than manor houses are investigated. Good work of this

12. Constance McLaughlin Green, *History of Naugatuck, Connecticut* (New Haven, 1949); Vera Shlakman, *Economic History of a Factory Town: A Study of Chicopee, Massachusetts* (Northampton, 1935).

13. For example, Edwin T. Freedley, *Philadelphia and Its Manufactures* (Phila., 1859); Harry B. and Grace M. Weiss, *Forgotten Mills of Early New Jersey* (Trenton, 1960).

14. Leo Marx, *The Machine in the Garden* (N.Y., 1964).

15. Daniel J. Boorstin, *The Image, or What Happened to the American Dream* (N.Y., 1962); John A. Kouwenhoven, *Made in America: the Arts in Modern Civilization* (Garden City, 1948).

16. Garvan explains his project in "Historical Depth in Comparative Culture Study," *American Quarterly*, 14 (1962), 260–74.

sort is currently being done under the direction of Ivor Noël Hume at Colonial Williamsburg and by the National Park Service at several sites. Much more investigation, directed specifically to technological objectives, is in order.

The aesthetic evaluation of technology and its products dominates several categories of writings which are useful to the historian of technology. Especially valuable are the many highly competent studies of architecture and of the history of architecture. Because architecture combines art and engineering in a peculiarly intimate fashion, good architectural historians are always conscious of the technological elements in their story and they usually write about them. Primary attention to the technology of architecture—that is, the study of building—is more rare. Marcus Whiffen's books include building technology within architectural history.[17] Condit's fine work is a pioneering isolation of the technology involved.[18]

The aesthetic approach is strong also in another, more diffuse, group of writings, the work of antiquarians and collectors of art objects. These range from erudite, highly disciplined evaluations by connoisseurs and museum personnel to loosely arranged catalogs by enthusiasts with little knowledge of history or of the technology represented in the products they collect. In total, these works contain mountains of data about one craft after another. Among them can be found specific details, fine photographs, and an understanding of both products and manufacturing techniques. The best treatment has been given to items that have retained a high monetary value with collectors; for example: silverware, glassware, pottery, furniture, and coins. Gun collectors, proceeding from similar motivation, have compiled material of more obvious technological value because of the nature of the product. The whole of this work ought not to be scorned because some is of poor quality and because technology is secondary in all of it. Much can be read back from the product into the manner of making it.

Leading art museums have played a central role in stimulating the study of craft products and their manufacture—the Henry Francis du Pont Winterthur Museum and the Boston Museum of Fine Arts are especially active in this work. Useful museum publications are voluminous but, because of their frequently scattered and occasional character, hard to find and use.

17. For example, Marcus Whiffen, *The Eighteenth-Century Houses of Williamsburg: A Study of Architecture and Building in the Colonial Capital of Virginia* (Williamsburg, 1960).

18. Carl W. Condit, *American Building Art: The Nineteenth Century* (N.Y., 1960).

The best libraries often do not hold this sort of pamphlet material or do not catalog it in a manner calculated to aid the historian of technology. Yet particularly where museums have turned to laboratory analysis of their holdings, acquaintance with their activities is essential. Indeed, Cyril S. Smith has urged the necessity of acquiring the insights of the art historian.[19]

A special plea should be entered for those collectors who are motivated neither by the aesthetic nor by the monetary value of the product. Such are the members of the Early American Industries Association who collect examples of and data upon the tools and products of primarily pre-machine technologies. Otherwise known as the "Pick and Shovel Club," this group has met from convention to convention at Dearborn, Shelburne, Mystic, and the Hagley Museum. It prints in its *Chronicle* a great variety of scattered information, and some of its members have published useful books. The various folk museums and local museums preserve similar "homely" artifacts which reflect very directly the way life was lived—how things were made and done. Specific details about technology are more elusive and more easily lost than may be imagined. All such information needs to be cherished.

The efforts of these varied collectors highlight a central problem in all history of technology: what can be done and what ought to be done with the artifacts and physical remains from the technologies under study? The physical things of technology in many ways remain the ultimate source for the history of technology. Preserved products and tools as well as other traces left by these technologies—including railroad cuts and canal segments, razed factories and mine shafts—constitute repositories of information poorly understood by the general historian. Furthermore, there are many barriers in the way of their optimum use. A start can be made by recognizing the primary problems, but it is not likely that many can be solved except by a rather considerable joint endeavor.

To begin with, there is no simple difference or conflict between words and things—words do not represent ideas alone nor things mere material accomplishment.[20] With technology as with all other aspects of man's life, words serve the function of describing and translating into a transmissible medium. When combined with engineering drawings, mathematical and chemical formulae, photographs, and sound recordings, they can render

19. Early American Technology Conference.

20. Cf. John A. Kouwenhoven, "American Studies: Words or Things?" in Marshall W. Fishwick, ed., *American Studies in Transition* (Phila., 1964), 15–35.

the artifacts and machines of technology in forms that can be stored compactly and multiplied without limit. Words can do more; they can be used to evaluate and compare physical objects and their functions. They can even follow the projections of technology into the abstractions of poetry and aspiration.

Nevertheless, the "things" of technology retain a primacy that does not adhere to the physical objects associated with science, religion, politics, or any intellectual or social pursuit. The means of technology are physical; the objectives of technology are also physical or material. Three-dimensional physical objects are the expression of technology—in the same way that paintings and sculpture are the expression of the visual arts. They call for some of the same attention and celebration that is accorded to works of art.

Moreover, although it is technically possible to reduce a steam engine to a variety of paper-and-tape records, this has not been done. At best, the scholar has available to him only rough descriptions, specifications, sketches, and photographs. These may very well suffice for his ordinary needs but they do not provide all of the transmissible information that can be gathered. Conversely, it often happens that absolutely nothing is available save the specimen itself, preserved in an out-of-the-way museum which lacks even a full-time staff. The scholar must then get what he can from a visual inspection, with inadequate light and space.

What can be gained from the visual examination of a technological specimen depends directly upon the knowledge and perceptiveness of the examiner. If he is well acquainted with similar objects, he can compare and evaluate as he looks. The surface inspection of a plow, a gun, or a loom may yield insights which did not appear at all when the student examined a very good description or photograph. Fundamental characteristics as well as the aesthetic character and the quality of workmanship often appear very different after a visual inspection. The historian is not likely ever to agree that the best paper-and-tape rendition of a specimen is a satisfactory substitute—and today most specimens do not boast even a poor description.

However, there is much information inherent in any specimen that will not yield to a casual, visual inspection. The nature of this information varies with each category of artifact. For a simple metal casting, a spectrographic or chemical analysis may be useful. Such a chemical analysis was run on many of the iron stove plates in the Mercer Museum collection, permitting the examiner to assign their production to specific furnaces; this fundamental information could only be acquired by taking a plug from each specimen, but the operation had to be done only once.[21] Charles F. Hummel, and Theodore Z. Penn, a graduate student who worked in the Andelot-Copeland Museum Science Project of the Winterthur Museum and the University of Delaware, have described the application of optical spectroscopy to the dating and validation of brass artifacts through the identification of trace elements.[22]

With machines, data on working characteristics are the obvious need. Operating steam engines should be run at different speeds with different loads, indicator diagrams taken, and the results analyzed. Another useful approach is suggested in the sound motion pictures of old English engines taken by the Shell Film Unit under the stimulus of the Cornish Engines Preservation Society.[23] Most of the sounds of history are lost beyond recall, but those of some machinery can be recovered. The smells and other attributes of the past, especially of the perishable products of chemical technology, are still more elusive, although John J. Beer pointed out that they might be re-created to give added depth to restorations.[24]

Study and analysis could create a new species of source material whose value might become great if uniform tests and analyses were agreed upon for each category and applied to most of the remaining specimens. If at the same time uniform patterns for describing and picturing these specimens could be applied, historians would have at their command comparable, two-dimensional renditions of the three-dimensional objects of early American technology.

The big problem is the magnitude of the task; nothing similar has yet been done for the collections of a single museum.[25] Clearly, the maximum effort in recording the characteristics of a specimen is not likely to be feasible except where destruction of the original is anticipated or duplication in a three-dimensional model is planned. When the Deutsches Museum made a copy of the Science Museum's Watt and Boulton "Lap" steam

21. B. F. Fackenthal, Jr., "Classification and Analysis of Stove Plates," Bucks County Historical Society, *Proceedings*, 4 (1917), 55–61.

22. Early American Technology Conference.

23. H. W. Dickinson, *The Cornish Engine: A Chapter in the History of Steam Power* (London, 1950).

24. Early American Technology Conference.

25. One of the best series of published technological catalogs is that of the Conservatoire National des Arts et Métiers in Paris.

engine, seventy sheets of engineering drawing were required. Similarly, where it is impossible to preserve all of the factories and plants noted in the British Industrial Monuments Survey, a large program has been projected to record the outstanding characteristics of those that will be destroyed; in some cases this work is already in progress.[26] The Historic American Buildings Survey, although not conceived primarily to preserve technological edifices, accumulated large files of drawings and photographs of many sorts of buildings of importance to historians of technology.

The historian does not require at his fingertips the maximum recordable information about each specimen; indeed, under most circumstances, this would be a burdensome supply of riches. By extended practice, numismatists and gun collectors have arrived at abbreviated descriptions and similar photographs and drawings which satisfy *their* needs in comparing, cataloging, and evaluating these technological products. Historians of technology could probably agree upon the minimum data *they* would need in studying each category of specimen. In arriving at standards, they should not disdain the example of the drawings offered in the ancient publications of Stuart, Strickland, Rees, and Nicholson.[27]

Even if such a program could be agreed upon, its fulfillment would require major financial support and continuing cooperation from individuals and institutions.[28] The need for it is manifestly greatest in describing and recording the collections of individuals and of marginal museums— the most difficult collections to handle. The major technological museums are not only friendly to scholarship but almost eager to serve it. They can be counted upon to support any efforts that clearly represent the desires

26. This survey, under the guidance of Rex Wailes of the Ministry of Public Buildings and Works, is an aspect of a large movement coordinated in part by the Council for British Archaeology. Individuals and schools of architecture are cooperating in making drawings of designated buildings. New industrial museums are arising in different parts of the country and old ones are finding new life. The publication of a series of regional surveys has begun with Kenneth Hudson, *Industrial Archaeology of Southern England* (Devon, 1965).

27. Charles B. Stuart, *The Naval and Mail Steamers of the United States* (N.Y., 1853); William Strickland, E. H. Gill, and H. R. Campbell, eds., *Public Works of the United States of America* (London, 1841); John Nicholson, *The Operative Mechanic, and British Machinist* (1st Amer. edn., Phila., 1826); Abraham Rees, *Cyclopedia,* 39 vols. and 6 vols. plates (Phila., 1810–24).

28. Nathan Reingold regarded this objective as attainable although in a shape not yet discernible, but Robert S. Woodbury doubted whether people competent to carry it through were available. Early American Technology Conference.

of the scholarly community. Whether or not means are developed for handling three-dimensional technological specimens with some of the ease with which manuscripts can now be handled, scholars must place more of their reliance upon the collections available to them in museums. Important assistance is now offered them in the accompanying Directory of Artifact Collections prepared by Lucius F. Ellsworth.

In approaching artifact collections, the academic scholar should appreciate some of the difficulties faced by museums holding technological specimens—and, in some cases, related manuscripts and publications as well. Like all museums, they must serve two functions. First, they display their specimens in attractive settings, interpreting them to viewers in broad historical perspective—often with learning and originality. In this function they serve by far the larger number of their visitors, and to this function their support is often primarily directed. Second, they make available to scholars the specimens they hold and the research information their staffs have compiled about them.[29]

These two functions are not wholly separate but they cannot be combined as readily as is sometimes imagined. Consider, for example, the presentation of dioramas, small working models, or full-scale restorations incorporating parts of the original. Even in a restoration, only a small part of which is conjectural—and more obviously in all reconstructions—a portion of the result is guesswork. It may be well-researched, imaginative guesswork of the sort which every historian values, but it must be called guesswork, nevertheless.

The imaginative insights familiar to the historian are something else again. In written history, the author may admit his lack of knowledge where it is inadequate, or he may omit all mention of episodes upon which information is defective. The builder of a model or a restoration does not have that option. If he has an original water wheel, he cannot erect it

29. There is remarkably little literature on the research function of a museum in relation to its display function. Three schools of thought are set forth by Wilcomb E. Washburn, "Scholarship and the Museum," Clark C. Evernham, "Science Education: A Museum Responsibility," and Katherine Coffey, "Operation of the Individual Museum," *Museum News,* 40 (1961), 16–29. For a brief appraisal in the field of architectural preservation, see Alan Gowans, "Preservation," *Journal of the Society of Architectural Historians,* 24 (1965), 252–53. Also see Ralph R. Miller, "Museum Installations," *The Museologist,* No. 95 (1965), 6–8, and Kenneth Hudson, "The Taming of Industrial Archaeology," *Museums Journal,* 65 (1965), 35–39.

without supports because those are lacking; he must study similar wheels and supply his best guess about the missing elements. Nor can he paint the guesswork orange to indicate its conjectural character; that would destroy the illusion. Besides, he may have a drawing of the supporting structure, and his only uncertainty may concern the kind of wood used. Then the very limited nature of his guesswork would be harder still to indicate.

It remains very difficult for a scholar to know how much of a restoration is based upon original parts or reliable information. He may be able to discover how the figureheads differed in successive restorations of the *U.S.S. Constitution,* but how can he be sure that the present exterior finish is that of the fighting frigate? In the magnificent restoration at Saugus he is informed that the location of the water wheels serving the forge is based on archaeological evidence. But, without access to manuscript records, he has no clue to the sources for the arrangement and interior appearance of fires, hammers, and furnishings. Even if he recalls the almost exact similarity of the *Encyclopédie* plates, he is left wondering how much of the interior might be painted orange and how deep should be the color of the dye. Must not every conceivable restoration be surrounded with some such questions and doubts?[30]

Is there not then a need to bring the scholar closer to the sources than the visible restoration? Behind several restorations and museums there are very fine unpublished historical and engineering reports—upon which exhibits and restorations or reconstructions are based. At their best, they show precisely what is known and what is not; they discuss sources and similar specimens; they explore the historical setting.

Understandably, even the most generous museums and restorations are somewhat reluctant to open wide their research reports. They have not been written for publication or to stand the scrutiny of historians. Much that they contain is tentative. Errors may have been detected in them, but

---

30. The interpretive function of museums and restorations has not attracted the attention it deserves. Perhaps the best statement, though restricted to anthropology, is H. H. Frese, *Anthropology and the Public: The Role of Museums* (Leiden, 1960). Osbert Lancaster discusses criteria for preservation of historic architectural artifacts in "Some Thoughts on Preservation," which will appear in *Historic Preservation Today* (1966), published jointly by the National Trust for Historic Preservation and Colonial Williamsburg. The appendix to that book also contains "A Report on Principles and Guidelines for Historic Preservation in the United States."

time may not have been available to correct them. The sponsoring institutions often encourage their authors to continue their researches in the line of the reports and to publish their results in finished form. But the published works often become broader and more significant—at the sacrifice of some of the specific information contained in the original reports. From such beginnings have come many valuable publications—especially by staff members of the Museum of History and Technology, of the Eleutherian Mills-Hagley Foundation, and of Colonial Williamsburg. However, all museum publications remain difficult to control bibliographically; one of the great needs is the compilation of simple finding lists.

The need for the scholar to get behind the exhibit and the restoration can already be satisfied at some of the leading institutions. The unpublished research reports of the Hagley Museum and those of Colonial Williamsburg, both upon buildings and upon crafts, are available to competent researchers. Yet helpful as they are, reports can be no substitute for the more basic sources—contemporary writings and artifacts. As Eugene S. Ferguson put it, "The historian should consider looking at artifacts to be so much a part of his trade that he will, over the years, develop a keen critical sense regarding the authenticity and significance of the artifacts and restorations that he sees. To be used effectively, the artifacts must mean something to the historian directly, not once removed through the mind and eyes of a curator."[31]

The scholarly handling of the artifacts of technology involves another question of the first magnitude: how should the historian approach those products of the crafts which have become the province of the art museum and the collector? The art museum and the technological museum have drawn lines between their areas of concern which have the tacit approval of many historians of technology but which ought now to be re-examined. Folk and craft products, which to the outsider seem to have limited aesthetic value, have been drawn into the realm of the art museum along with the products of men who are more obviously "artists" rather than artisans. The interest of the art museum in machine products and industrial design is more limited—especially in the early American period. The art museum and the art historian may give careful attention to craft products, but the direction of their interest is aesthetic and, to some extent, social. On the other hand, the technological museum and, often, the his-

31. Early American Technology Conference.

torian of technology tend to place their primary attention upon machine production, recognizing the hand-production phase of their studies but relegating it to a kind of prehistory or embryonic stage of development— to some extent relinquishing hand production to the art museum and the art historian. This division is well marked in the respective domains of the Eleutherian Mills-Hagley Foundation and the Winterthur Museum in Wilmington, Delaware. It is a division followed by the Museum of History and Technology in Washington, the Franklin Institute in Philadelphia, and the Science Museum in London. It is altogether desirable from the viewpoint of specialization and good direction.

For the historian who pursues craft technology into the art museum, this division has the additional advantage of opening the extensive and fundamental relationships between technology and aesthetics. It also offers insights into craft technology developed over a long period of time by men who pursued the truth about the objects of their work—in whatever category of study it might fall. On the other hand, the historian is hampered by the disregard of craft technology on the part of the technological museum and the historian of technology. The satisfactory understanding of early American technology demands that the whole technology of the period be studied by the historian of technology. This may seem to call for aggression or imperialism on the part of the historian of technology, but it is essential for the integrity of his subject. The whole continuum of early American history must be brought within the scope of his study and subjected to similar research and reflection.

This need becomes doubly apparent as soon as the chronology of early American technology is examined—even in the roughest terms. The terminal date of 1850 for this book is one of convenience. It is supported however by the convergence of many lines of technological development which seem to attain measures of fulfillment in the 1840's and by the industrial exhibitions of the early 1850's which marked American achievement in a spectacular fashion.

The one chronological divide that cuts across the early period of American history is the separation of hand production from the introduction of the new machine manufacture. It is inescapable. In Mumford's periodization, the technology with which the colonists built the first foundations was paleotechnic, while the new machine technology was neotechnic.[32]

32. Lewis Mumford, *Technics and Civilization* (N.Y., 1934), 151–267.

The division is the same with those who have used less esoteric terms, whether they write of the introduction of steam power, the introduction of the technology of the Industrial Revolution, or the use of machine or mass production.

This division is inescapable but it is not sharp, and the difficulty is to state precisely what separated the earlier technology from the later. The earliest gristmills and iron plantations represented powered, machine operation—while at the other end of the period, individual handcraft continued to be a part of much production, in the shop and on the farm. Steam power was used as early as 1750, but as late as 1850 water power still dominated much industrial production and was still advancing in technological efficiency. Yet certainly between those two dates the great technological divide was crossed, not only in America but in Britain and western Europe as well. This fact dominates the logical desire of some of our best current scholars to dissociate their efforts from the initial, pre-machine technology—sometimes with vehemence, sometimes with indifference.

The technological aspect of the Industrial Revolution has been subjected to much criticism. On the point of a great technological divide, Daumas has argued that there was no technological revolution at all. John U. Nef suggests that there was a revolution but that the significant elements of change should be placed in an earlier period, in the sixteenth and seventeenth centuries. This would not only move the technological divide to closer coincidence with the scientific revolution but would place it prior to and during the early phases of American settlement.[33]

If these questions remain in doubt, then surely the historian of early American technology must study the whole of his terrain, *even if* his only interest is in the later, machine technology. The depth and chronology of the European divide does not, in any case, determine the depth and chronology of the American. This requires specific study of specifically American experience. The more the history of technology is studied, the more its significance in all periods of history emerges. Hopefully, more and more scholars will regard colonial technology as a valid study, whether "paleotechnic" or "neotechnic." The way the colonists made and did things is

33. Maurice Daumas, "Le Mythe de la Revolution Technique," *Ithaca: Proceedings of the Tenth International Congress of the History of Science* (Paris, 1964), I, 415–18; John U. Nef, *The Conquest of the Material World: Essays on the Coming of Industrialism* (Chicago, 1964).

a fundamental part of their history—heightened in interest by the speed with which they transferred European techniques and the variety of adaptations and innovations they introduced.

The tentative identification of colonial technology as a hand-production, craft technology highlights one of the staggering coincidences of American history. The birth of the Republic coincided almost precisely with the first efforts to introduce the power and textile machines of the Industrial Revolution. Whatever terms we use to describe it, the language and character of American technology altered abruptly between 1776 and 1850; at points it even advanced to world leadership. The organ tones with which the American mechanic in the 1830's and 1840's proclaimed his success in building better locomotives, longer rail lines, more steamboats, and more automatic machinery were echoed by the slightly shocked engineers who came to the United States from Great Britain and France to view the achievements of this period.

This is an epic which deserves to be studied and recognized—not alone as a story within the history of technology but as a central thread in American history. In the perspective of the general historian, does not this define the era with far more justice than the westward-moving wagon trains, the Age of Jackson, or the battles over the tariff? Indeed, these episodes, which still constitute the warp and woof of traditional history, are understandable only in terms of the technology which rebuilt the floor under the pontificating senators even as they declaimed and which shaped and reshaped the tools required to conquer a continent and to erect a variant civilization.

The grandeur of this story is perhaps the most telling force behind the overemphasis upon republican technology. It introduces the historian of technology to the too frequent practice of the general historian in lopping off the colonial period from serious consideration and beginning his account of American history with the Revolution.[34] It is a practice not to be encouraged in either sphere. In its own right, and as the root of later growth, colonial technology demands study by those qualified to evaluate it.

Even the most elementary questions about the chronology of American technology immediately introduce the necessity of understanding the European chronology. In fact, every probe into American technology re-

---

34. Carl Bridenbaugh, "The Neglected First Half of American History," *American Historical Review*, 52 (1948), 506–17.

quires attention to comparable, contemporary European patterns. This not only provides a necessary standard of evaluation; it acknowledges a process which remained fundamental throughout the period—the transfer of technology from Europe to America.

The transit of civilization from Europe to America is one of the established elements of American history. Although it has probably been more acknowledged than studied in all of its aspects, it is a part of the general historian's outlook. The transfer of laws, governmental institutions, architecture, science, religion, and attitudes is so familiar that we think of much narrative history in these terms. Curiously, the transfer of technology has not had equivalent recognition. At the scholarly level it is now receiving some study, but at the popular level there is a conflicting image of American whittling boys inventing their way—through cotton gins, steamboats, mass production, telegraphs, and vulcanized rubber—from the colonial world of hand production to the new world of machine production. This picture is not without truth, but it is not a satisfactory vehicle for crossing the great technological divide or, indeed, for taking any extended trip. The transfer of technology is a more fundamental factor within the context of technological history. In the broader sweep of general American history, it is an essential element which must take its place beside the other aspects of civilization which were transferred. Study may, in fact, reveal that nothing imported from Europe was so important to the success of the colonizing endeavors or to the growth of the new nation as its techniques for making and doing things.

As with other elements in the transit of civilization, the transfer of technology was a continuing process which early provoked a counter-current carrying American innovations and alterations in the opposite direction. The westward current flowed increasingly as American needs grew, but before 1850 it had ebbed considerably in those lines of conspicuous American attainment. The eastward current was continuing to rise at the end of the period.

The entire process requires detailed investigation, but a preliminary view of the major current westward suggests that it was primarily a matter of the immigration of artisans, mechanics, and engineers with the required skills and techniques. This was the ancient means of acquiring new technologies, as Elizabeth had done in importing Flemish weavers. In early America, it was supplemented by manuals and how-to-do-it books, which sold very well, and ultimately by textbooks and treatises. It was further enriched by visits of Americans to Europe for specific investigation and

for training. Time after time, however, those who visited Europe to discover new techniques decided that the safest approach was to bring back
Europeans with the skill to do what they wanted done. This immigration
was overwhelmingly British, seasoned regionally in the colonial period by
Germans and Dutch, in the Revolutionary era by French, and at the end
of the period by Germans.

A generalization frequently made by Americans was that British technology was practical and successful while French technology was elegant
but theoretical. American experience usually sustained the implications of
this view, although their reliance upon the French was strong in military
engineering, in some related civil engineering, and in engineering education. This was true in large part because the United States fought its first
two wars against Great Britain, thus barring British military engineering
during the periods of greatest need. In drug and fine chemical production, and in certain other lines where the scientific element was significant,
American reliance upon Continental Europe was strong in the early nineteenth century. When American academic men, such as Jacob Bigelow,
Thomas Cooper, and James Cutbush, sought to apply science to technology, they reached more and more for French assistance. The evidence
suggests, however, that in the textile mill and the machine shop the British
remained the dominant European influence.

A contrary view is suggested in studies which apply the methods of
intellectual history to the problem of technological transfer, but the methods are inapplicable and the conclusions false. Timoshenko's fine history
of the strength of materials can be used only as a genealogical survey of
the key writings around which present theory evolved.[35] It has no relevance to eighteenth- and nineteenth-century American practice in this
area and very little relevance to British practice. The foundation theory
was largely Continental and heavily French, but it was not influential in
America until much later. More specifically misdirected is Shaw's demonstration that modern libraries hold proportionately more French than
British engineering works published before 1830.[36] This statistical measure, of course, means little unless we know when the books were acquired,
by whom, and—more pointedly—who used them. In all probability, the
French books were acquired at a later date and ultimately used effectively
by those who sought to find better theoretical and scientific foundations

35. Stephen P. Timoshenko, *History of Strength of Materials* (N.Y., 1953).
36. Ralph R. Shaw, *Engineering Books Available in American prior to 1830* (N.Y., 1933).

for technology. There is little indication that they were as important to practicing American mechanics and engineers at the time of publication as their numbers in the libraries of the present suggest.

Such studies do not challenge the view of Britain as America's primary European reliance in building and extending her technology, but they do suggest that some methods of study must be regarded dubiously. The "intellectual thread" approach may not serve the history of technology as effectively as it does other fields. Statistical measures, of course, must always be controlled carefully, but perhaps they require special attention in intellectual approaches to technology. Shaw's book-counting method seems a classic example of being misled by "non-significant survivals."[37]

The same hazards must be guarded against in using some of the largest concentrations of writing about technology which seem to offer so much promise; for example, patents and patent-related collections, government reports, engineering reports, manuals, and college textbooks. The question must continuously be asked whether the surviving written record tells us the most important things about the technology we study, a subject which was itself not fundamentally an intellectual or verbal construction. Must not the focus be upon the things built and the men who built them? This would include all of the direct evidence on and about the functioning technology and all of the evidence—much of it informal, scattered, and difficult to assess—about the men involved.

Some of the questions to be asked are suggested by the key problems about early American technology which are now attracting competent attention. Perhaps the most dramatic of these is the development of the "American system" of manufacturing by interchangeable parts. This has long been celebrated as an American achievement, but the written record, so far uncovered, is as limited in specifics on this development as it is at many of the other critical points in the history of technology. The focal question in most of the writing on this subject has been whether the successful accomplishment sprang, more or less full grown, from the brow of Eli Whitney. It was answered in the affirmative by most of the early writers but in the negative by Woodbury and Sawyer, who then move on to larger considerations.[38]

37. Daniel J. Boorstin, Anglo-American Historical Conference, London, July 10, 1965.

38. Robert S. Woodbury, "The Legend of Eli Whitney," *Technology and Culture*, 1 (1960), 235–51; John E. Sawyer, "The Social Basis of the American System of Manufacturing," *Journal of Economic History*, 14 (1954), 361–79.

This single question carries with it a heavy freight. It implies the assumption that the great question about technology is: at what points and by what process did it move from one level of development to a higher and more complex level? It also carries the generic answer that advance has come by individual, inventive ingenuity. Yet neither of these propositions is self-evident. Because they are so generally accepted, it is essential that efforts be made to discover precisely what happened in the critical episodes. The specific, written information will always be inadequate, so related questions which can be answered must be pursued. It may turn out that they, after all, are the more important.

In the case of interchangeable parts, one must know just what Bentham, Brunel, and Honoré Blanc did. What was their machinery? (This question is easily answered for the Brunel-Bentham efforts in making blocks for the Royal Navy.) In what places and how widely were Blanc's efforts to manufacture muskets and Bentham's 1793 patent known? What else of this sort was in the air? How did these and other contemporary efforts compare with Whitney's? Were his solutions, especially his use of jigs, different? How did the chronology of Whitney's accomplishments compare with the chronology of Simeon North's and the government armories' applications of the idea of interchangeable parts?

Merely to ask such questions raises some doubt about the comparative significance of a great, naked idea—such as interchangeable parts—and its technological fulfillment. The fulfillment is never instantaneous; in this case it was clearly a story of long growth, development, and extension. Inventive ingenuity was a continuing factor along the route. Historians must certainly use their maximum insight to discover all that can be known about this process.

However, they may have to take answers where they are available rather than where they would like to find them. John Fitch left a full and introspective record of his steamboat-building endeavors which appears to answer the questions of greatest moment. Yet the historian no sooner encounters Fitch's unqualified assertion that the concept of the steam engine came fresh to his mind, with absolutely no knowledge of previous engines, than he realizes that it does not matter much whether this was true or not. For the history of technology, the important thing is the account of the process by which Fitch's ideas of engines were gradually expanded as he built one after another, applying them through different linkages and different propulsive schemes to setting a boat in motion. Some of this he did mentally and some in three-dimensional form—always changing, adapting, and rebuilding as one element or another proved unsatisfactory. Throughout his manuscripts he reveals—almost as an allegory—a conflict between himself, as a man of naked ideas, and his partner, Henry Voight, as a brilliant and intuitive mechanic who translated dreams into working reality.

More often invention is chronicled only in the most formal records. The heroic, stroke-of-genius view of technology is given significant support by the greatest body of available sources on technological specifics— the patent records. The patent system is both a product and a cause of the heroic view of invention. For this and for a variety of other reasons technology continues to be viewed too often as a story of invention. The words "technology" and "invention" appear in some histories virtually as synonyms.

The difficulties in defining invention are monumental, but the concept is valid not only as a part of the technological process but as an element of all creative work: aesthetic, literary, and scientific. Even the patentability requirements of originality, novelty, and utility are usable restrictions, within limits. At the same time, mechanical invention and technological innovation have emerged from recent study as complicated processes, exceedingly difficult to tie to one man or moment of time.

To begin with, technology, like science, has an internal life with an integrity of its own which determines the fruitful routes of its own change and adaptation. At any given moment some "inventions" are possible and some are simply impossible until more elements have been added to the complex. The routes of change actually followed depend upon environmental circumstances, including political, social, and economic factors, and upon the impress of individual personality. Yet, at each point of development, certain changes and adaptations grow logically out of the state of the technology. This has been well demonstrated in Robert K. Merton's study of the prominence of "multiples" in scientific discovery and invention, but it is an old story to the patent attorney.[39] Simultaneous invention and multiple invention or reinvention are continuing parts of history.

Moreover, it may be useful to assume that each technological complex

39. Robert K. Merton, "Singletons and Multiples in Scientific Discovery: A Chapter in the Sociology of Science," American Philosophical Society, *Proceedings,* 105 (1961), 470–86.

operates, on a small scale, within bounds which Thomas S. Kuhn, writing of science, has called a set of paradigms.[40] The basic ingredients for a new set are always present some time before they are put into working order. Thus we might project the steamboat story: all the elements of a technically successful steamboat were at hand before one was built. They were even put together in a steamboat before technical success was achieved and certainly before the revolution in water transportation was accomplished. The same thing can be said of machine manufacture on the basis of interchangeable parts, or the cotton gin, or the reaper, or the electric telegraph. If this is so, we may have to limit the identification of inventors to very specific items (such as the Whitney gin or the Danforth throstle), and abandon efforts to tag the inventor of the steamboat or of interchangeable parts. Yet such eponyms are more a part of the history of technology—especially of early American technology—than of most other areas of human creativity.

Is it not necessary for the historian to come to terms with the internal life of technology before he seeks to assign title as inventor or to confer other accolades? This requires an understanding of the details of the technology under study. It requires an end to prejudice against the "hardware historian." It requires an effort to see through the eyes and feel through the hands of the craftsman and mechanic. The social relations of technology must not be neglected; they represent the historian's highest goal, but they are attainable only after the historian has developed a direct understanding of the men and their works.

One of the great questions that might yield to the central study of technology is the extent to which its internal character is deterministic. It is clear that at each stage of development, new possibilities are opened. When, before it was possible to travel in steam vehicles, Oliver Evans spoke of the exhilaration to be anticipated, he was not thinking of monetary profit at all. He was participating in the fulfillment of an unfolding technology. Norris built more powerful locomotives not merely to make money but perhaps more basically from a great human urge to do everything that developing means permit man to do. Were they so inner-directed that they failed to exercise needful restraint in moving as fast as they did toward dependence upon machines?

Perhaps technology was not so much a tool or a means as it was an

---

40. Thomas S. Kuhn, *The Structure of Scientific Revolutions* (Chicago, 1962).

experience—a satisfying emotional experience. When it became possible to make thread and cloth by machine, they were so made; when boats, trains, and mills could be driven by steam, they were so driven. These things could be accomplished only to the extent that economic needs and social attitudes permitted; but is it not possible that the more elemental force was within the technology itself? It is an idea often denied.

The men who constructed the elements of early American technology were themselves a part of it. Their words reflect an emotional satisfaction in reaching forward with each improvement. It is doubtful that the central role of technology in American history will ever be apprehended until this drive is appreciated. Its recognition alone will constitute a response to recent critiques emphasizing the evil results of technology—a response in the form of an explanation, not necessarily a denial.

The greatest need is to stand at the center of the technology—on the inside looking out. Instead, we have usually looked at technology from the outside: through the eyes of science, economics, political reflections, social results, or literary antagonisms. There is no telling what related factors and forces will appear once the historian develops his insights from the center; it has hardly been tried. As Daniel J. Boorstin put it, the history of technology is indeed one of the "dark continents" of historiography.[41]

Once the external view of technology is modified and the intensity of concentration on invention is relaxed, a number of pervasive themes within technology come to mind, some of them not even progressive in nature. One of these is the spread and conflict of different cultural traditions within the same technology. For example, Baron Hermelin wrote that iron production at the end of the colonial period was dominated in some regions by German methods, in some by Flemish, and in others by English.[42] If this was so, a fine comparative study in differences and effects is in order. Similarly, the influx of foreign artisans in glass production continued throughout the entire period. Some of the effects appear in the catalogs of glassware collectors, but no comparative study of the technology has been attempted. Again, Ewbank suggested that Philadelphia fire engines followed French and German models while New York

41. Early American Technology Conference.

42. Samuel Gustaf Hermelin, *Report about the Mines in the United States of America, 1783* (Phila., 1931), 63–65.

adhered to British patterns. Why should this have been so, and how did it influence development? [43]

Another comparative pattern is offered in the divergence between the southern shovel plow and the northern wooden plow; why the difference? A different sort of comparison is the lengthy competition between the Rhode Island and the Waltham systems of cotton textile production. This has been more noted by historians, but the voluminous records have never been adequately studied.

Major questions in almost endless number can be identified within the context of the interrelationships of technology and American society. Did educational patterns promote the acceptance of technological improvements as foreigners believed; and, in turn, did the new technology push American education along already determined routes? Was the traditional secrecy of the mechanic's "little black book" minimized by the openness of American society and the unusual mobility enjoyed by craftsmen and mechanics? [44] How in each episode did the several governments influence and how were they influenced by technology? Especially intriguing is a question suggested by Eugene S. Ferguson: did the Americans speed their technological development by a "doctrine of imperfectability?" Were they more ready than the Europeans to recognize that imperfection in machine parts was not merely acceptable but inevitable, and, if so, what in their background encouraged this difference? [45]

Among other pervasive themes that hold keys to the "Americanness" of American technology is the wooden age of our technological history. Here the Americans diverged sharply from European example in the use of material and, unexpectedly, in other directions as well. Architectural historians, agricultural historians, maritime historians, and historians of other aspects of technology have described portions of this story, but its dimensions have not yet been measured. Wood was used not only for houses and monumental buildings but for canal locks, for canals themselves, for road surfaces (the famous plank roads), for steamboats, for steam engine framing, for railroad ties (in place of stone blocks), for bridges, and even for compasses and mathematical instruments.

This reliance upon wood subsided from its peak before mid-century,

43. Thomas Ewbank, *A Descriptive and Historical Account of Hydraulic and Other Machines* (14th edn., N.Y., 1858), 343.

44. Lawrence W. Towner, Early American Technology Conference.

45. Early American Technology Conference.

but it had many results and ramifications and it continued to differentiate America from western Europe. It brought the Americans acknowledged leadership in wood-working tools, marked especially by the use of the muley saw, the circular saw, and the elegant American axe.[46] By promoting rapid obsolescence and replacement, it promoted the adoption of improved techniques or of techniques that were merely different. It may have facilitated the spread of machine production; handmade brass clockworks were followed by machine-made wooden works and then, in turn, by machine-made brass works. It offered a model which the Russians, conspicuously, found more to their taste than European models when they sought railroad engineers with solutions to problems of large distances and few resources other than wood. It played a large part in conquering the American continent (and theirs) with limited labor and capital.

The wooden age was closely related to another pervasive theme: labor-saving machinery. This concern may be stated most positively in terms of attitude, because the early nineteenth-century Americans wrote often of the merits and desirability of labor-saving machines. European observers acknowledged that the American view of this subject differed from theirs. Habakkuk, on the other hand, has recently found reason to question whether American labor was scarce in the sense believed and, if it was, whether labor-saving devices were the best answer.[47] The Americans did not have such doubts, yet they sometimes recognized that their capital was almost equally scarce. If solutions were what contemporaries believed them to be (reliance on machinery, cheap materials, standardization, poor finish, and lower levels of skill in labor), these characteristics are very significant in technological history.

When European commentators sought the pervasive characteristics of American technology, they often noted a "character of magnitude" plus a battery of differentiating American attitudes. All reported the fixation upon labor-saving machinery, resulting, as one account put it, in "the universal application of machinery with a rapidity that is altogether unprecedented."[48] The lack of fine finish and highly skilled work was related to the American's pride "in not remaining over long at any particular

46. For a stimulating suggestion for comparative studies of tools, see the comment by Whitfield J. Bell in Walter Muir Whitehill, Wendell D. Garrett, and Jane N. Garrett, *The Arts in Early American History: Needs and Opportunities for Study* (Chapel Hill, 1965), 19.

47. Habakkuk, *American and British Technology*, 8, 21–25.

48. Publisher's advertisement in Strickland *et al., Public Works*, n.p.

occupation, and being able to turn his hand to some dozen different pursuits in the course of his life."[49] Such restlessness, another commentator believed, inverted European attitudes toward changing techniques: "In England it is said, a custom so old 'must be right;' in America, that a custom so old 'must be wrong,' and needs revolution or change."[50] A French observer added that in America "one learns to appreciate the value of time: they employ it, while we use it." Even more perceptively, he noticed the manner in which American attitudes toward technology were woven into the fabric of their most cherished ideals; "Steam, with the Americans, is an eminently national element, adapted to their character, their manners, their habits, and their necessities. With them it is applied as much to extend their liberty as to augment their physical welfare. . . . The American seems to consider the words democracy, liberalism, and railroads as synonymous terms."[51]

Study of the chronological history, the internal nature, and the pervasive characteristics of early American technology are fundamental needs at this stage. All require a continuing sensitivity to the social relations of technology. No historian has to be reminded that the social, economic, and political relationships cannot be added as a gloss on top of the basic technological history; they are an integral part of the story. The big questions about social history must be kept in mind even to get the most out of an investigation of the shape of a screw.

The immediate difficulty is that the chronology and general character of early American technology can only be established through the detailed study of one specific technological episode after another. The great need is to get inside the men, their tools, and their works. Biographical studies of men will be useful, but despite the central importance of the human element, they are probably less valuable than "biographies" of crafts, of the various mechanical and engineering developments, and of the institutions within which technology was developed and assimilated. Unfortunately, this detailed study cannot be successfully pursued until the general questions have first been answered! This is a familiar dilemma. Tentative

49. Joseph Whitworth and George Wallis, *The Industry of the United States* (London, 1854), 2.

50. John Richards, *A Treatise on the Construction and Operation of Wood-Working Machines* (London, 1872), 125.

51. Guillaume Tell Poussin, *Chemins de Fer Américains* (Paris, 1836), xvi; *The United States: Its Power and Progress* (London, 1851), 345, 371.

generalizations must be accepted in order to permit the scholar pursuing detailed research to ask the proper questions. Then, as more satisfying details become available, the generalizations can be improved.

The importance of the questions to be asked cannot be overemphasized because the historian, like the inventor, will find little significant information except that for which he has searched. He must be prepared to ask the large questions of small episodes. He must also be aware of the questions asked by society—especially the rising number of critiques of technology in our own day. The opportunities are enormous and the principal qualification is an open mind stocked with good questions. Since we are not yet entirely sure what are the "right" questions or the "right" approaches, it is doubly necessary that no doors be closed and that the insights of all who have looked in this direction be sought. Most important of all is the enthusiasm of the early American craftsman, mechanic, and engineer. Unless the historian can catch some of that spirit he will render a better service by studying in some other field.

The ultimate objective is to raise technology to its proper place within the context of early American history. It belongs very close to the center as an expression and a fulfillment of the American experience.

SUSAN E. KLEPP

Lost, Hidden, Obstructed, and Repressed

Contraceptive and Abortive Technology

in the Early Delaware Valley

In 1936, Norman E. Himes published his now-classic work, *Medical History of Contraception*. Himes sought to demonstrate that all human societies have attempted to "control fertility by artificial means." All, that is, except western Europe. There he found that only folkloric methods of doubtful efficacy prevailed before 1800 and that even these few techniques were not widely diffused in the population. Himes concluded that most women for most of European history had little or no knowledge of contraceptive technology. Social and economic historians have generally followed Himes in asserting that techniques of fertility control were unavailable to Europeans or Americans before the nineteenth century.[1]

For advice and valuable suggestions, many thanks to Ava Baron, Cornelia Hughes Dayton, Judy A. McGaw, Robert Post, Charles E. Rosenberg, Billy G. Smith, and Stephanie Grauman Wolf.

1. Norman E. Himes, *Medical History of Contraception* (1936; New York, 1970), ix, 209–238. Two books have begun revising Himes: Angus McLaren, *A History of Contraception: From Antiquity to the Present Day* (Oxford, 1990); and John M. Riddle, *Contraception and Abortion from the Ancient World to the Renaissance* (Cambridge, Mass., 1992). McLaren is less concerned with the efficacy of traditional practice, and Riddle assumes that much of traditional contraceptive technology was lost by the 18th century.

It is true that there are few references to the use of barrier methods of contraception in western European societies before the nineteenth century. In addition, references to coitus interruptus (withdrawal) as a method of preventing conception often occur in the context of illicit sexual encounters and have led informed scholars to the conclusion that contraceptive practices were unthinkable for the majority of married couples. There is little evidence that sexual abstinence was ever widely practiced by married couples or that there was any comprehension of an infertile period during the menstrual cycle. Demographers have only recognized the evidence of prolonged lactation by married women as an effective and commonplace method of postponing births in Europe and North American colonies before the nineteenth century. Even in this regard, some historians have argued that variation in the duration of breastfeeding was a matter of local custom and not used by individual women to restrict fertility. Most American historians have argued, from "the few public accounts of abortion, infanticide, and contraceptive practice," that "control of birth was not significant in colonial America."[2]

Not every study agrees with this negative assessment. Demographic historians of the eighteenth-century Middle Colonies, unlike the majority of social and economic historians, have assumed that contraception best explains the marital fertility patterns of some eighteenth-century Pennsylvania women. Robert Wells found evidence of "deliberate family limitation" among Quaker families living in Pennsylvania, New York, and New Jersey. These women steadily reduced fertility levels and family size from the middle of the eighteenth century. Beverly Smaby suggested that "direct birth control" explained both the reduced fertility of the Moravians of Bethlehem, Pennsylvania, during the Revolutionary war and a subsequent period of crisis in their religious community and the higher fertility levels when times were more propitious. A study of Philadelphia

---

2. Catherine M. Scholten, *Childbearing in American Society, 1650–1850* (New York, 1985), 13–14; similar conclusions are arrived at by many other American historians.

Most demographers and economists have argued that before the 19th century a system of "natural fertility" prevailed. According to this view, the only certain control over fertility in preindustrial Europe was age at marriage and celibacy rates, but, once marriage occurred, biology, and not human volition, governed fertility. Étienne van de Walle, "De la nature à la fécondité naturelle," *Annales de démographie historique,* 1988 (Paris, 1989), 13–17, summarizes this argument; see also his "Fertility Transition, Conscious Choice, and Numeracy," *Demography,* XXIX (1992), 487–502.

found the upper and middle occupational groups restricting fertility during the Revolutionary war, whereas Philadelphia laborers raised fertility as the war increased employment. Generally, however, poorer Philadelphians had lower fertility than their more prosperous counterparts. All Philadelphians reduced fertility early in the nineteenth century. Pennsylvania Schwenkfelders had smaller families when first settling the frontier of Pennsylvania but then increased family size as their material conditions improved. How can these cases of restricted fertility be reconciled with the presumed absence of contraceptive and abortive technology?[3]

The scarce historical references to abortion, infanticide, and contraceptive practice are better indicators of contemporary standards of public decorum than they are trustworthy guides to private practices. Contraception, birth control, and family planning are not eighteenth-century concepts, and so it should not be surprising that these terms do not appear. Much of eighteenth-century fertility control is hidden behind euphemisms and now-lost medical definitions. Barrier methods of birth control were not used in the early Delaware Valley, and the technologies of fertility restriction practiced by women have been largely lost in the twentieth century. Public accounts and the press were masculine spheres, concerned largely with issues appropriate to men, while obstetrics and gynecology were women's concerns and considered private.

But a careful analysis of the language of eighteenth-century women, reinterpreting the now-hidden meanings of women's words in the light of twentieth-century understandings, can recover the lost technology of fertility control. It is the unique ethnic diversity of the Middle Colonies that opens a window on the hidden and lost world of eighteenth-century women's contraceptive technology. In the mix of cultures and practices, some of the customary reticence eroded as women and men found different techniques being used by Swedish, German, British, French, African, and native American women or commented on techniques similar to their own but practiced by other people. Philadelphia was an early medical

---

3. Robert V. Wells, "Family Size and Fertility Control in Eighteenth-Century America: A Study of Quaker Families," *Population Studies*, XXV (1971), 80–81; Beverly Prior Smaby, *The Transformation of Moravian Bethlehem: From Communal Mission to Family Economy* (Philadelphia, 1988), 75; Susan E. Klepp, *Philadelphia in Transition: A Demographic History of the City and Its Occupational Groups, 1720–1830* (New York, 1989), 138–224; Rodger C. Henderson, "Eighteenth-Century Schwenkfelders: A Demographic Interpretation" in Peter C. Erb, ed., *Schwenkfelders in America* (Pennsburg, Pa., 1987), 25–40.

center and the American hub of an international pharmaceutical trade, where more information was available to a wider audience than would have been possible in rural or more provincial areas. From this world the technologies of eighteenth-century fertility restriction can be recovered.

Gender and Gynecology in Private Life

Many English historians have used the anthropological term "taboo" to describe the strict division between English men's and women's roles in obstetrics and gynecology. Menstruation, childbearing, and fertility were under the control of women and appropriate only to women. Most men, with clergymen and physicians occasionally excepted, would not transgress these rigid boundaries of acceptable social roles. Barbara Hanawalt writes of the medieval period: "Folkloric sources are virtually the only ones with information on childbirth. They illustrate the taboo on having a man present and the anxiety of both sexes at breaking it." Patricia Crawford finds that "scriptural and medical taboo during the seventeenth century" warned against male contact with menstrual fluid. Angus McLaren labels the fertility-enhancing, abortive, and contraceptive activities of English women a "separate female sexual culture" in the sixteenth through eighteenth centuries. While these taboos lessened in the eighteenth century, considerable shame still attached to the transgressor of gender-appropriate spheres as late as the nineteenth century.[4]

These English ideas of proper male and female spheres were transported to the Delaware Valley. One consequence was that the available male techniques of contraception—condoms and coitus interruptus—seem to have been little used or known by eighteenth-century Pennsylvanians. Fertility

4. Barbara A. Hanawalt, *The Ties That Bound: Peasant Families in Medieval England* (New York, 1986), 216; Patricia Crawford, "Attitudes to Menstruation in Seventeenth-Century England," *Past and Present*, no. 91 (May 1981), 61–62; Angus McLaren, *Reproductive Rituals: The Perception of Fertility in England from the Sixteenth to the Nineteenth Century* (London, 1984), 111; William L. Langer, "The Origins of the Birth Control Movement in England in the Early Nineteenth Century," in Robert I. Rotberg and Theodore K. Rabb, *Marriage and Fertility: Studies in Interdisciplinary History* (Princeton, N.J., 1980), 271. Also pointing to male ignorance of contraception are the sources quoted in R. R. Kuczynski, "British Demographers' Opinions of Fertility, 1660–1760," in Lancelot Hogben, ed., *Political Arithmetic: A Symposium of Population Studies* (New York, 1938), 283–327.

and childbirth were women's concerns, and a man either passively observed his wife's pregnancies or, at most, attempted to avoid her. A French visitor to Pennsylvania in the 1790s was surprised that, "when a Philadelphia woman bears a child, her husband is never present." Childbirth was a female ritual from which even husbands were excluded. John Fitch, a Philadelphia watchmaker, felt that, in assisting at a birth in 1789, he had been "obliged to degrade the man and become a nurse." Childbirth and other gynecological subjects were shameful for men.[5]

Eighteenth-century Pennsylvania men seemed quite ignorant of contraceptive techniques. Fitch abandoned his wife in 1769, "fearing an increass of my familey urged my departure. And before I left that place my wife was with child with a daughter. Which had I known I should never have left her but worried thro' life as well as I could." Joseph Shippen wrote to his father in 1770, at the birth of a child, "If Events were always to correspond to our own particular Desires, I should have rather wished this to have been postponed a Year or two longer." In 1781, Robert Barclay wrote to a friend that he "hoped his wife would wait for awhile before having any more children," as if he had nothing to do with the process. Even a compulsive libertine, like the anonymous Philadelphia sawyer who kept records of his liaisons in an accounting book, did not use prophylactics. While he was certainly no naïf, he was surprised by his wife's pregnancy: "Molly informs me she is big again with child—oh my—how can I endure it." This Philadelphian did not use contraceptives in illicit encounters either. "She caused me much fright on asking me where I dwelleth and that in case increase from contact she would make known to me the fact—I said I lived in the Carolinas—and fled." Nor did he use a condom to prevent the spread of venereal disease, although condoms were being used by London prostitutes and their customers to guard against infection. A belated six months after the first diagnosis he resolved: "Clapp—Much itching in my flopper—must keep away from my Wife." Men could wish and hope, they could be frightened, but they had little control over fertility except to escape the consequences by fleeing. In colonial Pennsylvania, fertility decisions were left to women.[6]

5. Médéric-Louis-Élie Moreau de St. Méry, *Moreau de St. Méry's American Journey* [1793–1798], ed. and trans. Kenneth Roberts and Anna M. Roberts (Garden City, N.Y., 1947), 289; Frank D. Prager, ed., *The Autobiography of John Fitch*, Memoirs of the American Philosophical Society, CXIII (Philadelphia, 1976), 126.

6. Prager, ed., *Autobiography of John Fitch*, 46; Randolph Shipley Klein, *Portrait of an*

Childbirth, menstruation, and fertility were rarely discussed by men in the eighteenth century. When gynecological problems occurred, it was the women who administered the treatment, provided medical advice, and attended the patient, sometimes in consultation with physicians. When some male physicians began to take a more active role in obstetrics at the end of the century, their role was initially confined to that of an assistant at the birth.[7] The restriction of fertility was not a masculine prerogative in the eighteenth century.

Gender and Gynecology in Political Life

Pennsylvania men were similarly reluctant to intervene in childbirth and fertility in their roles as legislators and judges. Pennsylvania legislators never adopted or enforced the English statute on abortion, which was rarely invoked in England by the eighteenth century. The English antiabortion law of 1623 had not been concerned with the fetus before quickening (the point at which the pregnant woman first felt movement in the third to fifth month of gestation). This was believed to be the point at which a soul animated the fetus. The statute had postulated a tripartite scheme of human development—conception, ensoulment, and reason—that corresponded to impregnation, perception of movement, and birth. Interference in the first stage was no concern of the law, abortion after quickening was a misdemeanor, and only after birth could the murder statute apply. Quickening was increasingly untenable as a distinct developmental stage by the late seventeenth century. This changing interpretation of quickening may be a factor in the failure of Pennsylvania either to adopt or to enforce this particular English statute. Rather than invoke the English statute on abortion in cases where there were suspicious circumstances at birth, eighteenth-century prosecutors inquired into the viability of the

---

Early American Family: The Shippens of Pennsylvania across Five Generations (Philadelphia, 1975), 150; Judy Mann DiStefano, "A Concept of the Family in Colonial America: The Pembertons of Philadelphia" (Ph.D. diss., Ohio State University, 1970), 25; anonymous diary in the James Wilson Papers, entries for June 3, February 15, December 22, American Philosophical Society, Philadelphia. The author was probably a sawyer in Wilson's employ. Internal evidence places the year-long record between 1792 and 1798.

7. Judith Walzer Leavitt, *Brought to Bed: Childbearing in America, 1750 to 1950* (New York, 1986), esp. 58–60.

deceased infant. Where a fetus was born without hair or nails, up to the seventh month of gestation, it was assumed to be not viable by the panels of women called to examine the corpse. This was regarded as a normal stillbirth, not as an induced abortion. A Pennsylvania woman whose child was found to have been stillborn even at later stages of gestation was dismissed and not subsequently tried for abortion. A woman who failed to convince a jury that her child was stillborn might be held on a charge of infanticide. Neither the colony nor the state of Pennsylvania chose to prosecute under the provisions of the English statute concerning abortion.[8]

Infanticide was prosecuted. It was a capital offense to conceal the death of a bastard child. Should the mother "endeavour privately" to hide the body, then she "shall suffer death, as in case of murder; except such mother can make proof . . . that the child . . . was born dead." An anomaly in English law, this enactment overrode the presumption of the accused's innocence by making the discovery of a dead bastard child presumptive evidence of its murder by its mother. It was the responsibility of the accused to prove her innocence and not the duty of the prosecution to prove guilt. There were eight Pennsylvania women convicted under the concealment statute between 1718 and 1775. Only three of these women were hanged, however, an execution rate of 37 percent, compared to 74 percent for all murder convictions. Between 1779 and 1792, fifteen women were charged with the crime of concealment, but only three were convicted, two

8. For British laws enforced in Pennsylvania, see Samuel Roberts, *A Digest of Select British Statutes, Comprising Those Which, according to the Report of the Judges of the Supreme Court, Made to the Legislature, Appear to Be in Force, in Pennsylvania: With Some Others, with Notes and Illustrations* (Pittsburgh, 1817). When abortion was punished in other North American colonies, it was because the abortion was intended to cover up evidence of the crimes of adultery or fornication. Most criminal prosecutions concerned men who had forced abortifacients on apparently unwilling women. See Cornelia Hughes Dayton, "Taking the Trade: Abortion and Gender Relations in an Eighteenth-Century New England Village, *WMQ*, 3d Ser., XLVIII, (1991), 19–49; Julia Cherry Spruill, *Women's Life and Work in the Southern Colonies* (1938; New York, 1972), 325–326; Barbara Duden, "Quick with Child: An Experience That Has Lost Its Status," *Technology in Society*, XXIV (1992), 335–344; McLaren, *Reproductive Rituals*, esp. 109–110, 122; Lawrence H. Gipson, *Crime and Its Punishment in Provincial Pennsylvania* (Bethlehem, Pa., 1935), 5–6; William Bradford, *An Enquiry: How Far the Punishment of Death Is Necessary in Pennsylvania, with Notes and Illustrations* (Philadelphia, 1793), 39; G. S. Rowe, "Women's Crime and Criminal Administration in Pennsylvania, 1763–1790," *Pennsylvania Magazine of History and Biography*, CIX (1985), 335–368.

of whom were pardoned. As William Bradford, the attorney general of the United States and former attorney general of Pennsylvania and judge of the state supreme court, commented, "Where a positive law is so feebly enforced, there is reason to suspect that it is fundamentally wrong."[9]

There was little support for the punishment of infanticide as murder. In Chester County in 1697, an accused woman, her brothers, and her sister fought off warrant servers. Prosecutorial standards became increasingly lenient as standards of evidence changed. The woman would not be convicted if she told anyone of her pregnancy, if she had assistance in childbirth, or if she had collected baby linen. These were taken to be proofs that she had not intended concealment. In addition, it became accepted that the body of the infant had to show marks of violence sufficient to convict under the murder statute. Even with these qualifications, the law remained unpopular. A case from western Pennsylvania in 1791 involved a widow who first "denied, but afterwards confessed having had a child.—She said she had buried her child, it having been dead born. Afterwards owned she had taken it up, and thrown it in the river." The jury found her not guilty in spite of the confession, the marks of violence on the corpse, and the evidence of concealment.[10]

Reforming this unpopular law was a priority of the new state of Pennsylvania. The law was revised in 1786 and 1790 to require proof that the child was born alive before the woman could be convicted. The burden of proof shifted from the accused to the prosecution. In 1794, the legislature revised the law to require proof that the mother "did wilfully and

9. Article 8, Pa. chap. 247, in Alexander James Dallas, *Laws of the Commonwealth of Pennsylvania, from the Fourteenth Day of October, 1700* . . . (Philadelphia, 1797–1798), I, 135–136; J. W. Ehrlich, *Ehrlich's Blackstone, Part Two: Private Wrongs, Public Wrongs* (New York, 1959), 397; Leon Radzinowicz, *A History of English Criminal Law and Its Administration from 1750* (New York, 1957), I, 430–436; Bradford, *Enquiry,* 40; Gipson, *Crime,* 14; G. S. Rowe, "Women's Crime," *PMHB,* CIX (1985), 360, 365–366. Rowe, "Infanticide, Its Judicial Resolution, and Criminal Code Revision in Early Pennsylvania," American Philosophical Society, *Proceedings,* CXXXV (1991), 207–209, has uncovered cases unknown to Bradford for the period from 1682 to 1800: of 73 women prosecuted for concealment and infanticide, 57 were tried, 24 convicted, and 8 executed: executions declined markedly over time.

10. Herbert William Keith Fitzroy, "The Punishment of Crime in Provincial Pennsylvania," *PMHB,* LX (1936), 268; Radzinowicz, *History of English Criminal Law,* 430–436; Bradford, *Enquiry,* 39–40; Alexander Addison, *Reports of Cases in the County Courts of the Fifth Circuit* . . . (Washington, Pa., 1800), 1–2.

maliciously destroy and take away the life of such child," a belated legal recognition that newborn infants died of causes other than murder. Where the unwed mother simply concealed the death of her child, the new law fixed a maximum sentence of five years' imprisonment. After 1794, most Pennsylvania women suspected of infanticide were tried and convicted for concealment rather than for murder. In one indictment for murder, "the circumstances were very strong, and might reasonably have been thought sufficient, to satisfy the mind of the jury. . . . However, the jury did not find the murder, but found the concealment." In general, Pennsylvania legislators, prosecutors, judges, and jurymen did not equate infanticide with murder, choosing acquittal or a lesser charge over conviction under the statute.[11]

William Bradford argued that pity was the appropriate and usual response of jurymen as long as the accused woman preserved her proper role by appealing to the court as a "helpless woman" or as one of those "unfortunate creatures." In these cases, the male judiciary would view them "with compassionate eyes" and acquit. The few convictions occurred when the woman showed no remorse. Women operating within the expected bounds of their gender were not to be punished. This reluctance to intervene in the private endeavors of women at childbirth, even when there was a suspicion of murder, was part of a tendency in Pennsylvania to "view female sexual activity as a private matter," a tendency also apparent in the failure to adopt the British statute on abortion or to enforce its provisions.[12] In this legal environment the law had no interest in fertility regulation.

The Regulation of Fertility: Definitions

Through much of the eighteenth century, the *concepts* of family planning, birth control, fertility regulation, contraception, or abortion were little known or used. Women did not envisage an ideal number of children, and so they did not try to stop childbearing after achieving some intended family size. Pennsylvania women of the eighteenth century rarely

11. Dallas, *Laws of the Commonwealth*, I, 136, nn. p, q; Addison, *Reports*, 8.

12. Bradford, *Enquiry*, 39–40; Rowe, "Women's Crime," *PMHB*, CIX (1985), 360, 367. There is no evidence that any religious group took an active interest in fertility regulation in the 18th century.

spoke about restricting fertility, but they did regulate their health and well-being by employing medicines and procedures known as emmena-gogues, "which provoke and maintain the periodical occurrence of the menstrual secretion." Rather than planning the future size of their families or curtailing the span of their childbearing years, Pennsylvania women focused on their present circumstances, especially their current physical and emotional health. They were apparently most interested in regulating their menstrual cycles.[13]

The distinction between contraception and abortion must be based on a certain knowledge of pregnancy. But pregnancy was difficult to determine in the eighteenth century, and most women would have agreed with Dr. Samuel K. Jennings: "An entire suppression of the menses attends almost every case of pregnancy. But as suppressions may be brought on by other causes this cannot be an infallible mark." The absence of menstruation could be equally a sign of pregnancy or a sign of illness. So in 1778, Sally Logan Fisher of Philadelphia was "much dissapointed" with the onset of menstruation, since she had just miscarried and longed for another child. In 1805, Lydia Tallender of the same city defined her amenorrhea as a sign of illness, not pregnancy. She "had not had the female customs for 2 mos, and been unwell . . . pills given for effecting the return of the aforementioned caused puking and purgation, which after a while was checked." The lack of a necessary connection between amenorrhea and pregnancy allowed both women and doctors to view a missed menstrual period as a question of health up until quickening. While neither eighteenth-century women nor doctors believed in the old idea that quickening was the mystical moment when the fetus was endowed with a soul, quickening remained important as the first certain sign of a viable pregnancy. Before that point, amenorrhea (called lost, hidden, obstructed, or suppressed menses) might be diagnosed. If a woman saw her condition as pathological, then pregnancy was excluded as a possible underlying cause. A woman was either gravid or obstructed, and the two conditions bore no necessary relationship to one another.[14]

13. Joseph Carson, *Synopsis of the Course of Lectures on Materia Medica and Pharmacy Delivered in the University of Pennsylvania* (Philadelphia, 1867), 82–84.

14. Samuel K. Jennings, *The Married Lady's Companion; or, Poor Man's Friend* (1808; rpt. New York, 1972), 75. Dr. George DeBenneville wrote angrily that "Some woman are so Ignorant, They do not know whin she are Conceived with Child, and others so easy, They will nott Confess when they do know it," but he did not include amenorrhea as one

The two symptoms that distinguished the pathological condition of obstructed menses from pregnancy were, according to one doctor, "mental despondency" and hysteria. The emotional state of the woman was the primary clue to her physical condition. "Grief and distress" were considered the predominant symptoms in cases of amenorrhea and could be accompanied by stomach pains, headaches, and melancholy. Hysteria was, in the eighteenth century, a disease with both mental and physical symptoms. Despondency and physical symptoms resembling early pregnancy were thought to define hysteria. The most popular home guide to health stated, "A sudden suppression of the *menses* often gives rise to hysteric fits," beginning with fatigue, "lowness of spirits, oppression and anxiety." Additionally, hysteria was indicated if the woman felt "as if there were a ball at the lower part of the belly, which gradually rises towards the stomach, where it occasions inflation, sickness, and sometimes vomiting." Another medical source added a curious collection of symptoms: "an unusual gurgling of the bowels, as if some little animal were there in actual motion, with wandering pains, constituting colic of a peculiar kind." Anxiety, low spirits, and a feeling of oppression could, of course, have been caused by fears of an unwanted pregnancy. The physical signs thought to be specific to the disease of hysteria—an inflated belly, vomiting, and colic—mirror some of the early symptoms of pregnancy.[15]

In popular discourse, a missed menstrual period was called "taking a cold." This term was a survival of the Galenic system of medicine. Since menstrual blood was a "hot" humor, its absence must signify the unhealthy

---

of the signs of pregnancy. Rather, he listed some very dubious symptoms such as "coldness of the outward parts," the "Belly waxeth very flat," the "Veins of the eyes are clearly seen," or the appearance of a "small living creature" in the woman's urine after thirteen days of storage. See George DeBenneville, "Medicina Pensylvania, or, The Pensylvania Physician," 124, microfilm of MS, ca. 1760–1779, APS. This was the advice he passed on to his married daughter, Harriet DeB. Klein. See also Mary Beth Norton, *Liberty's Daughters: The Revolutionary Experience of American Women, 1750–1800* (Boston, 1980), 80; "Strangers' Burials," MS, Sept. 3, 1805, Gloria Dei burial records, Genealogical Society of Pennsylvania, Philadelphia. Tallender died, not of the emmenagogues, but from yellow fever.

15. Andrew S. Berky, *Practitioner in Physick: A Biography of Abraham Wagner, 1717–1763* (Pennsburg, Pa., 1954), 112; William P. C. Barton, *Outlines of Lectures on Materia Medica and Botany, Delivered in Jefferson Medical College, Philadelphia* (Philadelphia, 1827–1828), I, 104; William Buchan, *Domestic Medicine; or, A Treatise on the Prevention and Cure of Diseases . . .* , rev. Samuel Powel Griffitts (Philadelphia, 1795), 455–456; Jennings, *Married Lady's Companion,* 54.

dominance of a cold humor. In the empirical and eclectic medical practice of the eighteenth century, few strictly followed any one system of medicine, but the name persisted. Dr. William Buchan's *Domestic Medicine* warned, "More of the sex [women] date their disorders [amenorrhea] from colds, caught while they are out of order [expecting menstruation], than from all other causes." This merging of two physical conditions—the common cold and pregnancy—can especially be seen in medical tracts. Both rheumatism and pleurisy, thought to be caused by colds, were linked to obstructed menstruation through the "stoppage of customary discharges." All these conditions were conflated and the treatments merged so that emmenagogic medicines were considered cures for rheumatism. Riding on horseback was also recommended for rheumatism, presumably because it would dislodge any obstructed customary discharges. Catching a cold had layers of meaning for women in the eighteenth and nineteenth centuries.[16]

The emotional and physical state of the woman determined whether she would view her "suppressed menstruation" as a sign of pregnancy or as a sign of illness. The illness might be labeled obstructed menstruation, which was considered a separate disease, or the obstructed menstruation could be interpreted as one symptom of despondency, hysteria, colds, or rheumatism. Even pleurisy or intestinal parasites might be diagnosed. The cures for all these disease categories overlapped. Definitions of disease provided women with a vehicle for dealing with unwelcome pregnancies. There are few references to the deliberate prevention of conception or to induced abortion in the eighteenth or early nineteenth centuries, many to the regulation of the menstrual cycle through the treatment of one or more of these diseases.

English-language health manuals, diaries, and letters did not discuss the fertility effects of these emmenagogic practices and medicines, since the focus was entirely on the woman's health. But about one-third of Pennsylvania women were German, not English. In the German-language health guides published in eighteenth-century Pennsylvania, most recipes concerned the finding of "hidden or lost menses" and the subsequent restoration of menstruation and health, just as in the English guides. However, German women could also suffer from carrying "dead fruit." Christopher Sauer's *Small Herbal of Little Cost,* printed in Germantown from 1762 to 1778, gave women recipes designed "to expel the dead fruit." Later in the

16. James C. Mohr, *Abortion in America: The Origins and Evolution of National Policy, 1800–1900* (New York, 1978), 7; Buchan, *Domestic Medicine,* 395–396, 529.

century, midwifery manuals published at Ephrata, Pennsylvania, recorded the same cures as effective in expelling the fetus, "be it dead or alive."[17] No such explicit references to abortion appear in the English language.

Even among the English it must have been widely known, if little recorded for posterity, that restoring menstruation could abort. In the eighteenth century and well into the nineteenth century, descriptions of violent exercise or the ingestion of various herbs and minerals could be found in English books of household medicine both under sections describing behaviors to avoid in pregnancy and in sections dealing with cures for the disease of "obstructed menses." The strong association of emmenagogues with the destruction of "little animals" and their simultaneous use as insecticides and vermifuges indicates a conceptual linkage with contraceptive and abortive function. Women usually made only oblique reference to their use of these medicines and practices. Two months after weaning her son in 1757, Esther Edwards Burr was bedridden for two days and then rode from Princeton to Brunswick: "Found the Ride of service," she wrote to a friend. One month later she was "poorly." She did not remark that these illnesses came in monthly cycles. Margaret Hill Morris included no emmenagogues in her collection of cures. Yet her recipe for rheumatism encompassed ingredients and dosages that were usually used to restore menstruation. Euphemisms and now-quite-foreign perceptions of disease can obscure the intentions of eighteenth-century Pennsylvania women.[18]

Also blurring the abortive intentions of women was that these same remedies were used prophylactically. One preventive method was to employ an emmenagogue just before the menstrual period was due in order to ensure its arrival. A second prophylactic technique was that "after birth,

17. Christa M. Wilmanns Wells, "A Small Herbal of Little Cost, 1762–1778: A Case Study of a Colonial Herbal as a Social and Cultural Document" (Ph.D. diss., University of Pennsylvania, 1980), 464, 470–471. See also James Woycke, *Birth Control in Germany, 1871–1933* (London, 1988), 16–19.

18. For examples, see John Wesley, *Primitive Remedies* (Santa Barbara, Calif., 1975 [orig. pub. as *Primitive Physick; or, An Easy and Natural Method of Curing Most Diseases,* 5th ed. (Bristol, 1755)]), 23, 89–90; Jennings, *Married Lady's Companion,* 43–49, 77–80; Carol F. Karlsen and Laurie Crumpacker, eds., *The Journal of Esther Edwards Burr, 1754–1757* (New Haven, Conn., 1984), 267, 270, similarly 211. "For the Rheumatism—half an ounce of gum guiacum in fine powdr; ¼ os Valerian root in powder[,] 1 Dram of Camphor, mix them well together in a morter and with a little syrup make a mass for pills. Take 4, night and morning": Margaret Hill Morris (1736–1816), Receipe Book, 4, photocopy, Quaker Collection, Haverford College, Haverford, Pa.

nursing mothers took it [a tea of savin] to hold off renewed menses." This could have been especially useful, since, according to one visitor, "in Philadelphia a husband resumes conjugal relations with his nursing wife a month after the birth," making the contraceptive effects of lactation considerably less reliable than they would be where breastfeeding was accompanied by a ban on sexual intercourse.[19] Definitions of disease—obstructed menses, colds, rheumatism, even worms or internal parasites—along with a reticence in speaking or writing about private female topics have hidden the contraceptive and abortive technologies available to Pennsylvania women in the eighteenth and early nineteenth centuries.

The Regulation of Fertility: Techniques

There were a large number of cures for the disease of obstructed menses, and women combined herbal decoctions, bleeding, vigorous exercise, and "baths" to treat their amenorrhea. In 1828, Dr. W. P. C. Barton provided his medical students with an unusually inclusive list of emmenagogic medicines and procedures. He described thirteen classes of medical treatments. There were forty medicinal herbs and minerals designed to be swallowed in liquid form or as pills. He also advocated bloodletting, vigorous exercise such as dancing and horseback riding, and application of pressure to the abdomen with brushes and "frictions" (a forcible rubbing of the body). He recommended baths of various sorts. The methods described by Barton would today be considered attempts to abort, and some, like jumping rope, are still practiced popularly in order to terminate pregnancy. The use of footbaths and bathing would seem ineffectual, but among the Pennsylvania Germans a footbath was a euphemism for a vaginal douche. It was defined in German medical texts as "bathing inside from below upward." No such definition has been found in English, although perhaps this usage was understood both by physicians and lay practitioners. Similarly, the German health guides can explicitly describe contraceptive and abortive tampons, yet at times employ the circumlocution of medicines

19. Bradford Angier, *Field Guide to Medicinal Wild Plants* (Harrisburg, Pa., 1978), 155; Moreau de Saint-Méry, *American Journey*, 289. Many, but not all, books of domestic medicine advised against the resumption of sexual relations while nursing: see Paula A. Treckel, "Breastfeeding and Maternal Sexuality in Colonial America," *Journal of Interdisciplinary History*, XX (1989–1990), 25–51.

"laid upon the thighs" as a euphemism for tampons. When Barton recommended fomentations and cloths applied to the pubes, these may well
have connoted douches and tampons to his readers. The veiled language
of the day makes it difficult to recover the full range of contemporary
American practice. Unstated assumptions almost certainly contributed to
wide variations in individual women's knowledge of contraceptive and
abortive techniques.[20]

The emmenagogic, contraceptive, or abortive practices of the eighteenth
century were not associated with the act of sexual intercourse, but, as befits a technology largely controlled by women, focused on menstruation
and women's perception of a possible pregnancy or disruption of health.
Medications might sometimes be prescribed or prepared by men, but they
were administered by women. The treatments were taken orally or anally
or applied externally. There are explicit references to vaginal intrusion in
German, none in English. There are no references to uterine intrusion.
Vomiting, purging, and sweating as well as the removal of obstructions in
the womb were the expected results of all these treatments. The emmenagogic practices of the eighteenth century seem harsh two hundred years
later, as does much of eighteenth-century medical practice, but the intrusive methods of later contraceptive practice struck most Philadelphians as
"hideous."[21]

It may be that physical remedies for obstructed menses like jumping
rope and horseback riding were the ones most commonly employed. They
would have been readily available to women. It is possible that oral communication informed many women of tampons and douching techniques.
It is, however, the herbal remedies designed to be swallowed as liquids
or pills that prevail in the historical record, not these other practices.
Some herbs dominate the literature, recurring frequently in popular and
professional medical books, in the advertisements of druggists, and in
the writings of women. Savin (*Juniperus sabina* L.), juniper or red cedar

20. Quoted in Wells, "Small Herbal," 467. For a discussion of the distinctive childbearing patterns of individual women, based on birth intervals, see Stephanie Grauman Wolf,
*Urban Village: Population, Community, and Family Structure in Germantown, Pennsylva-
nia, 1683–1800* (Princeton, N.J., 1976), 271–273.

21. Moreau de Saint-Méry, *American Journey*, 314–315. On women's concern with men-
struation, see, in particular, Lucille F. Newman, ed., *Women's Medicine: A Cross-Cultural
Study of Indigenous Fertility Regulation* (New Brunswick, N.J., 1985); Mary Chamberlain,
Old Wives' Tales: Their History, Remedies, and Spells (London, 1991).

(*Juniperus virginiana* L.), rue (*Ruta graveolens*), aloes (*Aloe barbaden-sis*), pennyroyal (*Hedeoma pulegiodes*), madder (*Rubia tinctorum*), and seneca snakeroot (*Polygala senega*) were the most common drugs used in early Pennsylvania, among both the English and Germans. These substances were employed to ensure the onset of the menstrual cycle, to bring a return of menstruation, or to prolong amenorrhea during lactation.

The plant most commonly mentioned in the restoration of menstruation was European savin or the related, chemically identical American species of juniper or red cedar. As early as the seventeenth century, savin, a plant native to the Mediterranean and a known abortive agent since Roman times, was being grown in Pennsylvania. A 1702 description of the colony included notice of "a little tree, which looks like Juniper, and is called the *Savan;* it has the property of making a mare barren, or bring out her foal before the time. For that purpose you need only give her a handful of it." By 1750, both naturalized European savin and the native red cedar were growing in abundance along the banks of the Delaware River and its tributaries. The uses of these related plant species were widely known. Frederika von Riedesel had been in the United States for less than two months in 1777 when she heard something of the abortifacient properties of American cedar. Juniper was sold in various forms at the Philadelphia market at midcentury, and imported and native preparations of juniper—both savin and red cedar—were staples in drug stores and other shops. By the early nineteenth century, savin "was the single most commonly employed folk abortifacient in the United States." Drunk as a tea made from the berries or tips of the plant, it was used to restore menstruation and to prevent impregnation during lactation.[22]

22. Thomas Campanius Holm, *A Short Description of the Province of New Sweden, Now Called, by the English, Pennsylvania, in America,* trans. Peter S. Du Ponceau (Historical Society of Pennsylvania, *Memoirs,* III, part 2 [Philadelphia, 1834]), 163; Peter Kalm, *Peter Kalm's Travels in North America: The English Version of 1770,* ed. Adolph B. Benson (1937; New York, 1964), II, 635; William P. C. Barton, *Compendium Florae Philadel-phicae* . . . (Philadelphia, 1818), II, 200; Frederika Charlotte Louise von Riedesel, *Baroness von Riedesel and the American Revolution: Journal and Correspondence of a Tour of Duty, 1776–1783,* ed. and trans. Marvin L. Brown, Jr. (Chapel Hill, N.C., 1965), 49; Gottlieb Mittelberger, *Journey to Pennsylvania,* ed. and trans. Oscar Handlin and John Clive (Cambridge, Mass., 1960), 55; Joseph W. England, ed., *The First Century of the Philadelphia College of Pharmacy, 1821–1921* (Philadelphia, 1922), 110–112; Barton, *Outlines of Lectures,* II, 196; Mohr, *Abortion in America,* 9.

Druggists and physicians sold pills of imported aloes mixed with other ingredients as a treatment for amenorrhea. As a medical student in 1759, William Shippen successfully treated a case of "obstruct'd menses" with three pills containing aloes, myrrh, sulfate of iron, valeriana, rue, dittany, asafetida, and other items. In the early nineteenth century, the *Dispensatory of the United States of America* noted of aloes, "It is perhaps more frequently employed than any other remedy, entering into almost all the numerous empirical preparations habitually resorted to by females in that complaint [amenorrhea], and enjoying a no less favorable reputation in regular practice." Barbados aloes were always specified and were available from pharmacists in the form of Hooper's Female Pills, a popular patent medicine combining Barbados aloes with iron sulfate, hellebore, and myrrh as the major ingredients and widely advertised in Philadelphia newspapers from the middle of the eighteenth century.[23]

For women with kitchen gardens, rue, another plant native to the Mediterranean but naturalized first in England and then in North America, and the native American pennyroyal were favored remedies. Rue was used primarily as an emmenagogue and abortifacient, although it was considered an important remedy in rheumatism as well. Daniel Francis Pastorius grew it in his garden in Germantown early in the century and associated it with mothers, time, and age for its ability to regulate the periodicities of life. It was one of the "common herbs for domestic remedies" by midcentury. Among the Pennsylvania Germans it had the additional reputation of being able to "prevent impregnation," a rare reference to a contraceptive intent.[24]

---

23. Betsy Copping Corner, *William Shippen, Jr.: Pioneer in American Medical Education* (Philadelphia, 1951), 14, 37–38; George B. Wood and Franklin Bache, *The Dispensatory of the United States of America* (Philadelphia, 1833), 97. See, for example, the advertisement of chemist Christopher Marshall in the *Pennsylvania Gazette* (Philadelphia), Jan. 2, 1749/50. The ingredients in Hooper's Female Pills are 400 parts Barbados Aloes, 200 parts desiccated iron sulfate, 100 parts dark hellebore, 100 parts myrrh, 100 parts soap, 50 parts white cinnamon, and 50 parts ground ginger; see Roslyn Stone Wolman, "Some Aspects of Community Health in Colonial Philadelphia" (Ph.D. diss., University of Pennsylvania, 1974), 360.

24. Cristoph E. Schweitzer, ed., *Francis Daniel Pastorius' Deliciae Hortenses; or, Garden-Recreations and Voluptates Apianae* (1705–1711) (Columbia, S.C., 1982), 16, 28; Israel Acrelius, *A History of New Sweden . . .*, trans. William M. Reynolds (Historical Society of Pennsylvania, *Memoirs*, XI [1874]) (Philadelphia, 1876), 151; David E. Lick and

The other common garden drug was pennyroyal, "a popular remedy throughout the country for female (plaints. . . . It is chiefly beneficial in obstructed catamenia, and recent cases of suppressions, given as a sweetened tea." According to the *United States Dispensatory,* pennyroyal "excites the menstrual flux when the system is predisposed to the effort. Hence it is much used as an emmenagogue in popular practice." The native American pennyroyal, also called squaw mint both to distinguish it from the unrelated European plant and to indicate its function and origin, was domesticated in gardens and grew wild throughout the region, especially on newly cleared land and at the side of roads.[25]

Seneca snakeroot, or rattlesnake root, was another native American species, but was not domesticated. It became a commodity, however, and was gathered in the South and shipped to Philadelphia vendors from the early decades of the eighteenth century. Benjamin Franklin, or more probably Deborah Franklin, sold the newly discovered seneca root from the Philadelphia post office in 1740. It was valued for various medicinal purposes but became closely associated with emmenagogic functions. By the late eighteenth century, consumers were spared the trouble of steeping their own roots when it was sold as Snakeroot Cordial at local pharmacies. Snakeroot became an important export item as it entered into foreign pharmacopoeias. It continued to be considered an important emmenagogue through the first three-quarters of the nineteenth century.[26]

Thomas R. Brendle, "Plant Names and Plant Lore among the Pennsylvania Germans," Pennsylvania-German Society, *Proceedings and Addresses,* XXXIII, part 3 (1923), 62.

25. C. S. Rafinesque, *Medical Flora; or, Manual of the Medical Botany of the United States of North America,* 2 vols. (Philadelphia, 1828–1830), I, 233. The author adds that pennyroyal's usefulness in restoring menstruation "is proved by daily experience." See also Wood and Bache, *Dispensatory of the United States,* 365; Kalm, *America of 1750,* ed. Benson, I, 103, II, 632.

26. Snakeroot "is highly extolled by many practitioners in the treatment of amenorrhoea" (John B. Biddle, *Materia Medica, for the Use of Students,* 4th ed. [Philadelphia, 1871], 257). In 1740 it was sold from the post office, "with Directions how to use it in the Pleurisy, etc." The "etc." could cover the indelicate topic of emmenagogues and was a euphemism for menses in German. By 1789, no directions were needed for the cordial. See also *Pennsylvania Gazette,* June 12, 1740, and the advertisement of Seth and Isaac Willis, *Pennsylvania Gazette,* Feb. 1, 1789; Wells, "Small Herbal," 468; England, *College of Pharmacy,* 110. In 1791–1792, 5,300 pounds of snakeroot were exported from Philadelphia, 40% of the country's annual exports of this item. Tench Coxe, *A View of the United States of America . . .* (Philadelphia, 1794), 414.

Other emmenagogic medicines appear in the written record. Gum guaiacum was imported from the Caribbean, madder from Holland. Black hellebore was "very highly esteemed by some practitioners." There were some ethnic and regional differences in medical practice. Two important plant groups used by the Germans in Pennsylvania were smalledge (wild parsley or wild celery, *Apium graveolens* L.) and various species of the Artemisia family. German women also ingested ragwort (*Senecio aureus*) and called it the life herb for its ability to "correct female irregularities." The Shakers of New Lebanon employed scaly dragon claw (*Pterospora andromedea*) as an emmenagogue. In rural Pennsylvania *Cunila mariana* (American dittany), *Gautiera ripens* (mountain-tea), or *Hamamelis virginica* (winter witch hazel) were among the drugs used. One of the few plants listed solely as a contraceptive was wild ginger (*Asarum canadense*). It was used "by the Indian females to prevent impregnation." Some in the German community thought that wild ginger was not only a contraceptive but also an aphrodisiac. This dangerous combination of attributes earned it the name "thing of evil" in rural Schwencksville. This is not an exhaustive list of plants thought to have emmenagogic properties but gives some indication of the range of plants in use. There was no shortage of drugs designed to restore menstruation.[27]

The herbal remedies that dominated the emmenagogic and abortive literature are often portrayed as a traditional and largely static body of knowledge rooted in folklore and in medical authorities dating back to the classical Greeks. Even if it is true that these treatments had changed little in Europe, the attempt to transplant European medical practice to Pennsylvania brought change and encouraged experimentation. Some established remedies were carried over from the Old World: rosemary, rue, and savin, among others. The European plants most successfully naturalized and domesticated in Pennsylvania were plants that had first been naturalized in England and Germany, often from the Mediterranean region. Women also duplicated Old World remedies by seeking out closely related Ameri-

27. Wood and Bache, *Dispensatory of the United States*, 367; Wells, "Small Herbal," 461–469; Lick and Brendle, "Plant Names and Plant Lore," Pa.-German Soc., *Procs.*, XXXIII (1923), 62, 157. Ragwort is also called the life root, the female regulator, or squaw weed. See Joseph E. Meyer, *The Herbalist*, rev. Clarence Meyer and David C. Meyer (Glenwood, Ill., 1981), 119; Rafinesque, *Medical Flora*, I, 139, 204, 230, 165, II, 69; Frederick Pursh, *Flora Americae Septentrionalis* (London, 1814), rpt., ed. Joseph Ewan (Hirschberg, Germany, 1979), 596.

can species of dittany, madder, and savin. African women labeled several species of plants as snakeroot.[28] The botanists of the late eighteenth century and early nineteenth century got much of their information on local flora from women who had explored, identified, and experimented with a wide variety of plant species.

There seems to have been some selectivity in the transit of herbal medicine, since not all common English remedies were transported to the New World. Ergot of rye, for example, was not adopted as a drug in Pennsylvania. English pennyroyal failed to find advocates in Pennsylvania, since American pennyroyal, a plant with a somewhat similar "shape, smell and properties" was found to be "more efficient" as an emmenagogue. On the other hand, imported savin was preferred over locally available savin or juniper, in part because it was "scarce and dear," but largely because it was believed to be a stronger medicine.[29] Magical contraceptive practices, widespread in Europe, are not evident in eighteenth-century Pennsylvania.

Cross-cultural borrowing can be seen in the interest in native American emmenagogues and related plants. Squaw bush (*Vibirnum opulus,* cramp bark), squaw mint (*Hedeoma pulegioides,* pennyroyal), squaw root (*Cimicifuga racemosa* or *Caulophyllum thalictroides*), squaw vine (*Michella repens,* a native relative of madder), and squaw weed (*Erigon philadelphicum,* fleabane) were among the plants borrowed from or attributed to the medical practice of native American women. Snakeroot was important in both native American and African American practice. One native plant—pennyroyal—was domesticated in the eighteenth century, and black snakeroot and Seneca snakeroot were prepared commercially. The process of investigation and experimentation was not one-sided. European plants, including rosemary and rue, were adopted by some native American and African American women.[30]

---

28. Martia Graham Goodson, "Medical-Botanical Contributions of African Slave Women to American Medicine," *Western Journal of Black Studies* (1987), rpt. in Darlene Clark Hine, ed., *Black Women in American History: From Colonial Times through the Nineteenth Century* (New York, 1990), 473–484; William J. Simon, "A Luso-African Formulary of the Late Eighteenth Century," *Pharmacy in History,* XVIII (1976), 114; George Way Harley, *Native African Medicine: With Special Reference to Its Practice in the Mano Tribe of Liberia* (London, 1970), 61–62, 215–216, 224.

29. Rafinesque, *Medical Flora,* I, 232; Barton, *Outlines of Lectures,* II, 196.

30. Meyer, *The Herbalist,* rev. Meyer and Meyer, 18, 21, 33, 92, 119; Rafinesque, *Medical Flora,* I, 88, 99, 162–165, 234. Squaw root and squaw vine were adopted in England by

Investigation of new sources was encouraged by the fact that there were as yet no distinct lines between folk and professional medical practice. Women's medicinal recipes for household use were as likely to be copied from medical books as they were to be handed down orally. Oral communication did not necessarily differ from professional medical advice.[31] Neither women nor doctors were strongly bound by tradition in this period. Elizabeth Coates Paschall, a Philadelphia shopkeeper, proudly recorded many medical recipes as "my own Invention" and discussed the efficacy and safety of drugs with a wide range of people, including doctors, botanists, apothecaries, midwives, relatives, friends, servants, and customers in her effort to discover the best cures.[32] Doctors and medical botanists asked questions of women practitioners and sometimes adopted their advice. While emmenagogic usage contained some traditional aspects in the New World, contact with native American and African medicine, a changed environment, and a fascination with new remedies encouraged experimentation.

Experimentation in Pennsylvania did not diffuse all forms of medicine equally. English and urban practices came to dominate over German and rural medicine. With only a few exceptions, the experiments of rural

---

the mid-19th century, at a time when herbal abortifacients or emmenagogues are usually considered to have been abandoned. See Malcolm Stuart, ed., *The Encyclopedia of Herbs and Herbalism* (London, 1979), 262, 274; Virgil J. Vogel, *American Indian Medicine* (Norman, Okla., 1970), 243–244; Gladys Tantaquidgeon, *A Study of Delaware Indian Medicine Practice and Folk Belief* (Harrisburg, Pa., 1942); William N. Fenton, "Contacts between Iroquois Herbalism and Colonial Medicine," in Smithsonian Institution, *Annual Report of the Board of Regents, 1941* (Washington, D.C., 1942), 503–527; Goodson, "Medical-Botanical Contributions," in Hine, ed., *Black Women*, 473–484.

31. Oral tradition was not confined to women chatting over the garden gate. After abortion was made illegal in Pennsylvania, Dr. Joseph Carson's printed lecture outline on the drug savin warned of its "criminal use." However, one student wrote in his lecture notes, "one of the most powerful agents of abortion," and then proceeded to record other specific abortion techniques. Obviously, what was said in class could not be published (MS student notes in 3d ed. of Carson, *Materia Medica* [Philadelphia, 1863], 181 [copy in stacks, Van Pelt Library, University of Pennsylvania, Philadelphia]).

32. Elizabeth Coates (Mrs. Joseph) Paschall, Receipt Book (ca. 1749–1766), photocopy, Historical Collections of the College of Physicians of Philadelphia; Ellen G. Gartrell, "Women Healers and Domestic Remedies in Eighteenth-Century America: The Recipe Book of Elizabeth Coates Paschall," *New York Journal of Medicine*, LXXXI, no. 1 (January 1987), 23–29.

women with western Pennsylvania plant species were not incorporated into urban pharmacologies. Neither the explicit language nor the drugs specific to German practice seem to have had any direct influence on English writings. African and native American remedies were generally secondary to other drugs. After the turn of the nineteenth century, few new plant species were added to the lists of accepted drugs. The period of experimentation came to an end in the early nineteenth century.

Domestic emmenagogic practices, based on herbs gathered from gardens and fields, were gradually supplanted by commercial emmenagogic preparations. These were imported from England or made by local pharmacists and physicians and predominated in the city of Philadelphia by the middle of the eighteenth century and in the countryside by the end of the century. Emmenagogic drugs were an important part of a flourishing international pharmaceutical industry. Patent medicines and standardized preparations like Hooper's Female Pills, Hungary Water (an infusion of rosemary), and Dr. Ryan's Worm-destroying Sugar Plumbs ("highly serviceable to the Female Sex") were advertised in local newspapers, particularly after 1750. Bulk drugs, including savin, madder, guaiacum, and other emmenagogues, were imported from various parts of the globe. Preparations of herbs—gums, infusions, essences, tonics, waters, cordials—could be purchased from pharmacists, physicians, or druggists. Medical chests containing an assortment of simple herbs, patent medicines, and prepared drugs were regularly advertised for sale by Philadelphia doctors and pharmacists for rural families with no access to apothecaries. These included emmenagogues. Commercialization began standardizing and publicizing prepared emmenagogic medicines even in remote rural areas by the turn of the nineteenth century.[33]

33. An error-ridden compilation of the ingredients in commercial emmenagogic medicines can be found in "A Collection of Medicinal Preparations By Saml Fahnestock M D Together with the Virtues and Dosis York Town Pennsylvania June 22d 1798," MS, Historical Collections of the College of Physicians of Philadelphia. Fahnestock was almost exclusively interested in emmenagogic drugs. Suppressed menses was the most common complaint of Chester County women seeking professional medical advice in the early 19th century. See Joan M. Jensen, *Loosening the Bonds: Mid-Atlantic Farm Women, 1750–1850* (New Haven, Conn., 1986), 33; S. Stander, "Transatlantic Trade in Pharmaceuticals during the Industrial Revolution," *Bulletin of the History of Medicine*, XLIII (1969), 326–343; Harold B. Gill, Jr., *The Apothecary in Colonial Virginia* (Charlottesville, Va., 1972). Quaker influence in the pharmaceutical trade is noted in Roy Porter and Dorothy Porter, "The Rise of the English Drug Industry: The Role of Thomas Corbyn," *Medical History*, XXXIII

The Regulation of Fertility: Results

The herbal remedies most commonly mentioned in eighteenth-century sources are still classed as emmenagogues in some national dispensatories. Although these medicinal herbs usually have been dismissed by physicians in the twentieth century as useless or dangerously toxic, a number of plants have been shown to be effective in laboratory tests. For example, a wide range of herbs, including aloes, mints, and rue have been reported to have produced antifertility effects in rats and mice by acting on the smooth muscles of the uterus or by disrupting ovulation. In addition, chemical analysis finds that none of these plants, with the exception of hellebore, has a "very dangerous" level of toxicity.[34] These herbal remedies were often combined with vigorous exercise, "baths," and other techniques.

The success of aggressive medical treatment can be seen in Elizabeth Coates Paschall's recollection, about 1750, of an induced abortion. "I once was verry Bad with a violent pain in my Back and Bowells and three months Gone with Child." She immediately decided that the pregnancy was the cause of what she called colic, but on her own effort she "Could not be Delivered." She then called for professional assistance: "It was Judged Both By the Doctor and midwife that if I was not Speedily Delivered I Should Dye." The male doctor confirmed her own diagnosis but played no other role. "The midwife tryed to Deliver me butt found it Imposible." With the failure of professional help, Paschall turned to another source of expertise, "an Elderly woman proposed Giving me a Glister [enema]." This worked. "I took it and Lay Still Near an hour after it: Being presently

---

(1989), 277–295; see also Robert A. Buerki, "Caleb Taylor, Philadelphia Druggist, 1812–1820," *Pharmacy in History,* XXX (1988), 81–88; advertisement of Nicholas Brooks in *Pennsylvania Gazette,* May 24, 1780.

34. As late as 1955, many of the drugs used in 18th-century Pennsylvania were still listed, but as "no longer official" or "abandoned," in Arthur Osol *et al., Dispensatory of the United States of America* (Philadelphia, 1955). These were dropped from later editions. For toxicity, see James A. Duke, *Handbook of Medicinal Herbs* (Boca Raton, Fla., 1985), 517–522 (the author is head of the Germplast Resources Laboratory of the U.S. Department of Agriculture). For research on the fertility effects of plants, see N. S. Farnsworth *et al.,* "Potential Value of Plants as Sources of New Antifertility Agents," *Journal of Pharmaceutical Sciences,* LXIV (1975), 535–598, 717–754. Using these and other medical sources, John M. Riddle, "Oral Contraceptives and Early-Term Abortifacients during Classical Antiquity and the Middle Ages," *Past and Present,* no. 132 (August 1991), 3–32, has argued for the effectiveness of herbal remedies.

Eased and the Child Came from me in the after Birth all together and with verry Little pain." In Paschall's case, the intensive use of a variety of techniques succeeded. Abortion was conceived to be a necessary by-product of her treatment of "a violent Chollick Pain," and her health was the only apparent concern.[35]

More ambiguous is the case of Elizabeth Drinker's daughter. Molly Drinker Rhoads was under investigation by her local Quaker meeting for her irregular marriage, was temporarily estranged from her father, and was without a home of her own. Only Elizabeth Drinker's "discourse with the Doctor relative to Molly" revealed the pregnancy, and Drinker worried that her daughter was "very careless of herself which makes me much the more uneasy." Rhoads complained of colds in her third and fourth month of pregnancy. Her mother reported that "she looked very pale and was in great pain in her bowels. she has been disorder'd and in pain for a long time past at times, oweing to taking colds—I gave her mint water etc." Peppermint and magnesia were the ingredients in mint water, given for intestinal complaints accompanied by "morbid depression." It was also an emmenagogue. The hysteric symptoms reappeared two months later: "She then burst into tears and inform'd us how ill she had been most of the night—with pain, disorder'd bowels and a fluttering etc. and at times fainty—she took mint water and lavender compound, became rather easier towards morning—she was very much terrified." Lavender compound contained several items used as emmenagogues and was also used in nervous disorders. At six months' gestation, a fluttering in the bowels was not identified with fetal movement. In the next month, however, preparations were underway for the expected delivery, which occurred two months later. The full-term fetus would be stillborn.[36] For the greater part of this pregnancy, Molly Rhoads hesitantly treated her symptoms as an illness. Her condition was labeled a cold and medicated

35. Paschall, Receipt Book (ca. 1749–1766), 9, photocopy, Historical Collections of the College of Physicians of Philadelphia. Until the late 19th century, medical texts defined clysters (glisters, enemas) as the anal insertion of liquids, although perhaps vaginal or uterine administration was practiced in this case and others.

36. Biddle, *Materia Medica,* 164, 180; Farnsworth *et al.,* "Potential Value of Plants," *Journal of Pharmaceutical Sciences,* LXIV (1975), table 3, 550; Carson, *Materia Medica,* 180. Lavender compound included lavender, rosemary, alcohol, and red dye. The red coloring may be a bit of sympathetic magic. See Elaine Forman Crane *et al.,* eds., *The Diary of Elizabeth Drinker* (Boston, 1991), II, 873–874, 878, 897, 900, 904–932.

as if it were hysteria through the first six months of pregnancy. She did not, as she was expected to do, reduce her labor or watch her diet. Molly Rhoads seems to have been as ambivalent about her pregnancy as she was tentative in coping with her physical and emotional symptoms.

Eight cases of emmenagogic treatments administered by Drs. Thomas and Phineas Bond in Philadelphia in the 1750s offer further clues to contemporary practice. In six of the eight cases a single round of four emmenagogic pills sufficed. A second round of medicines was given in the other two cases. In one, the second dose was administered two days after the first, and in the other case the second dose was given along with a tonic six weeks later. In seven of the eight cases other symptoms besides amenorrhea were being treated. Four patients were simultaneously treated for hysteria. Several patients were given cathartics, emetics, or mercury (a common prescription for syphilis but also used for worms or amenorrhea). Some were treated for fever or pain. In these seven cases, emmenagogues were not the sole medicines given—other symptoms were thought to be involved. In two cases the cures worked too well, and the patients had to be treated with styptics to stop bleeding, including one who apparently lost so much blood that tincture martis was given—a salt of iron popular for female troubles and later found to be specifically useful in anemia.[37] The course of treatment for obstructed menses was not pleasant, but it seemed to work immediately in six of the eight cases.

By the end of the eighteenth century, increasing population density meant that fewer city women had access to gardens. The Philadelphia Dispensary, a charitable institution, provided outpatient services to the worthy poor who were unable to attend to themselves or to afford private physicians. In the first six years, from December 1786 to November 1792, the dispensary treated 83 women for amenorrhea; complete cures were achieved for 80 percent of these patients. Of the remainder, 6 percent

37. Thomas and Phineas Bond Co-Partnership Ledgers, Historical Collections of the College of Physicians of Philadelphia. Vol. I: account of Widow Robeson (of Chester), 1752, p. 151; of Richard Edwards, 1753, p. 385. Vol. II: two accounts of Robert Taigart (shopkeeper), 1753/4, p. 9; of Thomas Bottom, 1755, p. 347; of Ludowic Cosser (shoemaker), 1755, p. 358; of William Jackson, 1756, p. 498. Vol. III: account of Mr. John Wallace, 1759, p. 62. An unscientific sampling of cases was made by looking at the accounts on roughly every 60th page. Later accounts are not suitable for analysis, containing either vague statements like, "Visiting attendance and advice to his wife," or rough notes that make it impossible to follow specific cases.

proved unworthy of care and were discharged for disorderly behavior. Somewhat fewer than 5 percent were only relieved of their symptoms, and almost 10 percent were held for additional treatment as outpatients or in the House of Employment. A substantially larger number of impoverished women complained of hysteria, at a time when amenorrhea was a primary symptom of the disease. Of the 246 cases, 71 percent were cured, but 24 percent were only relieved. Nine women were treated for the aftereffects of abortion, and only one required prolonged care. No deaths resulted from any of these treatments. During the same six-year period, 84 women successfully delivered their infants under the care of the dispensary, many fewer than sought treatment for amenorrhea and hysteria.[38] Both the Bond casebooks and the dispensary records indicate that emmenagogues successfully restored menstruation in 70–85 percent of cases. It appears that eighteenth-century emmenagogues were reasonably effective and safe.

Demographic evidence reveals that eighteenth-century fertility rates were variable, not constant. If anxiety, grief, and distress were signs of pathology in married women, then it would be expected that emmenagogues would be most widely and persistently used during wars and other crises. The connection between crisis and lowered fertility was recognized at the time. A rumor of war in May 1706, fabricated in order to frighten the Quakers, had another effect, since "many women who were in delicate situations miscarried in consequence of their fright." The number of births recorded in the Friends' meeting records dropped from twenty-six to four between 1705 and 1706. It is impossible that fright alone could induce spontaneous abortion and result in so great a shortfall in births (emotion is not now considered a factor in fetal loss). Both lower fertility and its cause were recognized by Abigail Adams in December 1775, at the beginning of the Revolutionary war. "I believe Philadelphia must be an unfertile soil, or it would not produce so many unfruitful women. I always conceive of these persons as wanting one addition to their happiness; but in these perilous times, I know not whether it ought to be considered as an infelicity, since they are certainly freed from the anxiety every parent must feel for their rising offspring." That is, women's anxiety led to infertility. Dr. Benjamin Rush noted, "In 1783, the year of the peace, there were several children born of parents who had lived many years together with-

38. Compiled from Samuel P. Griffitts *et al.*, "A Return of the Diseases of the Patients of the Philadelphia Dispensary" College of Physicians of Philadelphia, *Transactions,* I, part 1 (Philadelphia, 1793), 1–42.

TABLE 1. *Mid-Atlantic Fertility*

Total Marital Fertility Rates
(Expected Number of Children)
Average Family Size
(Actual Number of Children)

	Period			
Group	Colonial: ca. 1680– 1775	Revolutionary: ca. 1750– 1800	Early National: ca. 1780– 1830	National: ca. 1800– 1870
Mid-Atlantic Quakers	8.8	7.5	6.2	—
	6.7	5.7	5.0	—
Philadelphia gentry	10.1	—	9.4	7.1
	7.5	—	7.9	5.3
Urban Jews	10.4	8.7	9.0	—
	6.0	5.1	6.4	—
Philadelphia merchants and artisans	8.7	7.9	8.9	7.9
	6.0	5.6	6.6	5.3
Philadelphia mechanics and laborers	7.8	8.9	7.2	—
	5.0	5.2	4.5	4.4
Lancaster County	10.6	10.5	10.6	—
	7.2	7.3	7.1	—
Bethlehem Moravians	8.0	6.5	7.6	7.6
Pennsylvania Schwenckfelders	5.5	6.0	7.3	—

TABLE 1. Continued

	Period			
Group	Colonial: ca. 1680– 1775	Revolutionary: ca. 1750– 1800	Early National: ca. 1780– 1830	National: ca. 1800– 1870
Mean				
Overall	9.2	8.4	8.4	7.5
	6.3	*5.8*	*6.4*	*5.0*
Without				
Lancaster	9.0	8.1	8.3	7.5
	6.1	*5.5*	*6.3*	*5.0*

Note: The total marital fertility rate measures the number of births expected if all women married on their 20th birthday and remained married for thirty years. Average family size encompasses the actual number of children born at both earlier and the more common later ages at marriage and in those marriages broken by death.

Sources: Quaker women married before age 25, with estimate of marital fertility at ages 45–49 added: Robert V. Wells, "Family Size and Fertility Control in Eighteenth-Century America: A Study of Quaker Families," *Population Studies,* XXV (1971), 78 (family size for all families, 75); Louise Kantrow, "Philadelphia Gentry: Fertility and Family Limitation among an American Aristocracy" *Population Studies,* XXXIV (1980), 24, 27. Jewish women married before age 25: Robert Cohen, "Jewish Demography in the Eighteenth Century: A Study of London, the West Indies, and Early America" (Ph.D. diss., Brandeis University, 1976) 127–128 (actual family size for all families, 130). Philadelphia women married before age 30: Susan E. Klepp, *Philadelphia in Transition: A Demographic History of the City and Its Occupational Groups, 1720–1830* (New York, 1989) 217–219; Rodger Craige Henderson, "Community Development and the Revolutionary Transition in Eighteenth Century Lancaster County, Pennsylvania" (Ph.D. diss., SUNY-Binghamton, 1983), 169, 176, 349, 366, 523, 530; Beverly Prior Smaby, *The Transformation of Moravian Bethlehem: From Communal Mission to Family Economy* (Philadelphia, 1988), calculated from data on 75; Rodger C. Henderson, "Eighteenth-Century Schwenkfelders: A Demographic Interpretation," in Peter C. Erb, ed., *Schwenkfelders in America* (Pennsburg, Pa., 1987), 39.

out issue." He also found that the war did not affect all groups equally, since wartime employment led to an "increase of births" and "favored marriages among the laboring part of the people." Women's childbearing experiences reflected their circumstances.[39] War, economic stress, and the dangers and deprivations involved in establishing European settlements on the frontier resulted in lowered fertility. Peace and prosperity favored higher birthrates.

Demographic information shows the variety of fertility experiences in Pennsylvania (see Table 1). Important differences in women's experiences are apparent. Wealthy women had more children and higher fertility than other groups. Fertility was lower where financial pressures were greater. Rural, agricultural Lancaster County had the highest marital fertility rates and the second-largest family size, according to a study by Rodger Henderson. (This finding may be a consequence of his methodology, since women exhibiting a large gap between births were credited with an unrecorded baby. This assumption could artificially raise fertility rates and family size and mask change over time.) Other groups of Pennsylvania women had lower fertility rates. The tendency was for marital fertility to be highest in the early colonial period and then to drift downward, except on the frontier, where family size rose as material conditions improved. Even in his instance it should be remembered that settlement was marked by war against the native people.

For the majority of women in the mid-Atlantic region (see Table 2), marital fertility was high in the early eighteenth century. Colonial women surpassed even their supposedly noncontracepting European contempo-

---

39. An account of the 1706 episode is in Gary B. Nash, *Quakers and Politics: Pennsylvania, 1681–1726* (Princeton, N.J., 1968), 258–259. Andreas Sandel is quoted in Andrew Rudman, "Transactions relative to the Congregation at Wicacoa," 5, microfilm of MS, Gloria Dei Church Records, Genealogical Society of Pennsylvania. The births are calculated from the records of the Philadelphia Monthly Meeting of Friends, Genealogical Society of Pennsylvania. Sam Shapiro *et al.*, *Infant, Perinatal, Maternal, and Childhood Mortality in the United States* (Cambridge, Mass., 1968), 47–77; Charles Francis Adams, ed., *Familiar Letters of John Adams and His Wife Abigail Adams, during the Revolution, with a Memoir of Mrs. Adams* (New York, 1876), 129. On lower fertility on the frontier, see Richard A. Easterlin, "Factors in the Decline of Farm Family Fertility in the United States: Some Preliminary Research Results," *Journal of American History*, LXIII (1976–1977), esp. table 1, 602. Benjamin Rush, "An Account of the Influence of the American Revolution on the Human Body" (1788), in Dagobert D. Runes, ed., *The Selected Writings of Benjamin Rush* (New York, 1947), 330–331.

TABLE 2. *Mid-Atlantic Age-Specific Marital Fertility Rates*

			Period		
Age	European "Natural" Fertility	Colonial: ca. 1680– 1775	Revolutionary ca. 1750– 1800	Early National: ca. 1780– 1830	National: ca. 1800– 1870
		No. of Children per 1,000 Years of Marriage			
15–19	—	434	414	387	431
20–24	435	503	453	441	476
25–29	407	427	396	422	453
30–34	371	387	344	367	294
35–39	298	308	250	280	213
40–44	152	151	154	153	112
45–49	22	26	29	26	6
		Ratio of Change from Previous Generation			
15–19			95	93	111
20–24		116	90	97	108
25–29		105	93	106	107
30–34		104	89	107	80
35–39		103	81	112	76
40–44		99	102	99	73
45–49		118	112	90	23

Sources: European "natural" fertility: Louis Henry, *Population: Analysis and Models,* trans. Étienne van de Walle and Elise F. Jones (New York, 1976), 90. The other rates are the simple mean of the following. Quaker women married before age 25 with estimate of marital fertility at ages 45–49 added: Robert V. Wells, "Family Size and Fertility Control in Eighteenth-Century America: A Study of Quaker Families," *Population Studies,* XXV (1971), 78; Louise Kantrow, "Philadelphia Gentry: Fertility and Family Limitation among an American Aristocracy" *Population Studies,* XXXIV (1980), 24. Jewish women married before age 25: Robert Cohen, "Jewish Demography in the Eighteenth Century: A Study of London, the West Indies, and Early America" (Ph.D. diss., Brandeis University, 1976), 127–128. Philadelphia women married before age 30: Susan E. Klepp, *Philadelphia in Transition: A Demographic History of the City and Its Occupational Groups, 1720–1830* (New York, 1989), 219.

raries, especially when they were newlyweds in their early twenties. Substantially fewer European women would have been married at so young an age, but when they were, they bore fewer children. The differences between the Old World and the New were less pronounced when women were between twenty-five and forty-four. Small numbers inflate the differences in fertility after age forty-five, but the differences in the experiences of women at the beginning and end of their childbearing years may indicate that European women were deliberating curtailing births. The crisis of the Revolutionary war led many American women to limit childbearing. Women in their thirties experienced the greatest decline in fertility. They may have become more skilled in their use of contraceptive or abortive methods than they had been when in their twenties, or this age group may have felt the full brunt of the war and therefore have been more motivated to reduce childbearing. For the majority of this generation, the war was over by the time they reached their forties. Fertility levels reverted to prewar levels and contributed, along with the higher fertility of the next generation of mid-Atlantic women, to a postwar baby boom. The third generation of mid-Atlantic women did have higher fertility than their mothers, but they did not replicate the experiences of colonial women. Women in the early national period had fewer children in their early twenties than either of the two previous generations of women. They also had lower fertility after age thirty-five than colonial women. The fourth generation of mid-Atlantic women, who married and bore children in the early nineteenth century, reversed the trend established by their mothers and grandmothers of steadily reducing fertility while in their twenties. Rather, these women had high rates of childbearing in their earliest years of marriage (as colonial women had done), but unlike eighteenth-century women, they severely limited childbearing after age thirty. Fertility and family size would fall to levels lower than any prevailing in the previous century. These changes in childbearing over four generations of women's lives would occur without any substantial change in contraceptive technology. It would seem from these studies that Pennsylvania women were not at the mercy of their biological natures, that they were not beguiled by magical and ineffective folk practices, but that they had some real influence over their fertility and could adjust childbearing as a rational response to changes in their situations.[40]

40. These fertility shifts in Pennsylvania are mirrored, 100 years later, in Utah. See Lee L. Bean *et al.*, *Fertility Change on the American Frontier: Adaptation and Innova-*

The emmenagogic techniques in use in eighteenth-century Pennsylvania allowed women to exercise some control over reproduction and also allowed both men and women to avoid, in reading, in writing, and in public and private conversations, any reference to indelicate topics. It should not, however, be seen as a golden age of women's power over reproduction. The failure rates were high, the methods painful, and the acceptable reasons for employing emmenagogues limited to conditions of emotional and physical ill health. These justifications for attempting to restore menstruation would need to be reexamined every month and weighed at the beginning of each pregnancy. Hysteria could be invoked by women in times of personal, political, or financial crisis in order to delay childbearing, but at a cost of succumbing to irrationality, overwhelming fear, and emotional excess—the very characteristics that were used to define women as inferior human beings, unfit for independence in marriage or society.[41] Other diseases like suppressed menses and the resulting colds, rheumatism, and pleurisy labeled women as physically weak and sickly by their very nature. In addition, notions of shame and propriety restricted access to such information as was available and could obscure appropriate procedures. Restriction of fertility as practiced in the eighteenth century gave women a limited ability to control their fertility, but, rather than liberating women, only confirmed their separate and unequal status.

The control that most eighteenth-century women sought over their fertility would, under ideal circumstances, allow them to delay childbearing until physical or psychological health returned, until the family's financial security improved, or until the end of a war. If all was well, then childbearing might resume. Family limitation was not yet a lifetime goal.

The Emergence of Family Planning, Contraception, and Abortion

The emmenagogic techniques were most effective in lengthening the intervals between births: their primary use during the eighteenth century

---

tion (Berkeley, Calif., 1990). My thanks to Robert V. Wells for bringing this work to my attention.

41. For a discussion of woman's character, see Vivien Jones, ed., *Women in the Eighteenth Century: Constructions of Femininity* (London, 1990). For a later period, see Carroll Smith-Rosenberg, "The Hysterical Woman: Sex Roles and Role Conflict in Nineteenth-Century America," in Smith-Rosenberg, *Disorderly Conduct: Visions of Gender in Victorian America* (New York, 1985), 197–216.

was, insofar as can be determined, to allow health and well-being to return. Reducing family size was of little apparent interest to most women, since fertility rates tended to rebound once a crisis in health, security, or finances had passed. This reversion happened after wars, as the frontier was tamed, as religious uncertainty ended. When, at the end of the eighteenth century, first women and then men began to express an interest in family planning, in stopping childbearing after a predetermined number of children had been born, new techniques were tried. (The social and economic reasons behind that change are, obviously, too complex to discuss here.)

Margaret Shippen Arnold wrote to her sister, Elizabeth Shippen Burd, near the close of the eighteenth century: "It gives me great pleasure to hear of your prudent resolution of not increasing your family; as I can never do better than to follow your example, I have determined upon the same plan; and when our Sisters have had five or six, we will likewise recommend it to them." The Shippen sisters made these decisions on their own, as women had throughout the century. Their husbands were not involved either in the decision or in the attempt to not increase the family. The Shippen sisters knew how to stop bearing children, although what method they intended to follow was not stated. Sixty years later, "prudent considerations" was the code phrase for abortion.[42]

In 1794, Moreau de St. Méry began selling syringes in his bookstore in Philadelphia. "I wish to say that I carried a complete assortment of them for four years; and while they were primarily intended for the use of French colonials, they were in great demand among Americans, in spite of the false shame so prevalent among the latter. Thus the use of this medium on the vast American continent dates from this time." Despite these grandiose claims, retailing syringes was hardly new in the city. Nathaniel Tweedy, druggist, was advertising "small ivory syringes," along with Hooper's Female Pills and Fraunce's Female Elixir, at his store on Second Street in 1760. It may be, however, that these devices were more acceptable and in greater demand by the end of the century. Moreau de St. Méry observed: "Syringes, when first imported by French colonists, seemed a hideous object. Later they were put on sale by American apothe-

---

42. Klein, *Portrait of an Early American Family*, 285 n. 105. A denunciation of abortion—delicately named "prudential considerations"—appeared in an article about the low fertility rates of native white Americans. See "Social Science," *Ladies' Repository: A Monthly Periodical, Devoted to Literature and Religion*, N.S., I (1868), 52.

caries." What does appear to be new, and perhaps a consequence of the French presence in Pennsylvania during the American and French revolutions, is that these syringes were probably being used, not to douche, but to dilate the cervix. Moreau de St. Méry's description of these devices, "ingenious things said to have been suggested by the stork," lends credence to this supposition, if the stork's method of feeding its young is pictured. This is the first suggestion of uterine-intrusive abortion or contraception in Pennsylvania. Another hint that abortion techniques were changing appeared in 1792, when Dr. William Currie listed the distinctive diseases of various occupational groups in the city. The "ladies" of "the opulent and fashionable class" were especially subject to hysteria, and Currie also included menorrhagia, spontaneous abortion, and sterility as diseases specific to this group of women. These latter three conditions could have been the results of botched uterine-intrusive abortions: abortions that had perforated the uterus, had failed to expel the fetus immediately, or had produced infection. In the 1790s, abortion techniques included uterine intrusion as well as emmenagogic techniques. A generation later, in the 1830s, there would be much more openness in advertising the availability of both uterine-intrusive abortion and abortion medicines.[43]

An interest in abstinence also coincided with the interest in family planning, and some women attempted to control births by limiting sexual intercourse. Mary Drinker Cope had her fourth baby in May 1798. In November 1800, her husband wrote in his diary: "A man of feeling has his wishes thwarted by childish opposition. His home is rendered uncomfortable by the peevishness and cold manners of his spouse. Indifference ensues. His spirits are broken and if he does not seek to forget his chagrin by dissipation and unworthy indulgences abroad, melancholy follows and some fit of maniacal desperation finishes the tragedy; and all perhaps for want of a little common prudence in his wife." Mary Drinker Cope gave birth nine months later to a stillborn son. Abstinence could not work without the cooperation of wife and husband. By 1805, Cope had borne six children, and the family was in severe financial straits, when her husband came home and found her nearly dead. Her paleness and weak pulse indicate massive loss of blood; the "paroxysms" and "spasms," uterine

---

43. Moreau de Saint-Méry, *American Journey*, 177–178, 314–315; advertisement, *Pennsylvania Gazette*, Nov. 27, 1760; Mohr, *Abortion*, 3–45; William Currie, *An Historical Account of the Climates and Diseases of the United States of America . . .* (Philadelphia, 1792), 107.

contractions. Cope's husband did not approve of her attempts to restrict fertility, and she had two more children.[44]

Most women in the early nineteenth century apparently preferred to intensify their use of emmenagogic techniques rather than turn to uterine-intrusive abortion or abstinence. There seems to have been less embarrassment in discussing breastfeeding, at least as a method of delaying births, by the early decades of the nineteenth century. Elaine Crane has found that, in line with their mother's advice, "all three of Elizabeth Drinker's daughters appear to have nursed their children for longer periods than their mother did. Even Nancy (who nursed her children for shorter periods than the others), breast-fed at least sixteen months." Margaret Izard Manigault told a married daughter in 1809, "I think it less fatiguing to the constitution to nurse this one, than to bring forth another." Another of her daughters had so imbibed the new ideas that she criticized an aunt for becoming pregnant. Their "family is sufficiently large," she wrote in 1814. Having too many children was objectionable, imprudent behavior.[45]

The local demographic reconstructions of marital fertility find that increasing intervals between births were the first indications of a sustained fertility decline. Only later did women stop bearing children at ever earlier ages.[46] Druggists continued to advertise drugs for women's "troubles." The Philadelphia Dispensary and other institutions offered emmenagogic

44. See Nancy F. Cott, "Passionlessness: An Interpretation of Victorian Sexual Ideology, 1790–1850," *Signs: Journal of Women in Culture and Society,* IV (1978–1979), 219–236; Eliza Cope Harrison, ed., *Philadelphia Merchant: The Diary of Thomas P. Cope, 1800–1851* (South Bend, Ind., 1978), 48, 78, 172–175, 292–293. Is the outcome of this pregnancy, like that of Molly Drinker Rhoads's above, a consequence of the pills and potions taken for colds and lost menses? Since the emmenagogues can affect the smooth muscles, their ingestion during pregnancy might have damaged lung, digestive tract, and heart development. A common cause of death among infants in the 18th century was the inability to suckle, perhaps also related to drug use during pregnancy and lactation. It may be that lowered infant mortality was in part a consequence of the switch to new methods of contraception and abortion.

45. Elaine F. Crane, "The World of Elizabeth Drinker," *PMHB,* CVII (1983), 12; Norton, *Liberty's Daughters,* 233; Virginia Armentrout and James S. Armentrout, Jr., eds., *The Diary of Harriet Manigault, 1813–1816* (Rockland, Maine, 1976), 51.

46. See the work of Wells, "Family Size and Fertility Control," *Population Studies,* XXV (1971), 80; Louise Kantrow, "Philadelphia Gentry: Fertility and Family Limitation among an American Aristocracy," *Population Studies,* XXXIV (1980), 28; and Klepp, *Philadelphia in Transition,* 187, for birth intervals and age at last birth.

pills to the poor. Home health guides proliferated in the early nineteenth century, offering more detailed advice on familiar emmenagogic practices. Newer techniques of fertility regulation may have been suspect because of their intrusive nature, because of their risk or discomfort levels, or because of their newness. In the meantime older emmenagogic techniques of regulation were gradually being redefined as contraceptive and abortifacient procedures as women began to articulate a desire to reduce fertility permanently.

A combination of an intensified use of older technologies with a gradual employment of newer techniques initiated a long-term decline in fertility levels as women sought to achieve an ideal family size. With detailed censuses available, fertility rates can be measured for far larger groups than is possible for the eighteenth century. Child-woman ratios (the number of young children of given ages per one thousand women of childbearing age) show the urban leadership in the reduction of fertility that is a characteristic of Pennsylvania and the United States in the nineteenth century. But they also show, as do the estimates of crude birthrates, that fertility was falling, albeit from a higher starting point, in the countryside as well as the city. The pace of the decline in fertility is similar for urban and rural women.

The steady decline in the birthrate masks conflict over who would now control fertility and what methods were appropriate. The old, strict gender roles concerning gynecology began to disappear in the new openness about fertility regulation and family planning. Men were increasingly interested in the process of contraception, but they too initially urged traditional practices. Dr. Anthony Fothergill wrote a male friend in 1810, "5 Children already! after a 6th appears—'Festina lente "ne quid nimis." Voluptatis commendat rarior usus.'" (Roughly: "Make haste slowly 'nothing in excess.' Less frequent exercise of one's appetite will be rewarded.") Partial abstinence was the only technique that the doctor could or would promote even in the privacy of the Latin language, but he now assumed that men would make the decisions about childbearing. Obstetricians were taught by 1825 "that she [the patient] may secure to herself the best possible chance for her safety, let her have no opinion of her own, that may clash with those of her medical attendant; let her therefore be passive, and obedient." The high incidence of stillbirths in the city caused one doctor to worry that it will "serve to cast a shade of reproach on the females and medical profession of Philadelphia." By the middle of the nineteenth century doctors perceived married women complaining of amenorrhea as

TABLE 3. *Child-Woman Ratios*

	1698	1784	1800	1810	1820	1830
Group			Children per 1,000 Women			
Children Age 0–4, Women Age 15–50						
Swedish Lutheran	1,218	723				
Children Age 0–10, Women Age 16–44						
Philadelphia Co.			1,201	1,200	1,177	
Pennsylvania			1,880	1,840	1,750	
United States			1,820	1,820	1,730	
Children Age 0–4, Women Age 15–39						
Philadelphia Co.					742	598
Surrounding counties						
1st ring					815	772
2d ring					896	824
3d ring					968	855
Children Age 0–4, Women Age 15–50						
Philadelphia City						417
Philadelphia Co.						508
Eastern Pennsylvania						668
Western Pennsylvania						865
Mid-Atlantic States						718
United States						781

TABLE 3. continued

	1698	1784	1800	1810	1820	1830
Group	Children per 1,000 Women					
	Relative Ratio of Female Children Age 0–1 per Woman (Estimate)					
Rural Pennsylvania	100					
Pennsylvania cities, 2,500–10,000	85					
Philadelphia	57					

Sources: Susan Klepp, "Five Early Pennsylvania Censuses," *Pennsylvania Magazine of History and Biography,* CVI (1982), 496; A. J. Jaffe, "Differential Fertility in the White Population in Early America," *Journal of Heredity,* XXXI (1940), 407–411; Tom W. Smith, "The Dawn of the Urban-Industrial Age: The Social Structure of Philadelphia, 1790–1830" (Ph.D. diss., University of Chicago, 1980), 111; unpublished work of John Modell (quoted in Diane Lindstrom, *Economic Development in the Philadelphia Region, 1810–1850* [New York, 1978], 225 n. 21), MS in the Archives of the Philadelphia Social History Project, University of Pennsylvania; Jim Potter, "The Growth of Population in America, 1700–1860," in D. V. Glass and D.E.C. Eversley, eds. *Population in History: Essays in Historical Demography* (Chicago, 1965), 674–676, with calculation of Philadelphia County rate by the author.

having "a design to entrap me into the administration of drugs that might remove the difficulty by procuring abortion." More and more doctors saw their professional integrity in terms of their authority over women and women's medical knowledge. However, pharmacists, druggists, and doctors of materia medica (pharmacology) clung to their knowledge of herbal medicines.[47]

47. Anthony Fothergill Letterbook, 241, American Philosophical Society. See also G. J. Barker-Benfield, "The Spermatic Economy: A Nineteenth-Century View of Sexuality," in Michael Gordon, ed., *The American Family in Social-Historical Perspective,* 2d ed. (New York, 1978), 374–402; Charles D. Meigs, *Obstetrics: The Science and the Art* (Philadelphia, 1848), 160; W. P. Dewees, "Recapitulation . . . [On the Prevention of Abortion]," *Philadelphia Journal of the Medical and Physical Sciences,* N.S., II (1825), 4; Gouverneur Emerson, "Medical Statistics: Being a Series of Tables, Showing the Mortality in Philadelphia, and

Many husbands began to assume control over fertility decisions. For example, Daniel Stein of suburban Pittsburgh in the 1860s, "determined upon having no more than five children"; without apparently consulting his wife, Stein fathered two more children after two died in infancy. Sidney Fisher of Philadelphia in the 1850s decided, "I am not rich enough to have children," although his wife wanted, and eventually had, a child. Women, meanwhile, supported the idea of "voluntary motherhood," emphasizing their right to make independent, rational choices. New forms of contraception were publicized that concentrated on methods linked to the act of intercourse, like douches, sponges, or cervical caps, or on methods controlled by men, like coitus interruptus (male withdrawal before ejaculation), coitus reservatus (entire suppression of ejaculation, promoted by some American reformers in the nineteenth century), or condoms.[48] These were very different technologies from those that had prevailed in the eighteenth century. Centuries-old notions of shame had to be discarded by both women and men. These new technologies required continual premeditated interference with sexual intercourse and far greater self-control but were not markedly more failure-proof. They do reflect men's interest in and increasing control over fertility regulation. This later history of fertility regulation took place under very different assumptions about appropriate gender roles, about the meaning and practice of medicine, about the goals of fertility control. It has often been assumed that this later history marks the beginning of contraception and abortion in America, but women in the Delaware Valley already had a long history of fertility regulation.

Its Immediate Causes, during a Period of Twenty Years," *American Journal of the Medical Sciences,* I (1827–1828), 122; James Reed, "Doctors, Birth Control, and Social Values," in Morris J. Vogel and Charles E. Rosenberg, eds., *The Therapeutic Revolution: Essays in the Social History of American Medicine* (Philadelphia, 1979), 159–175.

48. James R. Mellow, *Charmed Circle: Gertrude Stein and Company* (New York, 1974) 31; Nicholas B. Wainwright, ed., *A Philadelphia Perspective: The Diary of Sidney George Fisher Covering the Years 1834–1871* (Philadelphia, 1967), 242. The standard work on the 19th century is Linda Gordon, *Woman's Body, Woman's Right: A Social History of Birth Control in America* (New York, 1976).

TABLE 4. *Crude Birth Rates, Philadelphia and Rural Pennsylvania Counties, 1690–1870*

| | Births per 1,000 Population | |
	Philadelphia	Rural Counties
1690	41	—
1700	36	—
1710	39	—
1720	45	—
1730	52	—
1740	50	—
1750	53	—
1760	53	61
1770	39	61
1780	50	61
1790	46	60
1800	37	59
1810	36	58
1820	43	57
1830	39	54
1840	35	52
1850	32	48
1860	27	44
1870	25	39

Sources: Susan E. Klepp, "Demography in Early Philadelphia, 1690–1860," in Klepp, ed., *Symposium on the Demographic History of the Philadelphia Region, 1600–1860* (American Philosophical Society, *Proceedings,* (XXXIII, no. 2 [June 1989]), 103–106, as revised in light of comments by Billy G. Smith and P.M.G. Harris, based on Philadelphia Bills of Mortality and church records; Morton Owen Schapiro, "Land Availability and Fertility in the United States, 1760–1870," *Journal of Economic History,* XLII (1982), 599, back projection based on 19th-century census data.

From One Technology to Another:
Linking the Eighteenth and Nineteenth Centuries

What today would be classed as abortive and contraceptive activities were widely practiced by women in eighteenth-century Pennsylvania, although not usually for the purpose of family planning. Birth control was a by-product of women's attempts to improve their physical and mental health by increasing the intervals between births when necessary. The emmenagogic techniques these women used to restore "lost or hidden menses" have themselves been lost to later generations and hidden by euphemistic language. In eighteenth-century Pennsylvania, women's decisions were private, and discussion of issues relating to fertility were appropriate only to married women and shameful for men. Given the great reluctance to discuss gynecological subjects in public, the contraceptive and abortive intentions of women are often obscure. The results are, however, much clearer. When women were motivated by ill health, the threat of war, or financial difficulties, they were largely successful in temporarily delaying childbearing. This control was achieved at a cost: the methods were painful and were limited to the treatment of severe physical and emotional weakness and distress. Only at the end of the eighteenth century and in the early decades of the nineteenth century is there evidence that family planning had become the primary goal of fertility regulation. And only then is there a sustained decline in both family size and marital fertility rates. Prudent choice, and not illness or anxiety, would henceforth justify fertility decisions made over an entire lifetime. But whose choice was a question that would be debated by men, women, physicians, church, and state.

Control over fertility had been women's preserve in the eighteenth century, although the rationality of women's choices was hidden by reference to illness and hysterics. But as fertility goals shifted from a concern over immediate circumstances to long-range planning of an ideal family size, the separate spheres of women and men in fertility and gynecology collapsed. A new social norm of acceptable family size, initially defined as five or six children, replaced the more varied and more flexible attitudes on fertility of the eighteenth century. Men were increasingly involved, because they, too, had social and economic interests in reduced fertility and because male involvement made the new goals more attainable. Some women found the new technologies requiring male interference in gynecology to be offensive or oppressive. These women continued to stress the older, woman-centered practices, or they denounced all contraceptive and abortive activities as shameful, preferring to advocate sexual abstinence. Some men, too, favored sexual restraint, either for everyone or for the lower classes. Other women turned to their husbands for help in achieving an end to childbearing through the employment of coitus interruptus or barrier methods of contraception. Some husbands followed their wives' wishes; other husbands asserted control over yet another aspect of their wives' lives. Some women turned to the medical profession for aid. But most obstetricians saw female-dominated medicine as an obstacle to professionalization. There were other commercial and professional sources of aid, all requiring cash and most in the hands of men. The disappearance of the old rigid gender roles on gynecology produced a growing consensus on the desirability of limiting births, but there was no unanimity among men or among women on the appropriate technology. Still, where goals and methods were contested across a gendered divide, it was the men who had more power and authority. Women's direct control over reproduction eroded throughout the nineteenth and much of the twentieth century even as they were freed of the nearly constant childbearing of their grandmothers. By the turn of the twentieth century, eighteenth-century gendered roles had reversed in at least some segments of the population. Shame was attached to women's knowledge of reproduction, and men were the experts who managed fertility decisions and births.

The characterization of eighteenth-century fertility regulation as a traditional medical art rooted in ineffective, irrational folk practice is inaccurate. In the eighteenth and early nineteenth centuries, emmenagogic techniques were not simply passed by tradition from one generation of women to the next. Advice was also available in medical texts for home use and provided in professional medical care. Women experimented, and some European practices were adopted in the New World while others were not. African and native American plants were added to the existing materia medica. Women's demand for emmenagogic drugs produced an international trade in these medicines. These eighteenth-century methods successfully restored menstruation and delayed childbearing in the majority of cases so that fertility levels were responsive to the objective circumstances of women. Knowledge of these fertility-regulating practices and drugs seems to have been widespread among women and was an important element in both fertility variation and in the later sustained decline of fertility in Pennsylvania. Although new techniques of abortion and contraception would be introduced over the course of the nineteenth

century and would eventually come to supplant earlier practices, emmenagogic techniques would remain important well past the period covered
in this study. However, as the interest in and demand for fertility regulation increased, the body of knowledge that had been largely private,
domestic, and female became commercialized, professionalized, and public. The debates that first emerged in the nineteenth century over who
would or should control reproduction and by what means continue to the
present day.

Bibliographic Essay: A Note on the Literature

Norman E. Himes's wide-ranging survey of contraceptive knowledge
(*Medical History of Contraception* [1936; New York, 1970]) has had a
lasting impact on the study of fertility. Himes devoted comparatively little
space to early modern Europe, but he argued that, because of the teachings
of Saint Thomas Aquinas and the prevalence of magical beliefs, there were
few, if any, effective forms of fertility control available in Western Europe
before the middle of the nineteenth century. Only the discovery of barrier
and chemical methods of birth control signaled an end to a long period of
ignorance. Demographers and economists, building on Himes's research,
have developed the concept of "natural fertility" to explain the high levels
of fertility that usually prevailed before the eighteenth century in France
and the nineteenth century in much of the rest of the Western, industrialized world. According to this theory, the only certain controls over
fertility in preindustrial Europe were late average ages at marriage and
relatively high celibacy rates. Once marriage occurred, biology, and not
human volition, governed fertility. See Étienne van de Walle, "De la nature
à la fécondité naturelle," *Annales de démographie historique*, 1988 (Paris,
1989), 13–17, who summarizes this argument; also his "Fertility Transition, Conscious Choice, and Numeracy," *Demography*, XXIX (1992),
487–502; and Christopher Wilson, "Natural Fertility in Pre-industrial
England, 1600–1799," *Population Studies*, XXXVIII (1984), 225–240.
Studies of contraception have stressed religious opposition and have seen
the nineteenth century as the inaugural era of contraception in the West.
Examples include: Robert V. Schnucker, "Elizabethan Birth Control and
Puritan Attitudes," in Robert I. Rotberg and Theodore K. Rabb, eds.,
Marriage and Fertility: Studies in Interdisciplinary History (Princeton,
N.J., 1980), 71–84; Nancy Folbre, "Of Patriarchy Born: The Political

Economy of Fertility Decisions," *Feminist Studies*, IX, (1983), 261–284; and Edward Shorter, *Women's Bodies: A Social History of Women's Encounter with Health, Ill-Health, and Medicine* (New Brunswick, N.J., 1991). American historians and economists generally concur with these assumptions and begin their discussion of fertility regulation in the nineteenth century. See Paul A. David and Warren C. Sanderson, "Rudimentary Contraceptive Methods and the American Transition to Marital Fertility Control, 1855–1915," in Stanley L. Engerman and Robert E. Gallman, eds., *Long-Term Factors in American Economic Growth* (Chicago, 1986), 307–309; Michael A. La Sorte, "Nineteenth-Century Family Planning Practices," *Journal of Psychohistory*, IV (1976–1977), 163–183; Jan Lewis and Kenneth A. Lockridge, " 'Sally Has Been Sick': Pregnancy and Family Limitation among Virginia Gentry Women, 1780–1830," *Journal of Social History*, XXII (1988–1989), 5–19; John D'Emilio and Estelle B. Freedman, *Intimate Matters: A History of Sexuality in America* (New York, 1988); James C. Mohr, *Abortion in America: The Origins and Evolution of National Policy, 1800–1900* (New York, 1978); Carl N. Degler, *At Odds: Women and the Family in America from the Revolution to the Present* (New York, 1980); and Laurel Thatcher Ulrich, *A Midwife's Tale: The Life of Martha Ballard, Based on Her Diary, 1785–1812* (New York, 1990), 56.

Recently, there have been a number of studies challenging Himes's assumptions on the dominance of religious opposition and the lack of contraceptive knowledge in the preindustrial West. Angus McLaren, *A History of Contraception: From Antiquity to the Present Day* (Oxford, 1990), argues that contraceptive and abortive techniques were widely known and used. He uses evidence of the spacing of births and of small family size to argue for the effectiveness of these methods and to argue against the concept of "natural fertility." John M. Riddle, *Contraception and Abortion from the Ancient World to the Renaissance* (Cambridge, Mass., 1992), finds that classical medical practices persisted and evolved through the Renaissance and that these practices were largely effective; however, he concludes that this body of knowledge disappeared by the eighteenth century. C. P. MacCormack, "Biological, Cultural, and Social Adaptation in Human Fertility and Birth: A Synthesis," in Carol P. MacCormack, ed., *Ethnography of Fertility and Birth* (New York, 1982), discusses these issues from an anthropological vantage point. There have been several local and regional studies of Europe. See Dorothy McLaren, "Nature's Contraceptive: Wet-Nursing and Prolonged Lactation: The

Case of Chesham, Buckinghamshire, 1578–1601," *Medical History*, XXIII (1979), 426–441; P. P. A. Biller, "Birth-Control in the West in the Thirteenth and Early Fourteenth Centuries," *Past and Present*, no. 94 (February 1982), 3–26; Angus McLaren, *Reproductive Rituals: The Perception of Fertility in England from the Sixteenth to the Nineteenth Century* (London, 1984); Mary Chamberlain, *Old Wives' Tales: Their History, Remedies, and Spells* (London, 1981). A few demographers have found evidence of contraception in preindustrial Europe: E. A. Wrigley, "Family Limitation in Pre-Industrial England," *Economic History Review*, 2d Ser., XIX (1966), 82–109; Alfred Perrenoud, "Les transitions démographiques," and "Espacement et arrêt dans le contrôle des naissances," *Annales de démographie historique*, 1988 (Paris, 1989), 7–11, 59–78; Massimo Livi-Bacci, "Social-Group Forerunners of Fertility Control in Europe," in Ansley J. Coale and Susan Cotts Watkins, eds., *The Decline of Fertility in Europe: The Revised Proceedings of a Conference on the Princeton European Fertility Project* (Princeton, N.J., 1986), 182–200. Bobbi S. Low, "Occupational Status and Reproductive Behavior in Nineteenth-Century Sweden: Locknevi Parish," *Social Biology*, XXXVI (1989), stresses the importance of disaggregating data in order to uncover evidence of control over fertility. In spite of this research, the concept of "natural fertility" still dominates.

Little has been written on abortive and contraceptive practices in North America. Zoila Acevedo, "Abortion in Early America," *Women and Health*, IV (1979), 159–167, summarizes scattered information in some of the standard social histories of colonial North America. Cornelia Hughes Dayton, "Taking the Trade: Abortion and Gender Relations in an Eighteenth-Century New England Village," *William and Mary Quarterly*, 3d Ser., XLVIII (1991), 19–49, examines one case, but finds knowledge of abortifacients widespread. Marvin Olasky, *Abortion Rites: A Social History of Abortion in America* (Wheaton, Ill., 1992), uncovers a larger number of prosecutions of abortion in colonial America, most from the seventeenth century. Demographic historians following the pioneering work of Robert V. Wells, "Family Size and Fertility Control in Eighteenth-Century America: A Study of Quaker Families," *Population Studies*, XXV (1971), 73–82, have been more likely than demographers of early modern Europe to assume contraceptive use (see note 3 and Tables 1 and 2). See also the earlier suggestions of A. J. Jaffe, "Differential Fertility in the White Population in Early America," *Journal of Heredity*, XXXI (1940),

407–411. The need for more information is recognized in Daniel Scott Smith, " 'Early' Fertility Decline in America: A Problem in Family History," *Journal of Family History,* XII (1987), 73–84; and Maris A. Vinovskis, *Fertility in Massachusetts from the Revolution to the Civil War* (New York, 1981), 52n; and is implicit in the comments by contributors to "Roundtable: Historians and the Webster Case," *Public Historian,* XII, no. 3 (Summer 1990), 9–75.

Appearing as this book was in preparation are Barbara Duden, *Disembodying Women: Perspectives on Pregnancy and the Unborn,* trans. Lee Hoinacki (Cambridge, Mass., 1993); and Janet Farrell Brodie, *Abortion and Contraception in Nineteenth-Century America* (Ithaca, N.Y., 1994).

MICHAL MCMAHON

"Publick Service" versus "Mans Properties"

Dock Creek and the Origins of Urban Technology

in Eighteenth-Century Philadelphia

Dock Creek was a complex entity even before its civil history began. By the time of Philadelphia's founding in 1682, the natural state of the cove and streams of the watercourse had already been used by the native peoples and European settlers who moved along the lower Delaware River. What the Delawares called the Coocanocon provided a safe harbor when rough weather caught small boats nearby on the broad river.

Research for this essay was supported by grants from the New Jersey Department of Higher Education, the Hagley Foundation, Greenville, Del., and the National Endowment for the Humanities. I wish to thank the staffs of several of Philadelphia's great libraries: the American Philosophical Society, the Historical Society of Pennsylvania, the Library Company, and the Van Pelt Library of the University of Pennsylvania, as well as the Philadelphia City Archives. Versions of this paper have been rigorously discussed, much to its benefit, at a meeting of the Society for Environmental History, at the University of Houston, and by members of the NEH-sponsored Transformation Project at the Philadelphia Center for Early American Studies. This article gained greatly from astute readings by Steve Golin, Judith McGaw, Stephanie Wolf, Fredrika Teute, Michael Black, and Arlene Fracaro McKenna, as it did from the able research assistance of Gregory Schmidt, Donna Rilling, and Sean McMahon. For creating a map that shows so well what I am telling, I thank Jeffrey McMahon. My thanks also to my wife, Sarah Lynne, whose support and incisive commentary has once again strengthened my work.

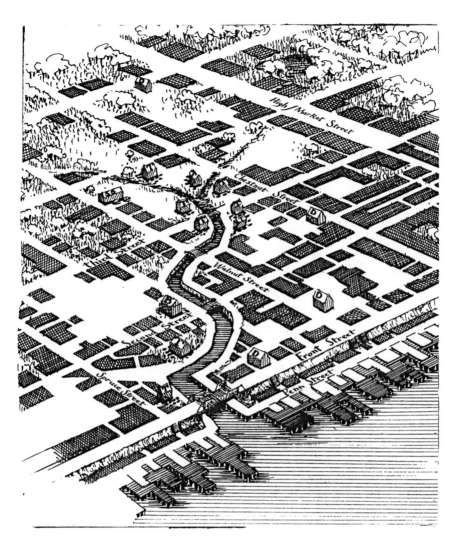

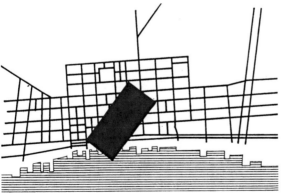

PHILADELPHIA AND DOCK CREEK. *Drawn by Jeffrey B. McMahon*

The tidal cove presented an ideal site to set fish traps and collect shellfish, gather seaweed to be used for fertilizer, and cut hay from the abundant *Spartina,* the marsh grass that filled the salt marsh bordering parts of the cove. So apparent were the cove's utilities that, before William Penn and his party arrived, Swedish settlers had built a tavern and a log cabin near the marshy area north of it.[1]

Along the one-mile eastern border of the townsite of Philadelphia, only the cove cut through the high banks of the river. It lay just above the extensive marshes that covered the bottom of the wedge formed by the convergence of the Delaware and its main tributary, the Schuylkill. Small hills around the cove nonetheless marked the townsite as the high, dry land Penn had sought. Located on the lower, tidal Delaware, the river fed the cove's waters as much as did the small watershed that extended to the ridge dividing the two rivers. Water entered from the land mostly by sub-terranean paths, yet surface waters flowed in from the watershed's three streams. The smallest, from the south, later called the Little Dock, entered where the cove cut through the riverbank; two larger streams flowed from the west, above the cove. What residents called the Dock, the swamp, and (sometimes) the creek was thus a hybrid body of water, part tidal cove and salt marsh and part running stream.[2]

The Dock's importance to Philadelphia began the day in October 1682 that Penn stepped onto the landing at Blue Anchor Tavern. Gary B. Nash has shown how Penn and the early settlers made the area "around the cove . . . the commercial center of town" and "the focal point of the im-port and export trade." Citizens reminded the Provincial Council in 1700 that the cove was "the Inducing reason . . . to Settle the Town where it now is."[3] This was so not only because of its usefulness to commerce but

1. See John and Mildred Teal, *Life and Death of the Salt Marsh* (Boston, 1969), esp. 19, for the utilities of the salt marsh. On the structures at the mouth of the cove, see Barbara Liggett, "Report on the Study of the Dock: Results of Archeological Examination, Phila-delphia, Pennsylvania, 1975," draft MS, on deposit at the Library, American Philosophical Society, Philadelphia.

2. The fullest account of the Dock is Liggett, "Report on the Study of the Dock." Earlier accounts are John F. Watson, *Annals of Philadelphia, and Pennsylvania, in the Olden Time . . .* (1830), rev. Willis P. Hazard (Philadelphia, 1909), I, 336–349; J. Thomas Scharf and Thompson Westcott, *History of Philadelphia, 1609–1884* (Philadelphia, 1884), I.

3. Gary B. Nash, "City Planning and Political Tension in the Seventeenth Century: The Case of Philadelphia," APS, *Proceedings,* CXII (1968), 60–64, *Minutes of the Provincial Council of Pennsylvania,* Oct. 18, 1700 (Philadelphia, 1852), 10.

because of its amenities, serving, as it did, as a parklike setting for fine residences and quasi-public gardens.

Although largely unrecognized by modern scholars, the cove and creeks served chiefly as a manufacturing and materials-processing center for the growing city. These activities strengthened Philadelphia's role as the vital center of a regional marketing system. Yet, to cove and city, the area's service as a manufacturing center held more concrete meaning, which contemporaries early recognized. As parts of the cove were walled and lined with docks, the same citizens who instructed the Provincial Council on the creek's centrality protested its treatment as a sump for domestic and manufacturing wastes. All of this, of course, stemmed from a single geological fact: that, in the words of a 1784 legislative act, the small watercourse was the conduit through which the city's "most populous and central parts" drained.[4]

The very hydrological and geological features that drew settlers to the banks of the watercourse would prove to be of the utmost importance for Philadelphia's inhabitants as they built over the land. Although founded more than a half a century after Boston and New York, Philadelphia passed New York in size by 1700 and by 1760 had drawn abreast of Boston. As the natural properties of the watershed mixed with rapid growth, a complex environment resulted.[5] The interrelationships of settlement pressures and ecological exigencies would shape the city.

Settlement and ecology did not peacefully converge. By 1730, land and city were locked in a contest that would determine the fate of each. As the Dock became so dangerously polluted as to be linked with deadly disease, it moved from being the useful and pleasant center of a town to being a center of civic dispute. For half a century, Philadelphia's leaders debated fundamental issues of urban form, environment, technical choice, and the politics of land use and development. In an emerging social order framed essentially by commercial and large-scale manufacturing interests,

4. James T. Mitchell and Henry Flanders, comps., *Statutes at Large of Pennsylvania, from 1682 to 1801*, 16 vols. (Harrisburg, Pa., 1896–1911), XI, 298 (hereafter *Statutes at Large*).

5. For population data, see James T. Lemon, "Urbanization and Development of Eighteenth-Century Southeastern Pennsylvania and Adjacent Delaware," *William and Mary Quarterly*, 3d Ser., XXIV (1967), 502; Gary B. Nash, *The Urban Crucible: Social Change, Political Consciousness, and the Origins of the American Revolution* (Cambridge, Mass., 1979), appendix, table 13, 407–408.

ecological questions formed at the edges of a conflict between competing social values, personal desires, and political and economic interests. The debate faded during the 1760s when community leaders devised a comprehensive infrastructural solution to the city's urban problems and when the onset of the Revolution diverted attention to other concerns.

Urban scholars have long acknowledged the national influence of Philadelphia's experience in city building, particularly through the grid design chosen by Penn and his associates in 1681. In his pioneer study of urban growth during the half-century after the Revolution, Richard C. Wade extended Philadelphia's influence beyond street plans to urban technologies and the design of infrastructures. In accounting for urbanization in the trans-Appalachian West, Wade found specific instances in which Philadelphia's choices led western cities to adopt urban technologies, ranging from street lighting and central watering systems to market house design and the construction of river landings. By 1818, a Pittsburgh editorial writer could declare Philadelphia to be "the great seat of American influence."[6]

Although watering systems were being adopted or considered by other towns and cities during this period, Philadelphia's reputation was launched, to a great extent, when the city's leaders built the nation's first, large-scale water system between 1799 and 1801. At a time when the water supplies of Boston and New York were equally polluted, Philadelphia's initiative justly earned the city a reputation as an influential pioneer in urban development.[7] Yet viewing the Philadelphia Waterworks as mark-

6. Richard C. Wade, *The Urban Frontier: The Rise of Western Cities, 1790–1830* (Chicago, 1959), 24, 27–28; Lawrence H. Larsen, *The Urban West at the End of the Frontier* (Lawrence, Kans., 1978), 3, who asserts the general influence of Eastern Seaboard cities on the urban West; John W. Reps, *The Making of Urban America: A History of City Planning in the United States* (Princeton, N.J., 1965), 173–174. See also Henry Wright, "The Sad Story of American Housing," in Lewis Mumford, ed., *Roots of Contemporary American Architecture: A Series of Thirty-Seven Essays Dating from the Mid-Nineteenth Century to the Present* (1952; New York, 1972), 326–327. Wade, Reps, and Larsen found in Philadelphia's grid a model for towns and cities along the moving frontier. Versions of the town plan—a one-by-two-mile rectangle broken by five squares, one located in the center and one in each of the quadrants—appeared in cities like Pittsburgh, Cincinnati, Lexington, Louisville, Kansas City, and Denver. The gridiron's influence extended, by Reps's account (125), even to the small, crossroads towns containing a central square intersected by a small grid of streets. On Philadelphia's wider influence, see Wade, *Urban Frontier,* 314, 316, and esp. 318–319.

7. For the waterworks, see Nelson Manfred Blake, *Water for the Cities: A History of*

ing the origins of urban technology in the United States misses the path that led to that event, a path that ran through the 1700s. To follow that route is to encounter not only Dock Creek but also an evolving drainage system that was finally to include the creek itself. This earlier story confirms Philadelphia's role as a model for much of national development, but in a far more complex way than Wade's post-Revolution model would suggest. The series of choices made by Philadelphia's inhabitants revealed a pattern of economic responses to settlement that would be repeated as the nation expanded across the continent.[8]

Dense settlement patterns and concentrated industrial activity early marked a settlement that virtually began its existence as a booming regional center. Long before the nineteenth century, when urban historians and historians of technology generally begin their studies, urban growth induced the need to establish technical means to support it. Environmental historians have long recognized the major role technology plays in environmental change, but they have tended to study rural environments rather than urban.[9]

To approach city building by uncovering the matrix of this earlier city expands the ways in which not only history of technology and urbanization but also environmental history can broaden our perspective. For such an investigation, no other colonial city can compare with Philadelphia as a model for urban and, by extension, regional development in the nation. Not only does Philadelphia's rapid rise to urban hegemony in the English colonies recommend it for study, but the role of the Middle Colonies themselves reinforces the city's significance. Entrepreneurial energy

---

the Urban Water Supply Problem in the United States (Syracuse, N.Y., 1956), 18–43; for an account of an earlier centralized system, see Brooke Hindle, *The Meaning of the Bethlehem Waterworks* (Bethlehem, Pa., 1977).

8. Besides Wade, *Urban Frontier,* see Larsen, *Urban West;* Eric H. Monkkonen, *America Becomes Urban: The Development of U.S. Cities and Towns, 1780–1980* (Berkeley, Calif., 1988); and Blake, *Water for the Cities.*

9. In his early assessment of the field of environmental history, Roderick Nash explained that technology had "placed [the] tools in man's hands that allow him to sculpt the physical world," in "The State of Environmental History," in Herbert J. Bass, ed., *The State of American History* (Chicago, 1970), 250. In a special collection of articles in the *Journal of American History,* Donald Worster placed technology and its interaction with the environment alongside nature and ideology as one of the three central facets of the new environmental history, in "Transformations of the Earth: Toward an Agroecological Perspective in History," *JAH,* LXXVI (1989–1990), 1090–1091.

and pragmatic self-interest make the civil history of Philadelphia and the Dock an exemplary tale for understanding city building and national expansion.[10]

At its narrative core, the Dock's story depicts an eighteenth-century city responding to the challenge of accommodating growth to natural utilities and amenities. Competing visions of the creek as advantage or problem, private property or public space, forced Philadelphia's leaders to grapple with demands for a healthy city environment. This interplay meant devising organizational structures and technical procedures for paving and cleaning streets, establishing drainage and water retrieval systems, and handling and disposing of solid and liquid wastes generated by households and materials-processing industries. Ultimately, it meant making the city livable.

The story of the Dock extends beyond its importance in demonstrating the intersection of ecology and the technological underpinnings of the American city. Whether perceived as a docking facility, a swamp, a cove and salt marsh, or a creek, the treatment of the Dock during the middle decades of the eighteenth century raised fundamental social and political questions. They concerned the relative weight of public and private values in the shaping of the city environment and the value of a community's store of scientific and technical knowledge. The adjudication of these values and the uses made of the knowledge can be seen in the decisions made about such matters as public works engineering proposals and the siting and regulation of industrial activities. And so, seen from the angle of the Dock, the story of the building of early Philadelphia resides in

10. The historiography of city growth is summarized in Josef W. Konvitz, Mark H. Rose, and Joel A. Tarr, "Technology and the City," *Technology and Culture*, XXXI (1990). For examples, see the special issues of the *Journal of Urban History* on technology and the city, V, no. 3 (May 1979), XIV, no. 1 (November 1987).

Michael Zuckerman has asserted that while New Englanders and Southerners were claiming primacy for their peculiar ideologies and principles, "men of the middle states were without incentives to such local consciousness. They simply held hegemony. Their principles and practices were increasingly inseparable from those of the emerging and expanding nation" ("Introduction: Puritans, Cavaliers, and the Motley Middle," in Zuckerman, ed., *Friends and Neighbors: Group Life in America's First Plural Society* [Philadelphia, 1982], 5). For a synthesis of this literature, see Jack P. Greene, *Pursuits of Happiness: The Social Development of Early Modern British Colonies and the Formation of American Culture* (Chapel Hill, N.C., 1988).

a social arena in which competing economic and political interests juggled questions of public rights and private purposes.

Pollution and the "Liberties of Tradesmen": The Political Response

In the middle of concerns about private interests and public order sat the graphic problem of pollution and the large-scale processing operations that clustered around Dock Creek. These included a brewery, tanyards, distilleries, and a slaughterhouse—all with vital links to the watercourse. Environmental problems had long existed alongside the positive attributes of Penn's green country town: In 1705, as in 1711, 1720, 1726, and 1750, grand juries recorded complaints about impassable roads, jammed sewers, and standing pools of water.[11]

The idea that filth and pollution led to disease appeared frequently in the formal complaints of citizens. In 1699, inhabitants had blamed the periodic visitations of epidemic disease on the two tanyards located on the cove. Seven years later, a group of residents charged the brewery above the drawbridge at Front Street with being "injurious to people." In 1712, citizens complained simply of "standing water" in the area. By the 1730s, the problems stemming from unregulated manufacturing and dense settlement were coalescing into poor drainage and inadequate waste removal procedures. At Third and Market streets, for example, "a deep, dirty place" existed "where the public water gathers and stops for want of a passage." Wastewater from residences and manufactories flowed overland into the Dock and the Delaware. Except for a few sewers covered over by the municipal government, most remained open culverts until after 1750.[12]

By midcentury, environmental problems presented the city's central, domestic challenge, and the Dock and its environs had become the center of pollution and controversy. Critics increasingly focused their complaints on the tanyards, which by 1730 numbered at least eight (counting separately the small yards scattered around the intersection of the creek's

11. Watson, *Annals,* I, 214–215.

12. *Ibid.; Minutes of the Common Council of the City of Philadelphia, 1704–1776* (Philadelphia, 1847), 412, 414, 422, 465, 500 (hereafter *MCC*); Judith Marion Diamondstone, "The Philadelphia Corporation, 1701–1776" (Ph.D. diss., University of Pennsylvania, 1969), 108–109. Ellis Paxson Oberholtzer, *Philadelphia, A History of the City and Its People: A Record of 225 Years* (Philadelphia, [1912]), I, 144.

branches above Third Street). Tanneries not only covered much of the land around the Dock but were scattered throughout the length of the stream. One sat below the Drawbridge at Front Street, and two sat just above Front on the western bank. Two were located at the top of the Dock, just below Third, and three literally surrounded the banks of the stream system between Third and Fourth streets.

Since the tannery pits and yards dominated the area, residents pointed to the tanneries as the primary source of the polluted waters of the stream and the foul air in the neighborhood. Most seriously, it seemed to residents that the stench of the tanyards must somehow be related to the nearly annual epidemics and to the endemic disease residing permanently in the urban core. Given the persistence of pollution, some epidemic diseases—tuberculosis, smallpox, and, with the aging settlement, a waterborne contagion like typhoid fever—might well have seemed endemic. Yet the citizens' complaints were problematic. If the illnesses they complained of were yellow fever or smallpox, then Philadelphia's busy seaport was likely responsible for introducing the vectors for these epidemic diseases. As contributors to the accumulating filth in the city, however, the tanneries might truly have been linked to diseases like tuberculosis as well as to endemic diseases such as the common cold and pneumonia.[13]

Not only the number of tanneries but the nature of the tanning process itself made the physical state of the watercourse a central and persistent issue throughout the century.[14] A tannery was generally a large operation, containing a number of structures necessary to the task of transforming hides into leather. Although hides sometimes came from farmers, city tanyards more often relied on nearby slaughterhouses because they were more reliable sources of hides.

Slaughterhouses, like the one on the south side of the Dock, were traditionally located near streams, along with cattle pens. Around them, hornworks, chandlers, and soap- and glue-boiling yards made use of the slaughterhouse's by-products. But the largest users of slaughterhouse leav-

13. Watson, *Annals,* I, 339; Richard Harrison Shryock, *Medicine and Society in America: 1660–1860* (Ithaca, N.Y., 1960), 86–87, 91.

14. The following discussion of tanning relies chiefly on Dorothy Hartley, *Lost Country Life* (New York, 1979), 107, 201, 254–255; Peter C. Welsh, *Tanning in the United States to 1850: A Brief History,* Smithsonian Institution, United States National Museum Bulletin 242 (Washington, D.C., 1964); Jared van Wagenen, Jr., *The Golden Age of Homespun* (Ithaca, N.Y., 1953), 182–189.

ings were the tanneries. A tannery included not only mills for grinding bark used in tanning but also vats and pits for soaking the hides during the stages of removing the hair, tanning the hide, and giving texture to the leather. The method of curing varied according to the intended use, whether for saddles or for women's gloves. Beyond the yard, then, the tanneries also supported a cluster of linked manufacturing activities, including shoemakers, saddlers, curriers, and glovers.

Water was needed throughout the processing of the hides. Initial treatment consisted of soaking the hides in lime pits. Measuring, on the average, six feet square, sometimes lime vats were twice that size. Vats were usually wooden structures made to function like "a small water meadow slightly sunk." [15] A retaining dike wall was built around the vat, and, in rural settings, water was flooded in from the adjacent stream. Whether a country tanyard or an urban yard with wells and a nearby tidal cove, the water always returned to its source along with the various substances that had been added, including acidic liquids resulting from the refuse of cider presses, sour milk, fermented rye, and alkaline solutions made up of buttermilk and some forms of dung. The process signified as well the strong connections between the industrial activities around the Dock and the countryside. Breweries, tanneries, slaughterhouses, and distilleries gathered the products of the countryside—animals and grain primarily—processed them, and discarded the unused by-products on the Dock's watershed.

Although the stages of tanning involved a series of chemical processes, tanners during the 1700s worked with traditional techniques without understanding the chemistry of tanning. Even before a chemical revolution began to transform the tanning process near the end of the century, tanners possessed a complex body of skills and lore that belonged to this ancient and critical materials-processing industry.

If the situation in Philadelphia had involved only craft lore and chemical knowledge, the issue of pollution might have been approached more dispassionately. But the perceived villain was a large, sprawling processing industry, responsive to growing export and domestic markets and already in conflict with shoemakers over charges of shoddy materials and with curriers who charged the tanners with threatening their livelihood by adding currying to the yard's activities. The precipitating factor turned out to be the long-standing suspicion that the odors and standing wastes around

---

15. Hartley, *Lost Country Life*, 254.

the cove caused the pestilence, a comprehensive term for any virulent and contagious disease. A series of virulent epidemics during the 1730s led the inhabitants once again to home in on the tanyards. Individual citizens, organized groups of petitioners, and Dock Creek craft manufacturers became involved; they in turn drew in the chief governing institutions of the province: the proprietor, the provincial Assembly, and the municipal Corporation, a self-perpetuating body of freemen created by Penn, which included a mayor, recorder, assessor, and a council of aldermen and common councilmen. The complaints became detailed attacks on pollution and the location of industrial operations. Finally, the debate broadened to include the polluted state of the city and intensified as the degraded environment appeared ever more dangerous.

As the central arena of debate in the province, the Assembly was the site of the first of a series of dramatic confrontations over the condition of the Dock and its environs. It began during the spring of 1739 when a "great number" of inhabitants petitioned the legislators to correct the conditions pervading the Dock's neighborhood. They specified "the great Annoyance arising from Slaughter-Houses, Tan-Yards, Skinner Lime Pits, *etc.* erected on the publick Dock, and Streets, adjacent." Besides forbidding the construction of new tanneries, the petitioners wanted existing yards "removed within such Term of Years as shall be judged reasonable." [16]

No one saw the issue as an abstract question of rights, as the petitioners made clear. They wanted not only to move the tanyards out of the city—to the Delaware, specifically—but also to regulate the use of the land on and around the Dock.[17] After noting that the watercourse had recently been navigable as far as Third Street, the petitioners charged the tanyards with diminishing the value of other properties in the area by destroying the amenities that formerly graced the watercourse and its banks—hanging gardens, grassy slopes, and clear waters.

Within a few days, the tanners responded with a counterpetition in which they asked for a copy of the original petition and time to respond. In their petition, the tanners admitted that the pits could give off offensive smells, but charged the residents in the area of the Dock with going to unreasonable lengths to remove the tanyards when only a regulation would

16. Samuel Hazard *et al.*, eds., *Pennsylvania Archives* (Philadelphia, Harrisburg, Pa., 1852–), 8th Ser., III, 2487 (hereafter *Pennsylvania Archives*).

17. Carl Bridenbaugh, *Cities in the Wilderness: The First Century of Urban Life in America, 1635–1742*, 2d ed., (New York, 1959), 318.

be necessary.[18] Apparently impressed with their arguments, the Assembly gave the tanners until the next session, to be held at the end of the summer, to provide a plan.

In August, when the Assembly turned again to the dispute, the tanners welcomed the petitioners' willingness to accept the tanyards if they were "so regulated as to become inoffensive." In a detailed scheme to clean up their yards, the tanners proposed that

> the Tan-yards be well paved between all the Pitts, and wash'd once every Day: Let the Watering-Pools and Masterings (which are the only Parts that afford offensive Smells) be inclosed on every Side, and roofed over, within which Inclosure may be a subterranean Passage to receive the Washings and Filth of the Yard into the Dock or River at High-Water: Let the whole Yard be likewise inclosed on all Sides with some strong close Fence, at least seven or eight Foot high, and every Tanner be obliged every Week to cart off his Tan, Horns, and such offensive Offals.[19]

The Assembly not only accepted the tanners' regulations and left the yards on the Dock: it charged the Corporation with enforcement while withholding authority adequate to the task. As if to acknowledge the inadequacy of its decision, the provincial legislators promised to consider a request for the power to force compliance should it become necessary.[20]

The legislative decision underscored the substantial and long-standing power of the tanners. The large number of names on their counterpetition attested to their social influence as well. All of the six tanners who defended their rights before the Assembly and in the pages of the *Mercury* stood among the town's elite. Perhaps lowest on the social scale, John Snowden, John Ogden, John Howell, and William Smith were nonetheless influential citizens. Two signers of the letter in the *Mercury* stood securely in the ranks of Philadelphia's Quaker elite. William Hudson, Jr., and Samuel Morris also brought extensive political connections to the tanners' cause. The twenty-eight-year-old Morris was a Quaker who had

18. *Pennsylvania Archives,* 8th Ser., III, 2490.

19. *American Weekly Mercury* (Philadelphia), Sept. 6–13, 1739; *Pennsylvania Archives,* 8th Ser., III, 2503.

20. Ernest S. Griffith, *History of American City Government: The Colonial Period,* I (New York, 1938), 24. *Pennsylvania Archives,* 8th Ser., III, 2501, 2503–2504.

married into the prominent Cadwallader family.[21] His tanyard lay on the western bank of the Dock, near the bottom of the curve extending from the stem to Second Street. Nearby was John Snowden's yard, at the base of Society Hill near Front Street. John Ogden's yard sat north of the Dock below Third, bordering the first of several tanneries owned by William Hudson, Sr.

This main yard of Hudson's lay on the north bank, below Third and between Dock and Chestnut streets, near the headwaters of the Dock. Besides the tannery, the elder Hudson's mansion and carriage house sat on the lot. Hudson Senior, who retained possession of the Hudson properties until his death in 1742, had arrived in 1685 and soon added to his standing in the community by marrying a daughter of a leading Quaker family. After becoming the owner of one of the two yards located on the Dock in 1700, Hudson established several additional yards during the early decades of the century. All of them were operated by his son, William Junior, and his son-in-law, John Howell, in 1739. Howell's yard bordered on Fourth, sitting at the highest point of the industrial area that followed the creek.[22]

The sheer physical presence of the tanners in the town must have contributed to their bold act of celebrating the Assembly's decision with a parade. At that point, the town's leading newspapers joined the debate. Although on opposing sides, articles in the *Pennsylvania Gazette* and the *American Weekly Mercury* reinforced the picture of the Dock presented by both the original petitioners and the tanners. Accompanying the rich description were ideas that clarified the political and economic issues being debated. The passionate exchange described a cluster of processing indus-

21. Morris's father-in-law, John Cadwallader, served on the Council in 1718 and in the Assembly between 1729 and 1734. After midcentury, Morris too served on the Council. See Diamondstone, "The Philadelphia Corporation," 298, 322, and 365; also Watson, *Annals*, I, 346.

22. The elder Hudson was 67 in 1739. A wealthy man with property in England and a large slaveholder, Hudson was named a councilman in the original city charter of 1701; in 1705, he served as an alderman—a select group chosen from the councilmen—and became mayor in 1725. Hudson's wife was a daughter of Samuel Richardson, a resident of the Dock Creek neighborhood and founder of one of the more considerable Quaker families in the city. See Diamondstone, "Philadelphia Corporation," 298, 322, 365; also Watson, *Annals*, I, 346. For the expansion of his tanyards above Third, see the unpublished map by Anna Coxe Toogood, Independence National Historical Park, Philadelphia.

tries on the banks of the watercourse. It documented the city's many uses of the cove and creek and added to the ecological understanding already present in fifty years of complaints from the community. Such concentrated attention from the newspapers to the physical state of the city—unusual in the presses of the day—responded to the critical level of pollution and the arrival of "autumnal fever" during the months of August and September.[23]

The involvement of the *Gazette* and the *Mercury* testified to the power of the contending interests, each of which drew a newspaper to its side. The petitioners had the power of numbers and of their residence in the area. Among their supporters were rising members of the community like the *Gazette*'s editor, Benjamin Franklin, who had long used his weekly to defend and advance public property ideas like those expressed in the initial petition. In this instance, the letter addressed to "Mr. Franklin," which appeared during the last week of August 1739 and covered the entire front page of the weekly, argued for alternative uses of the town center and for more controls over land use.[24]

The Assembly's failure to provide a means of enforcement left the tanners free to claim victory. As they pointed out in a joint letter to the *Mercury,* the Assembly had "upon the whole" rejected the petitioners' request and upheld the tanners' "right to follow their Trades within the City." Further, they insisted that the Corporation's right to approach the Assembly belonged no less to them.[25]

Through August and into September, even while admitting that their yards were "a nuisance" and "voided many unwholesome smells," the tanners continued to argue their rights. In their defense, they sketched a picture of a city in which sources of pollution were pervasive. A great amount of waste converged on the Dock, which served as "a Receptacle for all

23. The debate appeared in 1739 in the *Mercury,* Aug. 9–16, Sept. 6–13, and a Postscript dated Sept. 13, and in the *Pennsylvania Gazette* (Philadelphia), Aug. 23–30.

On "autumnal fever," see Martin S. Pernick, "Politics, Parties, and Pestilence: Epidemic Yellow Fever in Philadelphia and the Rise of the First Party System," in Judith Walzer Leavitt and Ronald L. Numbers, eds., *Sickness and Health in America: Readings in the History of Medicine and Public Health* (Madison, Wis., 1985), 356.

24. For a fuller discussion of Franklin's provincial career and public philosophy, see A. Michal McMahon, " 'Small Matters': Benjamin Franklin, Philadelphia, and the 'Progress of Cities,' " *Pennsylvania Magazine of History and Biography,* CXVI (1992), 157–182.

25. *Mercury,* Aug. 9–16, 1739.

kinds of filth from a very great part of the Town." Water no longer entered at a volume sufficient to carry off waste deposited at the head of the cove. Nor was the cleansing force of the tide adequate before the "abundance of Necessary-houses" that lined the Dock and drained directly into the watercourse.[26]

The industrial sources of pollution *were* comparably diverse, as the tanners strenuously argued. Critics of the tanyards failed to censure "Butchers, a Trade . . . much more offensive than Tanning," because it was convenient to have the slaughterhouse near the center of town. And why charge the tanners alone with the responsibility for "Tann, Horns, Dead Dogs, Country People losing their Dogs, Tanners Dogs biting People, a Dog mangled, [and] an other rescu'd from a Slaughter-House"? When the protesting citizens put these complaints before the Assembly, the tanners pointed out, the "Hon. House" found them so "impertinent to the point" it had not even replied.[27]

The Dock Creek tanners insisted that they were not alone responsible; "a gentleman . . . of some note" had cited the tanyards in Southwark, a settled area on the Delaware directly south of Philadelphia, as "truly offensive." To their critics' point that New York City had banned tanyards, the tanners countered that tanyards were allowed throughout the "Towns and thick settled Parts of *Great-Britain*" and existed in the wards of London with the approval of the lord mayor's government.[28]

"The Affair of the tanners had made a great deal of Noise" in the city, as *Mercury* editor Andrew Bradford wrote in mid-August. Yet the 1739 dispute revealed more than an understanding of the problems of the congested core. Questions of public rights and private land use were bared as well. To Franklin, the Dock and its utilities provided a "publick service" of the city; to Bradford, it was "mans property." The one account that appeared in the *Gazette,* which responded to the tanner's initial letter, answered in detail the arguments put forth then and later by Bradford and the tanners. Public rights to the Dock and the land around it rested on traditions beginning before 1700. Even Dock Street had been given "with the Dock for publick Service."[29]

26. *Ibid.*

27. *Ibid.*

28. *Ibid.*

29. *Ibid.,* Sept. 6–13, 1739; *Gazette,* Aug. 23–30, 1739. The charter is reprinted in *Ordinances of the Corporation of, and Acts of Assembly Relating to, the City of Philadelphia*

Removing the tanyards from the area around the Dock promised to enlarge public spaces within the city and thus allow smoother passage through the town center and make it easier to fight fires. Franklin admitted that improvements to the vacated grounds would make the land more valuable to the residents. As for the tanners' right to property, Franklin and the protesters suggested that the city could move the tanyards to a similarly "convenient" place.[30]

In a Postscript to the *Mercury* published two weeks later, Bradford insisted that no places existed near the city that were as convenient as the Dock. The tanners would have to bear the costs of tearing down and rebuilding the old structures that could not be moved, and the increased distance would raise the costs of carrying goods to market. Besides, the poor would be deprived of four to five hundred cartloads yearly of tanbark for fuel, "a material article to the City." Against the force of such arguments, the author of the *Gazette* article had already argued that the final issue went beyond agreeing on "the damage done" to actually assessing the relative weight of damage done to tanners and damage done "to the city."[31]

Everyone agreed on the sources of pollution and its impact on the Dock Creek area. Resolving claims of competing values called for balancing public costs and benefits against the costs and needs of private business. The supporters of a public Dock faced tremendous obstacles in a society that had gone far to privatize the possessions and purposes of civil society.[32] Franklin assumed the tanners would bear the expense of

---

(Philadelphia, 1851). See Diamondstone, "The Philadelphia Corporation," 71, and "Philadelphia's Municipal Corporation, 1701–1776," *PMHB*, XC (1966), 183–201.

30. *Gazette*, Aug. 23–30, 1739.

31. *Mercury*, Sept. 13, 1739; *Gazette*, Aug. 23–30, 1739.

32. The view of Sam Bass Warner, Jr., asserted in his study of the growth of Philadelphia after the Revolution, *The Private City: Philadelphia in Three Periods of Its Growth* (Philadelphia, 1968), that privatism existed "already by the time of the Revolution" (3) is amply supported in the specialized literature on Philadelphia. Warner errs, however, when he argues that, until the 19th century, "no major conflict [existed] between private interest . . . and the public welfare" (4). See Gary B. Nash, "City Planning," APS, *Procs.*, CXII (1968), 54–73; Nash, *Quakers and Politics: Pennsylvania, 1681–1726* (Princeton, N.J., 1968); Hannah Benner Roach, "The Planting of Philadelphia: A Seventeenth-Century Real Estate Development," *PMHB*, XCII (1968), 3–47, 143–194; and Frederick B. Tolles, *Meeting House and Counting House: The Quaker Merchants of Colonial Philadelphia, 1682–1763* (Chapel Hill, N.C., 1948).

relocation, although the Corporation might suggest a "convenient" location. This honored Penn's charter promise that the Corporation would recognize the rights and property of "persons."

But what of the tanners' charge that the petitioners acted more out of self-interest than out of concern for the city? The *Gazette* author had admitted self-interest in arguing that the removal of the tanyards would increase the value of the area around the Dock. Although Franklin would later move to within a block of the Hudson and Howell complex of tanneries, in 1739 he lived on High Street between Front and First. While located on the Dock's watershed, Franklin probably suffered more from the market buildings on High that fronted his property and from the closeness of Bradford's printing shop than from the sluggish waters of the Dock. Beyond the market shambles, a small alley opened into Letitia's Court, named for Penn's daughter, who owned the lot earlier in the century.

It would be easy to see a politically courageous act in a thirty-three-year-old newspaper publisher's willingness to cover his entire front page with a letter criticizing one of the most powerful groups in the city. A young printer still rising in the community, Franklin could not lightly squander the good will of wealthy Quaker tanners. Yet however much Franklin's concerns went beyond the value of his property, that value was nonetheless a factor. Even with his shops and home more than two blocks from the yards, the Dock's condition was seen as injurious to property values and the quality of daily life throughout the settled watershed.[33]

Nor did Bradford let the obvious personal investment pass: "Here is the Rabit! Good people observe what a concern they have for the promotion of lots!" The tanners' aims were no different, Bradford insisted. They wanted the opposition to show signs of "some tenderness to mans properties."[34]

The 1739 dispute had linked the issue of public rights and land use policy to the problem of pollution. It demonstrated also an emerging awareness from within the community of the connections between hydrological systems and settlement patterns. In Franklin's *Proposal for Promoting Useful*

33. See the map, "Franklin's Philadelphia, 1723–1776," showing his residences and properties, in Leonard W. Labaree *et al.*, eds., *The Papers of Benjamin Franklin*, II (New Haven, Conn., 1960), following 456. Although the evidence supports Franklin's authorship, the *Gazette*'s response represented the position of the original petitioners.

34. *Mercury*, Sept. 13, 1739.

Knowledge, written in 1743 to launch the American Philosophical Society, he included among the areas of knowledge needed by the colonials land drainage techniques and a knowledge of water-pumping machinery for the retrieval and distribution of water.[35] Franklin's growing understanding of technical design and scientific experimentation were manifest in the stove he designed and through his electrical investigations, all achievements of the 1740s. The same understanding could be seen in his approach to public works technology and his broad awareness of the city's infrastructural needs. It was the conjunction of these interests with his growing financial stake in his adopted city that involved Franklin in the affairs of the Dock for more than thirty years.

The other parties in the dispute over the Dock were similarly aware of links between technology and ecology. The original protesters connected the pollution of the cove with the state of the streams and watershed above. The tanners confirmed the importance of the cove's location when they described the upper parts of the Dock as being "without sufficient water to carry" off the filth that flowed into the cove "from a very great part of the town." They demonstrated their grasp of the properties of a hydrological system and the Dock's being situated such that pollution from throughout the city flowed into it.[36] Implicit was a sense of the relationship between dense settlement and cleared land and the amount of sediment that joined the garbage in the cove.

Arguments offered on all sides of the debate pointed to drainage and waste removal as the central urban challenges at midcentury. Bradford recognized this when he linked the state of the tanneries to the condition of the Dock and the town:

> And truly, as the Dock is now circumstanced, they must be fine Nos'd that can distinguish the smell of Tannyards from that of the Common sink of near half Philadelphia, that for want of Passage stands putrified, Carrion too frequently thrown there, and the Excrement from nearly Thirty Houses of Office communicating with the Dock, and lodged on the sides every Fresh.[37]

35. Benjamin Franklin, *A Proposal for Promoting Useful Knowledge among the British Plantations in America* (May 14, 1743), in Labaree *et al.,* eds., *Papers of Benjamin Franklin,* II, 381.

36. *Mercury,* Aug. 9–16, 1739.

37. *Ibid.,* Sept. 13, 1739.

Domestic and manufacturing wastes were often indiscriminately dumped or placed in pits, and many streets remained unpaved and covered with standing water after rains. Waste liquids from processing operations came from the oil and color shops of distillers and from soap-boilers and chandlers to mix with storm water runoff before filtering into an aquifer from which, through public and private wells, the city drew its water. In one way or another, nearly all wastes entered the Dock, the major exit point in the colonial city's natural drainage system.

With the decision of the Assembly in 1739, official consideration of the dispute had effectively ended. But the issue did not disappear as a matter of public concern. After all, the legislative action and the newspaper debate of 1739 had amplified complaints simmering since before 1700. And so, less than a decade later, the state of the Dock once more led to public consideration, in the form of a major study by a committee of the Corporation.

The immediate result was a report offering general policy recommendations and an engineering proposal for a public works project to revitalize the Dock. Its reception by the Assembly and the Proprietor, two potential sources of external revenue, and the response of the Common Council itself reveal even more about the commitments of the leadership to the ordering of space in the city.

The 1748 and 1763 Proposals to Restore the Dock: The Engineering Response

The Corporation's initiative came in the fall of 1747, following the decade's second harsh attack of epidemic disease. The descriptions of the Dock's condition that emerged from the 1739 dispute alone suggest that the Council's aim of reviving the Dock as a multipurpose wharving facility and an urban amenity presented a daunting task. For some, as city recorder and councilman Tench Francis put it, the Dock still held promise as an "ornament to the city." As late as the 1740s, a large home was begun on the south bank of the lower Dock, recalling the days in which the wealthiest settlers built homes along its banks.[38]

---

38. Watson, *Annals*, I, 341; Thompson Westcott, "A History of Philadelphia . . . to . . . 1854," 210. Completed only to 1829, the partial manuscript was printed without pagination

Yet the Dock's problems were beyond a stately new home, as might have been read into the building's destruction by fire before being completed. The presence in the area of "mansions" like William Hudson's at Third and Chestnut failed to alter the prevailing belief that the cove increasingly resembled "a foul, uncovered sewer." Most of the watercourse, from the walled portion from the Delaware to the bridge at Third Street to the twin streams that dissected the block above Third, constituted a major blight that literally split the southeast quadrant of the town plan. Nor could the problem expect to improve when a new market house had to be built in 1745 on Second below Pine in response to the dense, residential expansion south of the Dock, where short rows of two- and three-story tenements were replacing rural estates and cleared land.[39]

The slaughterhouses and the skinning troughs, lime pits, and large piles of dry and soaking bark of the tanyards lay on both sides of the Dock. Within the watercourse itself, industrial waste joined silt eroding from bare land and unpaved streets. Inhabitants regularly used the Dock as a receptacle for the city's domestic refuse, human and animal sewage, and dead animals.[40] Residents and visitors more and more referred to the town's once-prized watercourse as a common "sink" or sewer.

Epidemics were common. They had come in the form of smallpox three times during the 1730s, and, in 1741, five hundred people died of yellow fever thought to have been brought in on a ship. The next year, the pest-house was moved from Tenth and South streets to an island in the mouth of the Schuylkill. After yellow fever again came in 1747, arriving during the first hot days of July and lasting into September, the Corporation named a committee to assess the continuing problem of the polluted Dock and to propose means for cleaning up the environs. The committee was to consider especially the remaining marsh and to recommend "the best method of improving the sd swamp for the general use and benefit of the city."[41]

To carry out its charge, the Common Council had placed on the com-

---

and never published. Page reference is to the paginated facsimile in the Library of the American Philosophical Society, Philadelphia.

39. Watson, *Annals,* I, 341; Oberholtzer, *Philadelphia,* I, 144; Westcott, "History of Philadelphia," 210; Scharf and Westcott, *History of Philadelphia,* I, 212.

40. Scharf and Westcott, *History of Philadelphia,* I, 212.

41. Bridenbaugh, *Cities in the Wilderness,* 399; Oberholtzer, *Philadelphia,* I, 144; MCC, Oct. 19, 1747, 487–488.

mittee not only its own members but also citizens who were not on the council. The character of the half-dozen leading citizens who formed the ad hoc committee demonstrated the growing potential for a systematic, engineering response to Philadelphia's growth. Among them, only Franklin revealed a strong interest in science per se. Yet all but one were technically knowledgeable and experienced in public and industrial undertakings. The only nontechnologist, William Logan, shared Franklin's receptivity to new ideas about agriculture, and, like him, experimented in that area. The committee represented, in short, an emerging community of men involved in industrial entrepreneurship and the mechanic arts.

Two members, John Stamper and Logan, were distinguished chiefly as successful merchants. Stamper, a rising businessman, apparently gained from his work on the committee a reputation in the building arts as well. In December 1751, the council contracted with him to perform work on the south end of the drawbridge at Front Street and along the south side of the Dock extending east to the Delaware side of Water Street. Stamper agreed to do all "according to the plan of the city, in the best and workmanlike manner." When he became mayor in 1760, Stamper built a fine Georgian house on Pine Street south of the Dock.[42]

Logan derived his distinction in the city in large part from his father, James, a former governor who had served Penn in several capacities, whom he followed in trade. William joined the Common Council in 1743 and, in 1747, replaced his father on the governor's council. He held both positions until the councils were suspended in 1776. When his father died in 1751, William quit the mercantile business and moved to the family farm, where he gained a reputation as a progressive agriculturalist and wrote an essay entitled "Memoranda in Husbandry."[43]

Unlike Stamper, Edward Warner brought extensive experience in construction to the committee's deliberations. He described himself as a merchant and "of the city of Philadelphia, house-carpenter." A man of means and position, Warner's wealth came from his former master, a carpenter

---

42. Frank Willing Leach Collection, LXI, 71, Genealogical Records, Historical Society of Pennsylvania, Philadelphia; C. P. B. Jeffreys, "The Provincial and Revolutionary History of St. Peter's Church, Philadelphia, 1753–1783," *PMHB*, XLVII (1923) 337; *MCC*, Dec. 10, 1751, 553–554.

43. John W. Jordan, ed., *Colonial and Revolutionary Families of Philadelphia: Genealogical and Personal Memoirs* (New York, 1911), I, 30–31; Frederick B. Tolles, "George Logan, Agrarian Democrat: A Survey of His Writings," *PMHB*, LXXV (1951), 261–262.

and builder who died childless in 1737 and left his fortune to Warner and a fellow apprentice. Warner's master had been a founder of the Carpenters' Company, a trade association of builders and craftsmen that Warner later joined.[44]

Three committee members—Samuel Rhoads, Samuel Powel, and Benjamin Franklin—formed a core engineering group on the committee. All were born between 1704 and 1711 and spent most of their lives in Philadelphia. In their work, interests, and public service activities, they demonstrated an interest in manufacturing and infrastructural development and a knowledge of machines. Rhoads's activities indicated a knowledge of construction and pumping machinery. In 1744, he served the council by inspecting the ferry at Market Street on the Schuylkill, for which he recommended removal of the existing buildings and putting up new ones. A year earlier, he sat with Powel on a committee to inspect and recommend repairs to the city's fire engines.[45]

Although chiefly a merchant, Powel's father had won a reputation as a prominent builder and architect and a "carpenter of high ability." When the son was fourteen, the elder Powell erected a bridge over Dock Creek at Walnut Street, and, in 1734, father and son constructed a "public wharf and regulated the streets" on the south side of the Dock near Front Street. As they explained a decade later when seeking the council's permission to charge for its use, the area had become impassable and had been of no use to the inhabitants until their improvements. Powel also joined his father in 1726 in an enterprise called the Durham Company, which was intended to erect a furnace and other works for making and casting iron.[46]

Franklin shared characteristics with the entire committee: as a prominent craftsman, as a serious technologist, and as a well-off and influential citizen who promoted private and public undertakings. Although in 1748 he belonged to neither the Common Council nor the Assembly, Franklin had served as clerk of the Assembly for more than a decade.

By February 1748, the ad hoc public works committee had completed

44. Jordan, ed., *Colonial Families of Philadelphia*, I, 235; query, *PMHB*, I, (1877), 358–359; Anne H. Cresson, "Biographical Sketch of Joseph Fox, Esq., of Philadelphia," *PMHB*, XXXII (1908), 178. Of the six appointed members, only Warner failed to sign the report.

45. *MCC*, Jan. 12, 1743, Aug. 17, 1744.

46. *MCC*, Jan. 28, Feb. 14, Oct. 4, 1743; Jordan, ed., *Colonial Families of Philadelphia*, I, 110–111.

a public works proposal aimed at reclaiming the entire area of the original cove, extending to Third Street. It recommended development of a multiple-wharving system in the cove and creek, to include existing docks and landings. A sixty-foot dock would be built along the creek's main stem as far as it "extends westward." Two stretches of sloping beach of thirty and forty feet were to be left open on either side for the landing of flats and small boats. The lower bank between Front Street and the Delaware would be raised above high tide and walled in with stone, and the channel widened to between sixty and eighty feet. The "common sewer" that flowed from the southwest into the Little Dock would be continued to the main stream, covering over the small tributary. Finally, the cove was to be regularly dredged as far as Third Street so that the bottom would be covered with water even at low tide.[47]

Few precedents existed in Philadelphia for an actual public works project of such size and scope—although Thomas Budd's 1698 design for the Dock certainly equaled the committee's plans. Comparable projects had been undertaken in other colonial cities. Boston's Long Wharf, built between 1710 and 1713, measured sixteen hundred feet. Nearly thirty years later, Newport's merchants, with a grant of town land from the city government, constructed a fifty-foot-wide wharf that extended more than two thousand feet into the channel. Newport's project differed from Boston's purely public effort by joining private and public resources. Yet neither project entered the city in the manner of Philadelphia's Dock. The New England cities thus avoided chafing effects like those which followed the committee's attempt to reorder the very center of Penn's town.[48] Philadelphia's efforts differed also in being born out of conflict, not the needs of commerce: The committee responded to the widely held belief that a polluted Dock would continue to degrade the city.

The Dock Creek restoration plan foundered precisely upon the need to navigate between hesitant public authorities and entrenched private interests. The Assembly ignored the committee's proposal that "a tax be laid upon the city." With control of the Corporation's taxing authority and a lack of sympathy for its proposed solutions, the Assembly left the council with the necessity of approaching the proprietor.[49]

In March, after discussing the matter with the Assembly-appointed city

47. MCC, Feb. 24, 1748, 494–496.
48. Bridenbaugh, *Cities in the Wilderness*, 171–172, 326.
49. *Pennsylvania Archives*, 8th Ser., V, 244.

assessors, the Corporation named the mayor, recorder, and several aldermen to approach the proprietor, Thomas Penn, for assistance. Penn expressed his concern for the "sickly state of the city." Yet he questioned
whether the "mud" deposited on the sides and bed of the Dock was really
dangerous, since the Dock had been polluted for many years "and yet no
such fever was known until the year 1741." Penn advised that, if the mud
differed from that on the sides of the Delaware, then the tanyards should
be removed. As to how to remove the mud from the Dock, he left that
decision to the Corporation, which was "on the spot." If the Corporation or public wanted to restore the Dock, they should pay for it or give
up any rights to it and allow private owners to "clear and build upon it"
and to receive the profits from "the landing of wood and other things on
the bank."[50]

Having been rebuffed by both the Assembly and the proprietor, the Corporation declined the full responsibility for the area, requesting owners of
land adjoining the Dock to clean up each private site. The incentive offered
was the owners' right to profit from their improvements. The committee
recommended, in turn, that the Corporation agree to use its funds to build
floodgates at the several bridges and to establish regular dredging of the
watercourse.[51]

Students of Philadelphia's colonial city government have sometimes
cited the Corporation's lack of taxing authority as a reason for its failure
to restore the Dock.[52] The record suggests, rather, a specific unwillingness
to make a commitment to restoring the Dock, even if the financing scheme
relied chiefly on private funds. From the beginning, Philadelphia's Corporation had regularly dealt with infrastructural matters, arranging for
the repair of collapsed bridges, breaches in the streets, and overflowing
sewers. Rarely did a civic leader deny responsibility for maintaining and
regulating the several public wharves on the Delaware, the public landing
at Blue Anchor Tavern, the Dock, the powder houses, or the stalls and
storage buildings in the market areas along High Street and on Second
Street below the Dock. To pay for this work, the Corporation relied chiefly
on revenues from fees, fines, and rents. While small repairs could be made
with relative ease from these sources, the magnitude of the 1748 proposal
to renovate the Dock called for far greater sums of money.

50. *MCC*, Feb. 24, 1748, 496.
51. *MCC*, Feb. 24, 1748, 495.
52. See Oberholtzer, *Philadelphia*, I, 145; Liggett, "Report on the Study of the Dock," 52.

When additional means of funding are examined, however, and other commitments considered, the absence of taxing authority fails to account for the neglect of the Dock. In the same year that the ad hoc committee put forth its engineering proposal, the Corporation raised a substantial sum toward the defense of the city from the French by guaranteeing the purchase of two thousand lottery tickets. Defense efforts dominated extraordinary city and provincial expenditures for the next three decades, yet other projects received substantial sums drawn from money paid to the Corporation for renting and leasing public property. In 1750, the Corporation committed seven hundred pounds over several years for the newly established Academy. It was not unusual for the Common Council to consider funding a new wharf, as in 1748. During the years following, the Assembly even allowed the city to raise funds by lottery for a landing in Northern Liberties and to pave streets in Philadelphia.[53]

Not only did the Corporation possess these additional means of raising funds, but its income grew substantially between the 1720s and the 1770s. Its cash reserve amounted to more than £500 by midcentury, which, lent at interest, grew to £730 in 1757. Revenues rose so rapidly during these years that, by 1763, annual income was £640 and the reserve stood at £1,000. During the 1760s and early 1770s, the Corporation gave £300 for a bridge and £500 for a ferry across the Schuylkill north of Philadelphia country.[54] For some time before the 1748 proposal, in short, the Corporation had demonstrated that, rather than a lack of funds, a lack of interest in the urban values represented by the Dock controlled its administration of the city. Through petitions and committee service, a number of individuals had made clear their belief in a civic order in which public rights balanced private desires. By action alone, city leaders demonstrated their commitment to a policy of allowing the individual choices of private property owners to determine the shape and character of the city. In the face of private property rights and competing claims, the Corporation abdicated its public responsibility in favor of relying on private enterprise to deal, or not, with the consequences of development.

The need for some decision was made clear in the conclusion of the February engineering report:

53. MCC, Jan. 18, 1748, 491–493, July 31, 1750, 529–530; *Pennsylvania Archives,* 8th Ser., VI, 5397; *Statutes at Large,* VII, 163–164.

54. Diamondstone, "The Philadelphia Corporation," 223–230.

as the Nusance is of such a Nature, and should it continue, may be of fatal Consequence in preventing the Growth and Increase of this City by discouraging Strangers . . . from coming among us, or filling our own Inhabitants with Fears and perpetual Apprehensions, while it is suspected to propagate infectious Distempers . . .[55]

The defenders of the Dock had argued for amenities, open space through which to move more freely in the city and the ornamental qualities of the watercourse, and for utilities, a source of water to fight fires. Yet not even the association with disease was persuasive enough to force quick action. Having rejected the option of controlling land use by moving the tanneries outside the city in 1739, and thus a confrontation with private economic interests, the preparers of the 1748 report left the Dock's manufacturers in place, offering instead a pure technological solution. It was this response that made the 1748 proposal so prescient.

The prescience of a pure technological solution became clear during the 1760s. The committee's makeup had attested to a growing body of experience and knowledge in construction and civil engineering, knowledge that the legislators of the 1760s would incorporate into a systematic treatment of the city's problems. The stature and character of the 1748 committee's members confirmed the foresight of local leaders whose experience could be called on to respond to the city's environmental problems. Yet the committee's proposals became moot as buildings and streets continued through the 1750s to expand over the watershed. As settlement moved beyond Fourth Street, lots above the cove were sold and built over, both in the remaining marshy areas near Spruce Street to the southwest and along the creek that fed the cove from the northwest. Under orders from the Corporation, the bulk of the hills surrounding the headwaters were used to fill and level the area above the cove and north of Spruce Street.[56]

In the face of the obliteration of the natural features of the cove's watershed such that no open, undeveloped marsh remained by 1760, the Dock itself still attracted defenders. In late January 1763, a group of petitioners made a final attempt to preserve the Dock. They reiterated to the Assembly that "the public Dock or Creek" endangered the health of citizens because of its use as a dump for "Carcases . . . Carrion, and Filth of various kinds." As earlier, the petitioners wanted to restore the Dock, yet this time they

55. MCC, Feb. 24, 1747, 496.
56. Watson, *Annals*, I, 341.

combined the notion of land use restrictions from 1739 with the engineering approach of 1748. They also moved, as they had not done before, to save the town's water supply. By the 1760s, the growing number of distilleries in the area around the Dock led the protesters to cite the "Number of Still-houses erected of late." [57]

The petitioners thought that the distilleries, being constructed mostly of wood, constituted fire hazards. Beyond that, the industry's processing operations brought more polluting wastes to the neighborhood of the Dock. Distilleries deposited their liquid "dregs or returns" in "wells, dug for that purpose." Like the contents of privies, the liquid waste tended to mix with the water drawn from wells for domestic use. Some of the distilleries even emptied their wastes into the "public gutters" that fed the city's expanding drainage system. [58]

Clearly demonstrating their grasp of the character of water flow beneath the ground's surface, the petitioners wanted the manufacturers removed in an attempt to protect groundwater, just as they wished the Dock to be once again useful for transporting building materials and firewood by small craft and for supplying water to fight fires. Reminiscent of the 1748 report, the petition recommended that the Dock be "cleared out, planked at the Bottom, and walled on each Side." [59] The Dock would serve, in effect, as a combined canal and reservoir.

After receiving the petition, the Assembly appointed a new committee to investigate and report back. As before, the committee took a favorable view of the Dock's potential, recommending it as a "public Utility" that should "be cleansed, and properly walled." The committee explicitly claimed the modified watercourse for the public, since they were unable to find "any Persons [with a] just and legal Claims to the said Dock, or the Streets laid out adjoining thereto." [60]

Once again, the Assembly set aside the assertion of public ownership and charged private owners with making the necessary repairs on those sections of the walled watercourse adjacent to their properties. In 1763, a supplement to a comprehensive act of the previous year to regulate streets and sewers contained specific instructions for restoring the Dock as far as Third Street. Each property owner was to erect "a good, strong,

57. *Pennsylvania Archives*, 8th Ser., VI, 5384–5385.

58. *Ibid.*, 5384.

59. *Ibid.*

60. *Ibid.*, 5397.

substantial wall of good, flat stone from the bottom of the said Dock," thickness, height, and depth to be specified later. If the residents did not complete the work within three months of receiving directions in writing, the commissioners would have the work performed and bill the owners.[61]

So slowly and ineptly did the city move on the restoration of the Dock that another supplement was passed in 1765 acknowledging that the attempts "to open, cleanse, repair, regulate and make navigable a certain watercourse . . . known by the name of the Dock" had failed to "answer the good purposes that were expected." Indeed, the attempt to command improvements had apparently been abandoned, since the 1765 supplement directed that the Dock be filled in between Walnut and Third streets "over the arch *now erected.*"[62]

The fate of the Dock was fixed. The act of building an arched conduit through the section above Walnut Street and covering it over confirmed that the citizens no longer found the once-marshy, stream-fed cove a "commodious Dock." After a century of ill use and grudging maintenance, the Dock had become the filthiest site in a habitat that was generally polluted with discarded waste, poor drainage, and fitful regulation of obnoxious industries located in the middle of settlement. Residents later recalled that "few people at the time regretted the dock being arched over," since "at high water great patches of green mud floated on it" and the few fish that still entered the creek "soon floated belly up."[63]

As the city closed in, the creek had become no more than a link, though perhaps a major one, in a drainage system that was expanding along with the settled core. The cove and streams had become a contradiction in nature: a watercourse without a watershed. From the perspective of the comprehensive legislation passed in the Assembly during the 1760s, the Dock was an engineered structure, no more, no less.

The Urban Technology Acts of the 1760s: The Systematic Response

Between 1763 and 1765, the first step had been taken toward arching and covering the channel that had been under construction since the 1600s. The second step came in 1784, when city regulators extended the storm

61. *Statutes at Large*, VI, 238–240.
62. *Ibid.*, 409–410 (italics mine).
63. Watson, *Annals*, I, 347.

water drain from Walnut to Front Street, and the final in 1818, when the closed sewer was continued to the river. The initial step did not represent a new understanding of the crisis by city leaders, but, rather, capitulation to the consequences of long neglect. Of more consequence was the community's long-standing awareness that the polluted Dock was one problem among many. Private privy pits and manufacturers' waste pits leaked their contents into water wells, processing industries polluted the air, citizens habitually threw dead animals on the commons, and practically everyone dumped garbage in the streets, empty lots, open sewers, and watercourses of the city. The systematic extension and intensification of infrastructural networks and municipal service became the only logical choice.

These acts clearly represented an expansion of public responsibilities after eighty years of hesitating to take charge of the physical needs of the city. The response was in many ways modern, using technology to fix problems that were social and political. The 1760s legislation rejected not only the 1739 attempt to control and establish the location of industrial practices but also converted the Dock into the key element in a citywide drainage system.

The first comprehensive step appeared in 1762 with the passage of a seven-year act that, although it would be replaced in 1769, successfully codified much of the technical legislation of eighty years.[64] Its breadth was captured in the title:

> An Act for Regulating, Pitching, Paving and Cleansing the Highways, Streets, Lanes and Alleys and for Regulating, Making and Amending the Water-courses and Common Sewers within the Inhabited and Settled Parts of the City of Philadelphia, and for Raising of Money to Defray the Expenses Thereof.[65]

After dealing with specific problems regarding streets, drainage, and waste removal, the Assembly assigned responsibility for enforcement to the Corporation and to a group of commissioners to be elected by the freeholders. In its main regulatory sections, the act mandated programs to pave the streets, lanes, and alleys and to lay out and maintain storm water sewers and watercourses. Although unmentioned in the 1762 act's title, and recognized in the 1769 title with the added words, "and for Other

64. Charles S. Olton, "Philadelphia's First Environmental Crisis," *PMHB*, XCVIII (1974), 90–100, discusses in detail the legislation of the 1760s.

65. *Statutes at Large,* VI, 196–214, VII, 277–307.

Purposes Therein Mentioned," the longest section established a system for removing solid waste and filth from the city.

Both the method and materials for the streets—usually stone surfaces laid with sand and gravel—were to be established by the commissioners, who were to give priority to streets "most used by the country in bringing their produce and effects to market." Similar instructions were included for sidewalks, or footways, the expense for which was to be borne by owners of property adjoining them. Along with the regulations already established for the Dock, subterranean and surface storm water drains were to be built through the settled parts of the city. Enabling authority was granted to a committee of the mayor or recorder, four aldermen, and four elected commissioners, both to lay out sewers on private land and to compensate the owners fairly. The regulators were to determine the degree of descent of watercourses and to fine persons who obstructed the passage of waters through the sewers.[66]

A section on solid waste removal followed the several paragraphs that dealt with streets and the drainage system. Most of the work of carrying off the "mud, mire, dirt, and other filth" found on streets was to be performed by hired scavengers, or street cleaners. Porters, church sextons, and the caretakers of other institutional buildings were to sweep all waste dirt and soil into the street on Fridays for scavengers to haul away on that day or the day after. Special wastes like shavings, ashes, and dung were to be set aside for the scavengers, with stiffer penalties set for noncompliance. Scavengers' fees were set by city administrators and were to be paid by businessmen and tradesmen, with no charges attached to the removal of waste material "incident to common house-keeping." The final ten pages of the 1762 act dealt exclusively with sources of general funding, including the remains of previous lotteries and fees assessed on property—"houses, lands, tenements, rent-charges, bound servants and negroes."[67]

The waste removal regulations of the comprehensive act were considerably advanced in a 1763 act "to Prevent and Remove Certain Nuisances in and near the City of Philadelphia"—that is, to regulate the disposal of contaminating, or toxic, wastes. A list of nuisances included parts of buildings—porches, windows, and spouts and gutters—that interfered with passage on streets and sidewalks, but the act focused on industrial wastes: foul liquids, wastes from stills and boiling vessels used by distill-

66. *Ibid.*, VI, 198–202.
67. *Ibid.*, 200–201, 204.

ers, renderings from soap-boilers and tallow-chandlers, and fat and grease from slaughterhouses. In addition, organic wastes generated by the general community such as the contents of privies and necessary houses and dead animals (cattle, sheep, hogs, and dogs) were designated. All such wastes were to be buried "a full or sufficient depth" either on the commons or "on or near any of the streets." Nothing was to be thrown "into the public water-course of the said city, called the Dock." In addition, the act set the depths of privies so as "to preserve the waters . . . wholesome and fit for use."[68]

Although the lawmakers provided no specific labels, the 1762 legislation had dealt mostly with inorganic waste, including sweepings of harmless material and household refuse, along with ash and manure, but not food waste. The 1763 act dealt instead with the disposal of potentially unhealthy or toxic, and definitely unpleasant, waste material, specifically carrion and unwanted and unused by-products from manufacturing processes. In contrast to its predecessors, the 1763 act held the businesses and individuals that produced waste responsible for removing it. The 1769 act incorporated all of these concerns and added others. Besides regulating the size and placement of signs and sign poles, a final paragraph provided that the Dock below Second Street continue to be cleaned and repaired. The act also established two public landings on the Delaware to complement those on the Dock below Front Street.[69]

Philadelphia's leaders had at last responded in a comprehensive fashion to the long environmental crisis—in which the polluted Dock served to focus the problems with pollution that occurred throughout the city. After thirty years of makeshift engineering, the provisions of the 1760s acts reflected a now-established understanding of the links between pollution, infrastructure, waste removal services, and funding arrangements.[70] Again, the connection between ecological systems and settlement patterns appeared in a warning contained in a January 1769 petition. After reiterating the problems of economic losses to landowners in the area of the Dock and the greater chance of disease, the petitioners explained that these

---

68. Ibid., 230–234.

69. Ibid., VI, 230–232, VII, 295–296, 304.

70. See Michal McMahon, "Makeshift Technology: Water and Politics in Nineteenth-Century Philadelphia," Environmental Review, XII, no. 4 (Winter 1988), 20–37, on the repeated use of the term "makeshift engineering" during another environmental crisis in late-19th-century Philadelphia.

problems would worsen "as the vacant Parts of the City, from which the Water is conveyed to the said Sewer, become improved and built upon." [71]

Despite this understanding of the hydrological properties of the land, the city's leadership had turned to mechanical solutions. Earlier in the 1760s, city leaders began what, within fifty years, would convert the entire Dock into a sewer. At the same time, they tightened the administration of city services and took steps to systematize an infrastructure of technology and services. Not only did the new comprehensive approach borrow from past legislative efforts, but the legislation relied less on individual landowners to initiate the task that would create a healthy and pleasant city. These public initiatives did not so much overturn the relative standing of public and private rights as accept the need for public authorities to work within a world shaped by private interests and individual acts. The shift came with the community's recognition that piecemeal fixes were inadequate when pervasive conditions called for a centralized, systematic response.

These steps were slowed and even halted by the distractions of the Revolution and the transformation of local and provincial governments during the 1770s and 1780s. But this procrastination does not belie so much as underscore their importance. Seen from the path leading to the Revolutionary years, the disastrous epidemics that struck Philadelphia at the end of the century appear as a response to past neglect as well as a stimulus to the building of a centralized waterworks. Endemic and epidemic disease killed from a fifth to a fourth of the population between 1793 and 1797; possibly, if the city had applied technological insights from the thirty years prior to the Revolution, such a death toll might not have been exacted.

Conclusion

Compared to the end of the century, Philadelphia's experience of urban technology between the 1730s and the 1760s was more complex because public use of natural advantages was in open contest with private rights to property development. Issues of technological choice and settlement design were complicated by ecological systems, intense environmental pollution, and the absence of established governing arrangements. During the nineteenth century, the situation became simpler with the acceptance of a

71. *Pennsylvania Archives*, 8th Ser., VII, 6308–6309.

fully mechanized habitat. Eighteenth-century Philadelphia offers a richer context for the study of settlement patterns, for it displays the process of making choices that resulted in the triumph of mechanization. With mechanization, the possibility of incorporating ecological systems into technological systems was gone.

In their tight focus on the hardened patterns of the nineteenth- and twentieth-century city, historians of urban technology tend not to appreciate organic perspectives of city development like Lewis Mumford's. This disdain has led some historians to find in Mumford's work "overarching but difficult-to-establish notions." [72] Observing a concrete terrain—served by a pervasive mechanized urban infrastructure complete with water-carriage sewage networks directed toward vast sewage treatment facilities and chemically based water treatment plants—understandably obscures the organic issues raised by the early city-building experience.

Yet how difficult would it be to demonstrate Mumford's statement that "a characteristic feature" of the industrial era involved the "transformation of the rivers into open sewers"? Or the accompanying results: poisoned aquatic life, destroyed food, and unswimmable waters? Is it really overarching to state that some of the chief contributions of the nineteenth-century United States city—"water closets, sewer mains, and river pollution"—represented a "backward step ecologically, and so far a somewhat superficial technical advance"? [73]

A close look at Philadelphia during its formative century, when vital organic elements remained active forces within the city, renders Mumford's holistic approach not only entirely relevant but more helpful to gaining an understanding of the problems of modern settlement patterns. Because seventeenth-century capitalism "treated the individual lot and the block, the street and the avenue, as abstract units for buying and selling, without respect for historic uses, for topographic conditions, or for social needs," early cities were flattened and built over. [74] Obscured along with the topography was an alternative, environmentally sounder vision of the city.

The importance of the story of Dock Creek lies in its model explanation

72. Konvitz, Rose, and Tarr, "Technology and the City," *Technology and Culture*, XXI (1990), 288.

73. Lewis Mumford, *The City in History: Its Origins, Its Transformations, and Its Prospects* (New York, 1961), 14, 459.

74. *Ibid.*, 421.

of the way that a natural cove and stream system became a quasi-natural, quasi-mechanical entity: a planked, walled canal. This in turn became an element in an evolving urban technological system: a brick conduit for transporting sewage, liquid wastes, and storm water. As an artifact of the process of mechanization, early Philadelphia raises fundamental questions with a clarity directly rooted in their seminal character. Eighteenth-century Philadelphia possessed an urban core not yet built over, which provoked a long struggle, in which the disposition of a natural utility and urban amenity was determined by the inability of a predominantly private economic order to accommodate natural utilities.

The very persistence of the values behind mechanization illustrates how the extended struggle over the Dock was no simple act of filling in a tidal cove and leveling a watershed in order to install a comprehensive drainage system. The loss of the Dock and creek reflects the narrow set of values that have shaped American settlement patterns. Public ends are transformed by the aims of a private order, social and environmental goals are submerged in mechanistic solutions, and natural entities like the Dock are relegated to memory and to the pages of antiquarian histories.

Patrick W. O'Bannon

Inconsiderable Progress

Commercial Brewing in Philadelphia before 1840

Until a generation ago, the notion of technological determinism informed the approach of historians of technology. An outgrowth of the industrialization of the American economy, technological determinism views technology as operating largely outside the bounds of societal influence, with one advance leading inevitably and inexorably to the next. Since then, historians of technology have developed a more complex understanding of the interaction between technology and society, recognizing that human choice has a major impact on the shape technologies take. Historians of technology now study societal influences, pressures, and goals to help explain the dynamic between human agency and technological systems.[1] The incorporation of technology into society removes the aura

Many people helped this essay reach its present form. Mark Wilde recommended me to Michael Zuckerman and Judith McGaw at the University of Pennsylvania Center for Early American Studies. My then employer, John Milner Associates, granted me half-time status to join the Center, whose scholars offered insightful commentary and criticism. Michael, Judith, Billy Smith, Peter Thompson, and the various anonymous readers all helped focus the argument. Hollis Payer read every draft and endured long talks about barley, hops, and malt.

1. John F. Kasson, *Civilizing the Machine: Technology and Republican Values in America, 1776–1900* (New York, 1976); Leo Marx, *The Machine in the Garden: Technology and the Pastoral Ideal in America* (New York, 1964). Some of the most important works

of inevitability from the process of technological change and increases the importance of human selection.

When confronted by a new technology, humans have a choice of accepting, rejecting, or modifying it. Change is not inevitable, and the continuation or adaptation of established or traditional methods and techniques is often more attractive than innovation. Most historical studies that address the issue of choice in the process of technological innovation and change examine industrial technologies of the nineteenth and twentieth centuries. The factory settings within which much of this technology functioned often favor an emphasis upon machinery, which can reveal more radical departures from past usage, rather than upon subtle forms of technology, such as knowledge and techniques.

The history of commercial brewing in Philadelphia prior to 1840 offers an opportunity to examine the dynamic relationship between technology and societal choice in a preindustrial setting. During this period Philadelphia's producers and consumers of beer clearly demonstrated the role of human choice in the development and transfer of a new technology. This technology, a new type of beer known as porter, necessitated no investment in new and untried machinery; indeed, the basic process used to brew porter differed little from that used to produce traditional beers and ales. However, porter required a substantial departure, on the part of both brewers and beer drinkers, from traditional notions of what constituted a malt beverage.

Commercial brewing in Philadelphia displayed a remarkable degree of continuity within a slow process of adaptation in terms of methods, markets, and product line over a 150-year period. Not until the introduction of lager beers in the 1840s did the patterns of brewing and beer consump-

---

investigating these dynamics include Reese V. Jenkins, *Images and Enterprise: Technology and the American Photographic Industry, 1839 to 1925* (Baltimore, 1975); Merritt Roe Smith, *Harpers Ferry Armory and the New Technology: The Challenge of Change* (Ithaca, N.Y., 1977); David J. Jeremy, *Transatlantic Industrial Revolution: The Diffusion of Textile Technologies between Britain and America, 1790–1830s* (Cambridge, Mass., 1981); David A. Hounshell, *From the American System to Mass Production, 1800–1932: The Development of Manufacturing Technology in the United States* (Baltimore, 1984); Judith A. McGaw, *Most Wonderful Machine: Mechanization and Social Change in Berkshire Paper Making, 1801–1885* (Princeton, N.J., 1987); and the essays collected in Marcel C. LaFollette and Jeffrey K. Stine, eds., *Technology and Choice: Readings from Technology and Culture* (Chicago, 1991).

tion in Philadelphia reveal a significant degree of departure from past practices.

Porter did indeed constitute a new technology, not simply a modification of or innovation on an existing product. Developed in England in the early 1720s, it formed the cornerstone of a vast English brewing industry that by the mid-eighteenth century boasted gigantic London breweries capable of producing 200,000–300,000 barrels annually. The London porter brewers not only introduced an entirely new product; they also developed innovative methods of mass production and distribution.[2]

Porter differed from the traditional beers and ales brewed before the 1720s principally in its use of dark, slightly scorched barley malt that provided a characteristic dark color and bitter taste. The brew's robust color and taste tolerated rougher treatment during the brewing process than lighter, more delicate ales and beers, allowing brewers to exercise less care in brewing. Brewers also discovered that the quality of porter did not suffer from rigorous processing, which enabled them to extract the maximum value from their raw materials. As brewers gained experience with porter, they also discovered that it could be successfully brewed using second-quality raw materials. Dark malt did not require as fine a quality of barley as the lighter-colored malt used for ales and beers, and the bitter taste imparted by the dark malt allowed the use of lesser-quality hops without harming the final product.[3]

Porter proved more resistant to spoilage than beer or ale, largely because of its higher alcohol content and greater tolerance of heat. Porter's tolerance of heat also allowed brewers to continue production further into the warm months of summer than was possible with more sensitive brews. Spoilage resistance and heat tolerance also produced a beer capable of surviving the rigors of transportation to markets a considerable distance from the brewery in a drinkable condition.

In many regards porter appeared an ideal industrial beverage: extracting the utmost value from second-quality raw materials and capable of being transported to distant markets with a minimal risk of spoilage. English historian Peter Mathias describes it as the first beer technically suited to mass production at contemporary standards of control. Mathias argues, "The appearance of the new beer should be seen . . . as an event of

2. Peter Mathias, *The Brewing Industry in England: 1700–1830* (Cambridge, 1959), xxiii, 12, 14.

3. *Ibid.*, 15–17.

the first importance, or as an invention exactly equivalent in its own industry to coke-smelted iron, mule-spun muslin in textiles or 'pressed-ware' in pottery."[4]

The characteristics of porter lauded by Mathias—particularly the brew's tolerance of second-quality raw materials, harsh processing methods, and temperatures sure to spoil traditional beers and ales—also suggest a product tailor-made for production in the American colonies or the new American republic. Ideally, American brewers could have produced porter without stringent control over either their raw materials or the production process and could have successfully marketed the beer in the West Indies or southern mainland ports, where the warm climate virtually precluded the production of beer. Porter, however, never became a notable component of the American economy, despite its theoretical profitability.

The techniques and methods—the technology—of porter production reached the North American continent in the 1770s, some fifty years after the new brew first appeared in London pubs. A London porter brewer bankrolled his son, Robert Hare, who arrived in Philadelphia in 1773 and by 1775 produced the first porter brewed in North America.[5] Hare's brewery prospered, but the introduction of porter failed to touch off a pattern of rapid growth and expansion similar to that which characterized the English industry. Instead, Philadelphia's brewing industry continued to produce chiefly traditional beers and ales until the introduction of German lager beer in the 1840s.

As late as 1810 former assistant secretary of the Treasury Tench Coxe expressed regret at the "inconsiderable progress" of the domestic brewing industry. Coxe cited the difficulty and expense associated with procuring a ready supply of strong bottles and the public's "peculiar taste for lively or foaming beers" as the principle causes of the industry's stagnant condition.[6] Behind the apparent lack of progress, however, were pragmatic responses to market limitations and gradual adaptations to changing circumstances.

Coxe's explanation hints at the complex array of factors that shaped

---

4. *Ibid.*, 13.

5. Robert Hare and Co., bill to General John Cadwalader, Oct. 9, 1775, Cadwalader Papers, Historical Society of Pennsylvania, Philadelphia; Stanley Baron, *Brewed in America: A History of Beer and Ale in the United States* (Boston, 1962), 114.

6. Tench Coxe, *A Statement of the Arts and Manufactures of the United States of America, for the Year 1810* (Philadelphia, 1814), xl.

Philadelphia's commercial brewing industry prior to 1850. The lack of suitable bottles constitutes a supply constraint that restricted production and limited output. However, most beer sold before 1850 appears to have been marketed in wooden containers, half-barrels, barrels, tierces, hogsheads, and puncheons of varying size. Only the finest and most expensive beers were sold in bottles, so a lack of these containers had little effect upon most of the industry's output.[7]

The public's seemingly perverse taste for "foaming" beer (as opposed to more traditional noncarbonated, or "flat," brews) perhaps played a more significant role in hampering the industry's development. Public taste represents a demand constraint. Without demand no product or industry will long endure in a capitalist society. Beer does not appear to have been a commercial product in high demand in Philadelphia taverns. Peter Thompson's detailed study of the drinking behavior of Philadelphians prior to 1800 suggests that all social and economic strata consumed wine and spirits, largely rum and brandy, in much larger quantities than they did beer. After 1800 whiskey replaced rum as the hard liquor of choice, and the trend continued.[8]

There are several possible explanations for consumers' choice of spirits over beer, all of which, because of the shortage of firm evidence, must be considered tentative. Hard liquor was plentiful and inexpensive throughout the colonial and early national period, though not as inexpensive as beer. Beer sales may be underrepresented in tavern account books that record retail sales, simply because beer drinkers may have been less likely to run a tab than those drinking spirits. The consumer demand for beer may also be underrepresented because no account books and few documentary records survive for alehouses. In contrast, several account books kept by tavernkeepers survive. Unlike the owners of alehouses, who were legally restricted to the sale of beer, tavernkeepers could serve spirits and other forms of alcoholic beverages.[9] Finally, the prevalence of home

---

7. James T. Mitchell and Henry Flanders, comps., *The Statutes at Large of Pennsylvania, from 1682 to 1801*, II (Harrisburg, Pa., 1896), 96; Peter John Thompson, "A Social History of Philadelphia's Taverns, 1683–1800" (Ph.D. diss., University of Pennsylvania, 1989), 72.

8. Thompson, "Philadelphia's Taverns," 27, 429; Leland D. Baldwin, *Whiskey Rebels: The Story of a Frontier Uprising* (Pittsburgh, 1939), 26, 57.

9. Thompson, "Philadelphia's Taverns," 146, 429. Thompson identifies two alehouses in his study. Virtually all the information related to the operation of these establishments,

brewing, using either malt or malt substitutes such as molasses, was a simple process requiring no specialized equipment or knowledge and may have seriously reduced the demand for commercial beer. Home brewing may have satisfied most of the daily demand for beer, leaving commercial brewers to fill the market represented by the export trade and the city's alehouses and taverns. Indeed, Thompson's research demonstrates that many of the city's largest commercial brewers sold a significant part of their production in their own taverns or alehouses.[10]

Confronted by urban markets a fraction the size of London and no great demand for their product, American brewers chose not to embrace the new technology that porter represented and establish huge breweries comparable to those in England.[11] Instead they adhered to long-familiar patterns of brewing established before the introduction of porter. In Philadelphia, brewers incorporated porter into their existing product lines and adopted those technical innovations most readily adaptable to their experience, but not until the second half of the nineteenth century did the

---

including data on their beer purchases that Thompson uses to extrapolate possible sales and profit figures, is derived, not from the account books of the publicans, but from the accounts of a brewery that supplied both establishments. No other details of the publicans' operations are known. *Ibid.*, 413–416.

10. Thompson, "Philadelphia's Taverns," 279–280. It is impossible to document the extent of home brewing that occurred within Philadelphia. Given the simplicity of the process, there is little reason to believe that brewing was markedly less prevalent than in rural areas. Indeed, in 1840 John Wagner brewed the first lager beer in the United States at his house in Philadelphia, utilizing home-brewing techniques little different from those employed by the earliest settlers. See Baron, *Brewed in America,* 176. For an indication of the commonplace nature of home brewing, albeit in a rural rather than an urban setting, see Laurel Thatcher Ulrich, *A Midwife's Tale: The Life of Martha Ballard, Based on Her Diary, 1785–1812* (New York, 1990), 73, 264, 313.

11. London's population in the mid-18th century is estimated at between 600,000 and 700,000. Philadelphia, the largest urban market in the American colonies, numbered only 25,000–30,000 during this period. E. A. Wrigley and R. S. Schofield, *The Population History of England, 1541–1871: A Reconstruction* (Cambridge, Mass., 1981); John K. Alexander, "The Philadelphia Numbers Game: An Analysis of Philadelphia's Eighteenth-Century Population," *Pennsylvania Magazine of History and Biography,* XCVIII (1974), 314–324; Gary B. Nash and Billy G. Smith, "The Population of Eighteenth-Century Philadelphia," *PMHB,* XCIX (1975), 362–368; Sharon V. Salinger and Charles Wetherell, "A Note on the Population of Pre-revolutionary Philadelphia," *PMHB,* CIX (1985), 369–386.

market prospects of American brewers permit construction of breweries approaching the size and sophistication attained in London a century
earlier.

Despite the apparent lack of progress reported by Tench Coxe, brewing
played an important role in early American social and economic life, and
nowhere more so than in Philadelphia. Commercial brewing began shortly
after the first harvests of barley and hops, the two essential ingredients
of English beer and ale that did not grow naturally in the area. By 1698
Gabriel Thomas could report, although probably with a degree of hyper-
bole, that the city boasted "three or Four Spacious Malt-Houses, [and] as
many large Brew-Houses." These establishments apparently produced for
both local and export markets, since Thomas claimed that Philadelphia
beer "hath a better Name, that is, is in more esteem than English Beer in
Barbadoes, and is sold for a higher Price there." [12]

The amount of commercial brewing conducted in Philadelphia during
the late seventeenth and early eighteenth centuries is not recorded. Legislation in the first decades of the eighteenth century placed imposts on
hops grown outside the Delaware Valley and forbade the manufacture of
beer using molasses and other adulterants. These political acts provide
little evidence regarding the scope of the industry, but they do indicate
that brewing was important enough to attract the attention of local interest groups and politicians and that these interest groups and politicians
wished to provide an environment within which such an industry could
develop. Several commercial breweries appear in the written record, but
it is not known how many others escaped notice.[13]

12. J. Leander Bishop, *A History of American Manufactures, from 1608–1860* . . . , I
(Philadelphia, 1861), 259; Albert Cook Myers, ed., *Narratives of Early Pennsylvania, West
New Jersey, and Delaware: 1630–1707*, Original Narratives of Early American History
(New York, 1967 [orig. publ. New York, 1912]), 241, 267–268n, 286, 327; Baron, *Brewed
in America*, 45.

13. J. Thomas Scharf and Thompson Westcott, *History of Philadelphia, 1609–1884*, III
(Philadelphia, 1884), 2278. Similar restrictions existed in most of the colonies where brewing could be conducted commercially. The imposts were intended to protect local brewers
and the farmers who produced their raw materials against competitors from other colonies.

Virtually all evidence regarding commercial brewing in Philadelphia prior to the 1730s
is circumstantial. It is known that several prominent citizens owned brewhouses and malt-
houses, but no account books, written records, or other documentary evidence detailing
the extent of these or any other brewing operations appear to survive. Thompson uses the
records of liquor dealers and tavernkeepers to document the drinking habits of Philadel-

The earliest detailed information pertaining to commercial brewing consists of the mid-1730s account book of an unnamed Philadelphia brewery. The account book entries clearly indicate that commercial brewing required, even before the introduction of porter, a considerable capital investment. The eight investors, mostly young men from prominent Philadelphia families, each contributed £250 to the venture. The £2,000 capital served to secure a long-term lease on a building and wharf, adjacent to the Sun Tavern, convert and equip the building for use as a brewery, hire a trained brewmaster, obtain a work force of indentured servants, and purchase raw materials. The first sales of beer occurred in September 1732.[14]

The Eight Partners Brewery produced five types of beer, differentiated by their strength and priced accordingly. Small beer, the weakest brew, sold for 4s. 6d. per barrel. Strong (or double) beer, the most potent of the brewery's products, commanded a price of 30s. per barrel, more than six times that of small beer. In addition to beer, the brewery also produced small quantities of ale, probably a traditional English unhopped brew, some of which, packaged in glass bottles, commanded a premium price of 3d. per bottle.[15]

Most sales recorded in the account book are for small purchases of the weaker brews. Few individual transactions exceeded a single barrel, and most were for half-barrels. The preference for regular half-barrel purchases of small and middling beer holds across social and economic lines. The accounts indicate that the brewery sold to a wide range of customers, from tailors and tavernkeepers to sea captains and merchants. Both the proprietor, Thomas Penn, and the governor, Patrick Gordon, patronized the brewery. Their accounts, while more extensive than those for laboring Philadelphians, list numerous half-barrel purchases of small and middling, with only an occasional barrel of ale or strong.[16]

Customers preferred the small, sixteen-gallon half-barrels because beer tended to spoil fairly quickly. Like most other foodstuffs, it was pur-

---

phians during this period, but these records provide no evidence regarding the scope of the commercial brewing industry, its methods, or its significance within the local economy. See Thompson, "Philadelphia's Taverns."

14. [Eight Partners Brewery], "Account Book," 1733–1735, HSP; *Pennsylvania Gazette* (Philadelphia), Dec. 2, 1736.

15. [Eight Partners], "Account Book." Statutes dictated that barrels contain 31.5 gallons (Mitchell and Flanders, comps., *Statutes at Large*, II, 96).

16. [Eight Partners], "Account Book."

chased frequently, generally each week, and consumed quickly. Processed liquids, such as beer or hard cider, reduced the risk of sickness associated with drinking fouled or polluted water and served as the "common table drink of every family in easy circumstances." The fairly constant level of imbibing may explain the greater consumption of less alcoholic brews.[17]

Barley, hops, and malt, along with fresh water and fuel (the latter consumed in the brewing process), served as the principal ingredients for all the various types of beer manufactured by the Eight Partners Brewery. The brewery obtained barley in small quantities from many sellers, principally during the brewing season, which extended from September to April. Between December 1733 and April 1734 the firm made sixteen separate purchases of barley from eleven different sellers. The average purchase amounted to only fifty-three bushels. A much smaller number of suppliers provided the hops, with a single seller furnishing 91 percent of the hops purchased in September and October 1733.[18]

The firm apparently sent its barley out to commercial maltsters for malting. Malting prepares the grain for brewing by steeping it in water, allowing it to sprout, and then drying it in kilns. This process converts a portion of the grain's starch into fermentable sugars. The accounts indicate that in 1733 the partners purchased nearly equal amounts of barley and malt; the similarity in these figures, combined with the lack of any mention of malting equipment, strongly suggests that the barley was sent out for processing. Several maltsters had accounts with the brewery, frequently receiving beer in payment for their services.[19]

The account book provides no information on the brewing process. However, its list of the various utensils in the brewhouse permits the basic outline of the process to be established with some certainty. The malt was

17. W. J. Rorabaugh, *The Alcoholic Republic: An American Tradition* (New York, 1979), 107; Israel Acrelius, *A History of New Sweden . . .* , trans. William M. Reynolds (Historical Society of Pennsylvania, *Memoirs,* XI [1874]) (Philadelphia, 1876), 160–163; Thompson, "Philadelphia's Taverns," 31, 37; Baron, *Brewed in America,* 5–6; Mark Edward Lender and James Kirby Martin, *Drinking in America: A History* (New York, 1982), 4.

18. [Eight Partners], "Account Book," fols. 61, 179.

19. William L. Downard, *Dictionary of the History of the American Brewing and Distilling Industries* (Westport, Conn., 1980), 115. The Eight Partners work force briefly included a maltster, but he "ran away" without settling his account in April 1733, an act that may have convinced the partners to place their trust in commercial maltsters. See [Eight Partners], "Account Book," fols. 18, 64, 72, 179.

ground between the stones of a horse-powered mill before being placed in a pair of mashing tuns, where it was mixed with liquor (the brewers' term for water). Workers armed with mashing oars thoroughly mixed the liquor and malt into a thick syrup known as wort. The wort drained from the mashing tuns into an underback, located below the mashing tuns, where the dregs of the mash were skimmed off. From the underback the wort drained into a copper, the most expensive piece of equipment in the brewery, where it was boiled with hops. After boiling, the wort was pumped or drained into three cooling backs, large shallow vessels that permitted the wort to cool rapidly to the temperature required for successful fermentation. The cooled wort flowed from the cooling backs into one of three working tuns, where it was pitched with yeast and allowed to ferment. After completion of the fermentation process, which required anywhere from two to six days, the young beer was drained into large, forty-two-gallon wood tierces for a short period of conditioning. From the tierces the beer was racked into barrels and half-barrels for marketing.[20]

The barrels, half-barrels, and other containers used to deliver the beer to the brewery's customers constituted a significant capital investment. Nevertheless, since beer and ale could be marketed after only a brief period of conditioning, this investment proved only a fraction of that required for porter brewers, who cellared their beer for as long as a year and often had one-third of their capital tied up in barrels and casks. In January 1735 nearly 450 casks, barrels, tierces, and other containers cluttered the premises of the Eight Partners Brewery. During 1736 the supply of barrels and casks at the brewery varied from 388 to 490, with a maximum value of more than £168. The brewery actively sought to retain ownership of these containers, charging customers three shillings if they neglected to return their empty barrels. This amounted to a significant penalty—two-thirds the cost of a barrel of small beer.[21]

The account book includes production figures for only 1735. During that year the brewery produced 1,511 barrels of beer. Small beer, the weak

20. [Eight Partners], "Account Book"; *The Complete Family Brewer* . . . (Philadelphia, 1805); Joseph Coppinger, *The American Practical Brewer and Tanner* . . . (New York, 1815); M. L. Byrn, *The Complete Practical Brewer* . . . (Philadelphia, 1867), 24–44; *One Hundred Years of Brewing* (New York, 1974), 76–115 (orig. publ. as *A Supplement to the Western Brewer,* 1903 [Chicago, 1903]); John P. Arnold and Frank Penman, *History of the Brewing Industry and Brewing Science in America* (Chicago, 1941), 87.

21. Mathias, *Brewing Industry,* 53–62; [Eight Partners], "Account Book," fols. 3, 175.

brew that constituted the majority of the firm's sales, accounted for 54 percent of the year's production. The only other beers brewed in significant quantities during the season were strong and double beer, which accounted for 29 percent and 16 percent, respectively, of the total output.[22] These production figures emphasize the small scale of the operation, particularly in comparison to the major English breweries of the period, which produced 200,000–300,000 barrels annually.

Their venture into commercial brewing failed to earn a profit for the Eight Partners. At the close of 1733 the business had incurred expenses totaling £2,905 while the credit side of the ledger, which included £570 outstanding from customers, totaled only £1,698, a loss of more than £1,200. Subtracting the cost of the brewhouse, its fixed works, and the brewing utensils (a total of nearly £850) still yields a book loss of nearly £360, or 18 percent of each partner's original investment. By December 1736 the investors had given up on the venture, selling the brewery and its utensils at auction.[23]

The few surviving account books and financial records from other Philadelphia breweries suggest that the Eight Partners Brewery was fairly representative of commercial breweries throughout the colonial and early national period. Commercial brewing required a heavy capital investment in brewing utensils, raw materials, and containers. Returns on this investment proved both elusive and small. Confronted by these prospects, most commercial investors chose to place their money elsewhere. Commercial brewing in Philadelphia remained largely the calling of families with a long commitment to brewing as a vocation. Transfers of malt, hops, and other raw materials between brewers occurred fairly frequently. These transactions appear to have functioned as short-term loans, since the amounts involved were generally small and cash seldom changed hands. The recipient simply returned the amount borrowed at a later date.[24]

Under these circumstances, the industry displayed a logical adherence to conservative practices over the course of nearly one hundred years. The changes wrought by the introduction of porter in the English brewing industry had little effect upon Philadelphia's commercial brewers because market conditions would not support such a large-scale innovation. Con-

22. [Eight Partners], "Account Book," fol. 19.

23. *Ibid.*, fol. 183; *Pennsylvania Gazette*, Dec. 2, 1736.

24. Reuben Haines and Godfrey Twells, "Account Book," 1767–1775, fols. 125, 174, 224, 235, 239, American Philosophical Society, Philadelphia.

tinuity, not change, characterized commercial brewing during the century following the failure of the Eight Partners Brewery.

Commercial brewers in Philadelphia long continued to secure their supply of raw materials in much the same fashion as the Eight Partners Brewery, purchasing barley in small quantities from a multitude of suppliers while buying hops in bulk from one or two sources. In 1815 Melichor and John Larer bought nearly seven thousand bushels of barley from seventy-three separate sellers in seventy-nine individual transactions. The average purchase amounted to only eighty-eight bushels. That same year the Larers bought virtually their entire supply of hops from the Boston firm of Grant and Stone, a broker or commission merchant for farmers throughout New England, where hops formed an important element in the local agricultural economy.[25] These patterns parallel those of the Eight Partners Brewery, which failed nearly eighty years before these transactions.

The Larers malted their own barley, unlike the Eight Partners, who sent their barley out to maltsters. The Larers' practice appears to have been the norm among Philadelphia's commercial brewers and suggests a degree of vertical integration within the industry. Fire insurance surveys for ten other breweries indicate that each had a malthouse associated with the brewhouse.[26]

As an alternative to malting, or shipping barley to maltsters, some brewers avoided this aspect of the business entirely and simply purchased malt. Reuben Haines, who inherited his father-in-law Timothy Matlack's brewery in 1752, followed this practice during the late 1760s, despite the fact

---

25. Melichor and John Larer, "Receipt Book," 1815–1820, HSP; Thomas Wilson, *Picture of Philadelphia, for 1824, Containing the "Picture of Philadelphia, for 1811, by James Mease, M.D."* . . . (Philadelphia, 1823), 78.

26. Tench Coxe lamented the "absence or infrequency" of malting as a separate trade. Coxe apparently believed that a separate malting industry, rather than the integration of malting into brewers' existing operations, would better serve the brewing industry. Coxe, *Arts and Manufactures of the United States, for the Year 1810*, xl.

On fire insurance, see the Philadelphia Contributionship: "William Abbott Survey No. 3891," Sept. 6, 1819; "Mordecai L. Dawson Survey No. 4692–4693," July 3, 1829; "William Dawson Survey No. 2358," Dec. 29, 1788; "William Gray Survey No. 1645–1646," Nov. 5, 1772; "Reuben Haines Estate Survey No. 2612–2615," Mar. 4, 1794; "Robert Hare Survey No. 2173–2174," Oct. 30, 1784; "Isaac Howell Survey No. 1930–1931," Dec. 5, 1775; "Anthony Morris Survey No. 1935–1936," Jan. 2, 1776; "Thomas B. Pritchett Survey No. 4212," Oct. 7, 1823; "Adam Seckel Survey No. 3919," Jan. 29, 1820.

that his brewery property included a malthouse. During these years Haines and his partner, Godfrey Twells, purchased malt in bulk from commercial maltsters as far away as Wilmington, Delaware. During a single week in May 1770 the firm purchased nearly thirty-seven hundred bushels of malt from two commercial maltsters at between four and five shillings per bushel. These purchases amounted to nearly one-third of the brewery's annual supply of malt.[27]

Well into the nineteenth century the physical plant of most commercial breweries in Philadelphia resembled that of the Eight Partners Brewery. Inventories of brewery vessels and utensils indicate little change in either number or type, although lack of information regarding the size of utensils may conceal significant differences over time. Before the 1830s no brewery in the city appears to have exceeded three stories in height, with the third floor frequently given over to grain and malt storage bins. This situation required Philadelphia brewers to rely upon pumps to move materials through their plants. Several breweries pumped the wort from the coppers to coolers located on upper floors, a practice largely eliminated in the large London breweries of the period, which fully exploited the economies offered by allowing gravity to carry materials down through the building during the course of the brewing process.[28]

The capacity of individual Philadelphia breweries is not recorded. It appears, however, that before 1830 productivity did not rise to levels significantly greater than those achieved at the Eight Partners Brewery a century previous. In 1810 Tench Coxe reported eleven commercial breweries in Philadelphia, more than in any other city in the country except New York, which boasted fifteen. Philadelphia's breweries produced forty-eight thousand barrels of beer annually, an average output of less than forty-four

27. "Inventory of the Goods Etc. in Timothy Matlack's Brew House and Malt House," ca. 1752, Society Miscellaneous Collection, box 2A, fol. 1, HSP; the Philadelphia Contributionship, "Reuben Haines Estate Survey"; Haines and Twells, "Account Book," fols. 179, 283, 382, 407.

28. "Inventory of Matlack's Brew House"; the Philadelphia Contributionship: "William Abbott Survey"; "Mordecai L. Dawson Survey"; "Frederick Gaul Survey No. 4079," Aug. 27, 1822; "Thomas Morris Survey No. 3895," Oct. 11, 1819; "Thomas B. Pritchett Survey"; "Adam Seckel Survey"; and Mathias, *The Brewing Industry*, 40–42. At least two Philadelphia brewers placed their liquor back, the cistern that held the water used in the brewing process, on a bridge or platform spanning an alley adjacent to the brewery (the Philadelphia Contributionship: "Thomas Morris Survey"; "Thomas B. Pritchett Survey").

hundred barrels each. Since Coxe's statistics probably exclude a number of the city's smallest commercial breweries, it seems likely that this figure resides at the outer limit of actual productive capacity.[29]

The output of Philadelphia's breweries constituted merely a fraction of the 200,000–300,000 barrels produced annually by London's largest porter brewers. The extraordinary disparity between English and Philadelphia brewery capacity masks, however, the true size of the domestic breweries. In 1819 Ludwig Gall, a German immigrant, marveled at Philadelphia's "enormous breweries, the likes of which in all Europe only England can boast."[30]

By the time Gall arrived in Philadelphia, the city's commercial brewers had cautiously introduced several technical innovations pioneered in the London porter breweries into their own plants. Porter itself had been readily incorporated into the repertoire of several local brewers. In 1816 the new brew constituted 28 percent of the total value of malt beverages exported from Philadelphia as part of the city's coasting trade. Stationary steam engines operated in at least three Philadelphia breweries by 1819, and by 1822 Frederick Gaul had installed a mashing machine in one of the two mash tuns he operated at the former Robert Hare and Company brewery.[31]

Technological innovations such as steam engines and mashing machines coexisted in Philadelphia's commercial breweries with time-honored traditional practices and methods. A widely read brewing handbook, first published in 1815, described brewing as a branch of commercial chemistry and advised brewers on the proper use of thermometers and hydrometers, scientific instruments introduced by English porter brewers that provided a degree of control and certainty over the brewing process. This same handbook also recommended that "should weevils at any time get into,

---

29. Philadelphia's breweries appear, on average, significantly larger than those in New York. The 15 New York City breweries produced an average of nearly 102,000 gallons per year, less than 3,200 barrels each. Coxe, *Arts and Manufactures of the United States of America, for the Year 1810*, 36, 59.

30. Frederic Trautmann, "Pennsylvania through a German's Eyes: The Travels of Ludwig Gall, 1819–1820," *PMHB*, CV (1981), 45–46.

31. Diane Lindstrom, *Economic Development in the Philadelphia Region, 1810–1850* (New York, 1978), 194–197; Trautmann, "Pennsylvania through a German's Eyes," *PMHB*, CV (1981), 46; the Philadelphia Contributionship: "William Abbott Survey"; "Thomas Morris Survey"; "Frederick Gaul Survey."

or generate in your malt, which is common when held over beyond twelve or eighteen months, the simplest and easiest way of getting rid of them, is to place four or five lobsters on your heap of malt, the smell of which will soon compel the weevils to quit the malt, and take refuge on the walls, from which they can be swept with a broom into a sheet or table cloth laid on the malt, and so taken off." [32]

Philadelphia's commercial brewers adjusted their processes and adopted low-risk innovations as economies of scale and available methods warranted. Not until the introduction of lager beer by German immigrants during the 1840s did the industry take a significant departure from accustomed manufacturing techniques. Even after the introduction of lager, traditional brewing, marketing, and consuming patterns remained powerful, as evidenced by the fact that traditional beers and ales remained the most important products of Philadelphia's commercial breweries until the Civil War. Philadelphia brewers brought to bear the pragmatic approach exhibited from colonial times to the mid-nineteenth century to the introduction of porter.

Confronted by the new technology that porter represented, Philadelphia's brewers made a series of choices over a period of approximately fifty years that enabled them to introduce selectively porter itself (as well as the most useful technical innovations associated with the new brew) into their existing product lines. Brewing in Philadelphia in 1840 relied upon traditional methods and techniques, augmented by the use of steam engines and scientific instruments in the city's largest and most progressive breweries. The productive capacity of individual breweries remained modest throughout the first half of the nineteenth century, but this output proved more than adequate to meet the demands of both the local market and the coasting trade. Philadelphia's commercial brewers adopted those elements of the package of technological innovations in machinery, knowledge, and methods constituting the new technology of porter that best enabled them to respond to their social and economic environment.

By 1860, and the onset of the Civil War, Philadelphia's brewing industry stood poised on the threshold of full-scale industrialization. The developments of the preceding half-century had introduced important technical innovations into the traditional industry, laying the groundwork for more dramatic changes that took place after the Civil War. These changes occurred not simply because of the introduction of a new technology but

32. Coppinger, *The American Practical Brewer*, 22–23, 74–75, 82, 99 (quote on 22–23).

because of the conjunction of a variety of factors at a single point in time. The increased availability of investment capital, the growth of urban markets, and the development of improved transportation facilities all made brewing on a massive scale a more likely source of profits than previously. These developments—combined with a popular acceptance of the new German lager beers that originated with the troops of the Union Army, many of whom were exposed to these new malt beverages for the first time while in uniform—led commercial brewers largely to reject traditional English beers and ales and choose to shift production to the new lagers, pushing the domestic brewing industry to levels of production similar to those attained by London porter brewers a century earlier.

Sarah F. McMahon

Laying Foods By

Gender, Dietary Decisions, and the Technology

of Food Preservation in New England Households,

1750–1850

Thursday 12th [March 1807]   A charming Morning. Self Seated . . . [in] a pleasant parlour with fine windows, . . . a good fire, a clean hearth   money in my pocket, the house well stored with bread, meat, pies, cake, milk, cream, tea sugar coffee, chocolate, bacon, chickens, mutton, meal, flour, corn, honey, chees, apples, po[ta]toes, wine, brandy, gin, beer, cyder, preserves of various kinds—a good cow
—Ann Bryant Smith, Portland, Maine

Between 1750 and 1850, rural New England households began to modify the pronounced seasonality of their customary diet as they secured

The main lines of this essay were first presented as a work in progress to the Transformation of Philadelphia Project of the Philadelphia Center for Early American Studies in 1988; the comments and suggestions of that research group clearly shaped the development of the project. Earlier versions of this essay were presented to the annual meetings of the Social Science History Association (1989), the Maine Women's Studies Association (1990), and the Society for the History of Technology (1990). The author is particularly indebted to Ed Hawes, Becky Koulouris, Judy McGaw, Kidder Smith, Jr., Tom Dublin, Bob Post, and Fredrika Teute for close readings and advice.

a more ample and varied food supply throughout the year. The daily fare of colonial New Englanders had alternated between fresh foods in warmer seasons and a more limited variety of stored provisions during the colder months. In combination, two efforts led to the gradual deseasonalization of the diet after the mid-eighteenth century. First, changes in agricultural decisions and practices increased the quantity and variety of staple foods that composed the yearly fare. For example, the stocks of swine and cattle on rural homesteads offered an ample source of meat, and families slaughtered and barreled quantities sufficient to last into the summer. Many households switched their fermented beverage consumption from beer to cider; doing so released the land previously sown in barley to breadgrain production and ensured a plentiful supply of grain, in spite of declining crop yields from increasingly exhausted land. Second, by experimenting with their methods of food preservation and storage, householders extended the seasonal availability of a wider variety of staple foods. After 1750, farmers filled cellar bins with their crops of potatoes, other roots, and hardy vegetables in addition to the traditional stores of peas and beans. Together, these provided a variety of vegetables—what they called "sauce"—to flavor their winter salt-meat pottages and stews. And while seasonal dairying routines continued to govern the production of butter and cheese, by the 1790s stores of dairy products were beginning to last from autumn to spring.[1]

As they endeavored to store a greater variety of foods for longer periods of time, rural New Englanders increasingly found that they were coming up against the limits of their traditional technologies of food preservation. In their attempts to overcome these limitations, some innovators sought mechanical solutions. By the mid-nineteenth century, the new technologies of commercial canning and icebox refrigeration had been developed. However, for reasons of risk, cost, and feasibility, these commercial products and methods did not meet the needs of rural families in preserving supplies of food in bulk for the winter.

For most rural households, the traditional technologies of food preservation and cold-cellar storage still offered the best methods for keeping the large quantities of foods that they produced and gathered during the

---

1. Sarah F. McMahon, "A Comfortable Subsistence: The Changing Composition of Diet in Rural New England, 1620–1840," *William and Mary Quarterly*, 3d Ser., XLII (1985), 26–65; "'All Things in Their Proper Season': Seasonal Rhythms of Diet in Nineteenth Century New England," *Agricultural History*, LXIII (1989), 130–151.

autumn harvest. Early nineteenth-century farmer's almanacs and cook-books continued to recommend the "economical" advantages of preserving and storing foods in quantity. Seeking better ways to secure their yearly subsistence, most families devoted more of their efforts to "taking care" of their various stores of food. In addition, some households experimented with their preservation methods; those alterations were primarily adaptations in the knowledge and skills that informed their traditional strategies of food preservation.[2] Yet, even as they sought to improve their methods, rural households had cultural investments in their customary practices. Architecturally, the layout of many New England houses was designed around cold cellars; perhaps this reinforced the investment in that method of food preservation. The complex structure of household work roles also supported the continuation of those practices. Food preservation is the nexus in foodways between men's work in agricultural production and women's work in meal preparation. While most tasks and responsibilities were gender-specific in rural New England, women's and men's efforts overlapped in food preservation; in that shared domain, the designation of certain tasks was ambiguous. The way in which households accomplished those tasks, defined roles, designated priorities, and allocated responsibilities and decision making affected the diet, the domestic economy, and the gendered culture of values and expectations in rural New England.

Thus, an examination of the methods, strategies, decisions, and expectations of food preservation reveals more than the role of food storage technology in the deseasonalization of the diet. Food preservation provides an optic for examining the application in daily life of the gendered ideology of "separate spheres."[3] That model of two opposing spheres, first

2. Both historians of technology and anthropologists have explored the concepts of "care" and "strategy" as means of broadening our understanding of technology and technological evolution. David Pye, *The Nature and Art of Workmanship* (New York, 1968), 4, argues that "in workmanship care counts for more than the judgement and dexterity." He also suggests an interesting gender contrast: most of male craftsmanship is workmanship of certainty, but women's craftsmanship is workmanship of risk.

Leslie White, *The Science of Culture: A Study of Man and Civilization* (New York, 1949), 364, first proposed that the technological system of any culture is composed of the material, mechanical, physical, and chemical instruments, together with the techniques of their use. Thus, "strategy" would be the effective use of the materials and instruments, which involved, at least in part, timing, scheduling, and taking care.

3. Food preservation offers evidence of ambiguity in a second ideological dualism.

proposed in the early nineteenth century and then revived as twentieth-century historians began to examine and explain women's experience, is now recognized as an ideological construction or metaphor rather than an accurate description of women's (and men's) daily lives. In rural households, the gender division of certain tasks may well have been more apparent than real. In a study of rural households in upstate New York, Nancy Grey Osterud shows that "work itself was gender marked; certain activities were regarded as the responsibility of men, and others as the responsibility of women. At the same time, women and men routinely coordinated their labor. Not only did husbands and wives synchronize their work on a daily basis, but farm processes often began in men's domain and culminated in women's." Similarly, there is compelling evidence that women had their own physical spaces that were separated from men's spaces. Yet, in rural households, while some work spaces and hence some responsibilities (such as women's kitchens and men's fields) were clearly separated, contiguous work spaces (such as gardens and barnyards) "involved a complex sharing and overlapping of activities" and tasks.[4] In

---

Claude Lévi-Strauss described a dichotomy between the raw and the cooked—a relationship that he linked with the fundamental dualism between nature and culture. He divided different techniques of cookery into "natural" techniques and "cultural" techniques. In food preservation, however, methods that he designated as natural might produce results that he characterized as cultural, and vice versa. Perhaps this ambiguous relationship between process and product occurred because food preservation was the intermediate stage between food production (the raw) and meal preparation (the cooked). See Claude Lévi-Strauss, *The Origin of Table Manners: Introduction to a Science of Mythology: 3*, trans. from the French by John Weightman and Doreen Weightman (New York, 1978), 478.

4. Nancy Grey Osterud, *Bonds of Community: The Lives of Farm Women in Nineteenth-Century New York* (Ithaca, N.Y., 1991), 147; Linda K. Kerber, "Separate Spheres, Female Worlds, Woman's Place: The Rhetoric of Women's History," *Journal of American History*, LXXV (1988–1989), 9–39, provides an excellent review of the use of the metaphor of two opposing spheres both to describe women's domain and to explain women's experience.

Thomas C. Hubka, *Big House, Little House, Back House, Barn: The Connected Farm Buildings of New England* (Hanover, N.H., 1984), 150–151, created a "gender map" of work on a connected farm. Similarly, in their introduction to women's work, Leslie Parker Hume and Karen M. Offen (introduction to Part 3, in Erna Olafson Hellerstein *et al.*, eds., *Victorian Women: A Documentary Account of Women's Lives in Nineteenth-Century England, France, and the United States* [Stanford, Calif., 1981], 274–278) describe a "geography of women's work" in a preindustrial rural economy.

many instances, the structure of gender roles and responsibilities in this shared space was easily determined. But, for some tasks, the distinctions may well have been less clear.

The ambiguous gender orientation of certain tasks and the ambiguous spaces that were occupied by both women and men raise questions about the way that women's and men's responsibilities, decisions, and domains were perceived when their work overlapped. Inherent in the structure of gender roles in rural society is the question, not just of who made decisions, but how that authority was determined. In a synthesis of the literature on rural households, Allan Kulikoff argues that in rural households, where presumably cooperation would lead to an overlap of women's and men's work that would be even more pronounced than in urban households, "the presumption of household unity precludes the possibility of conflict or tension within households, especially between husbands and wives, over authority, the sexual division of labor, and the distribution of goods produced by members for consumption, exchange, or sale." Similarly, Osterud argues that "responsibility for particular operations was divided more clearly and consistently than was the actual performance of tasks." If this division occurred in rural New York, where men and women did share responsibility for a wide range of dairying tasks, then the gender division of labor and responsibility in rural New England households might have been even more pronounced.[5]

Although historians have focused much of their attention on the ideology of gender division and cooperation, daily practices—and the decisions and choices that shaped those mundane tasks—might have been more important for rural families. The methods and strategies of food preser-

---

5. Allan Kulikoff, "The Transition to Capitalism in Rural America," *WMQ*, 3d Ser., XLVI (1989), 137; Osterud, *Bonds of Community*, 150, 284: "In New England, the gender division of labor on dairy farms was as sharp and rigid as on arable farms, while in the mid-Atlantic region women and men shared both the responsibility for and the labor of dairying."

Toby L. Ditz, "Ownership and Obligation: Inheritance and Patriarchal Households in Connecticut, 1750–1820," *WMQ*, 3d Ser., XLVII (1990), 257, argues that, in rural households, the patriarchal household organization ultimately determined the locus of authority: "The economic vitality of households . . . depended greatly on women's skilled labor, and, as many have suggested, the importance of women's work may have enhanced the respect accorded them in practice. The cultural emphasis on affection and respect was not, however, at odds with patriarchal household organization. Mutuality in marriage was almost always coupled with another theme—the wife's obligation to obey her husband."

vation may offer one of the clearest examples of the intersection between rural women's domestic domain and men's agricultural domain. It can suggest how women and men divided daily responsibilities and whether the lines between women's and men's work were always drawn in the same way.[6] The division of the tasks and responsibilities of food preservation may indicate how dietary decisions were made in rural households, and in particular the decisions that determined the quantity and variety of the food stores that created the daily fare throughout the year. The way in which families allocated the final authority in dietary decisions had important implications for the resolution of a potential conflict between women's domestic goals in meal preparation and men's economic goals in agricultural production for both household and market.

Finally, the blurring of the boundaries of the gendered division of labor that occurred in the techniques and responsibilities of food preservation suggests that some ambiguity was inherent in that gender structure. In particular, such ambiguities smoothed the transitions between gendered domains; the overlap created a realm of shared labor in an otherwise clearly demarcated system. Indeed, rather than attempt to resolve the ambiguities, families may have sought to preserve them. The concern in rural households was not whether the separation of spheres reflected reality, but rather whether the areas of ambiguity in that system allowed for the perpetuation of clearly defined ideological structures— which ordered relationships within the household—by providing some flexibility in their application.

The evidence from probated inventories and wills from Middlesex County, Massachusetts, 1653–1835, suggests improvements over time in the quantity and variety of foods produced on rural homesteads and indicates a gradual extension in the seasonal availability of those staple and supplementary foods. The changes in the patterns of food storage in New England households that occurred in the second half of the eighteenth

---

6. If some preservation tasks offered ambiguous guidelines for dividing responsibilities, those guidelines might be related to the ambiguous position of food preservation in Lévi-Strauss's nature/culture and raw/cooked dichotomies (above, n. 3). While he viewed women as symbolically closer to nature and equated culture with men, in rural New England raw foods (nature) were generally related to men's work in agricultural production, and cooked foods (culture) were women's responsibilities. See Sherry B. Ortner, "Is Female to Male as Nature Is to Culture?" in Michelle Zimbalist Rosaldo and Louise Lamphere, eds., *Woman, Culture, and Society* (Stanford, Calif., 1974), 67–87.

century were accompanied in the late eighteenth and early nineteenth centuries by a proliferation of advice manuals (farmer's almanacs and cookery books) that were part of a growing popular culture of information. That advice literature, apparently both responding to and encouraging new expectations about the composition of the diet, detailed refinements in the methods and strategies of the old technology of food preservation. Seven men's journals, or farm daybooks, and thirteen women's diaries confirm, at least in a general way, the use of those methods of food preservation, the compensating strategies to which households resorted, and the concerns and expectations that households held about their yearly food supplies.[7]

7. On probate inventories, see McMahon, "A Comfortable Subsistence," *WMQ*, 3d Ser., XLII (1985), 52–59, tables 1–7; "All Things in Their Proper Season," *Ag. Hist.*, LXIII (1989), 131, 134–137, tables 1–5. The evidence about the composition of the diet in rural New England from the late 18th to the mid-19th centuries is confirmed by the testimony in the autobiographies of more than 100 New Englanders born between 1733 and 1865.

According to Albert Lowther Demaree, *The American Agricultural Press, 1819–1860* (New York, 1941), 6, "The agricultural books available to the farmer before 1819 were few in number and difficult to obtain." Various almanacs were published in the last half of the 18th century. In 1774, *The New-England Almanack . . .* , by Edmund Freebetter [Nathan Daboll], included a best method of fattening pigs. Daboll's almanac, published in New London, Conn., continued to offer preservation recipes and agricultural and husbandry instructions. In 1786, *Thomas's Massachusetts, Connecticut, Rhode-Island, New-Hampshire, and Vermont Almanack* (published by Isaiah Thomas) offered an expanded monthly calendar as well as a "concise calendar for young farmers and gardeners." By the 1790s an increasing number of almanacs offered information and advice for farming families. In 1793, Robert B. Thomas began to publish his almanac, specifically designated as *The Farmer's Almanac . . . for the Year 1793*. The first cookbook by an American author appeared in 1796. Until 1818, a new volume of recipes appeared every few years. Between 1818 and 1824, a new volume appeared every year; and, after 1824, no fewer than two new cookbooks joined second and third editions of previous volumes on the shelves of bookstores. See Eleanor Lowenstein, *Bibliography of American Cookery Books, 1742–1860* (Worcester, Mass., 1972). In both cases, the quantitative change in the number of advice manuals can be taken as evidence of a qualitative change in the popular culture of information.

The men's journals and daybooks span the years from 1722 to 1865. The different accounts overlap between 1791 and 1865, and they offer evidence from southern, midcoast, and inland Maine. The women's journals cover the years from 1760 to 1868. Although these individual accounts were less likely to span the same number of decades as the men's accounts, still they do offer some overlap throughout the period. The earliest accounts are from Massachusetts (Salem and Hadley); the later accounts are from Vermont and coastal and inland Maine.

The various methods in the technology and strategy of food preservation in rural New England between 1750 and 1850 were limited in their effectiveness. The problems of food preservation, and the ways in which households attempted to overcome those problems, suggest that food preservation was indeed an uncertain link in the food process. That uncertainty heightened the importance of the tasks involved. As households divided and shared responsibilities in food preservation, they found that, while some tasks were clearly gendered, the gender orientation of other tasks was more ambiguous. The ways in which families allocated or shared responsibilities were determined in part by the way that women and men viewed and understood male and female spheres. The uncertainties and ambiguities of the process and methods of food preservation led to particular strategies for dealing with these decisions. In turn, those solutions reinforced certain priorities that determined the way that decisions about the diet were made.

The Technology and Strategy of Food Preservation

Sabbathday 15th [February 1807]   The rain pours down in torrents   our
   Cellar is covered with water, one can not get anything out unless they wade
   half leg—potatoes, onions, bread, all all Spoiled. . . . Tuesday 17th   The
   water has left our Cellar—I, our boy, & Eliza have been all the fore part of the
   day to work on apples, Eggs, onions, & potatoes, trying to [save] them
   they have all been drenched in water, but I hope to save some of them
   —Ann Bryant Smith, Portland, Maine

December 16, 1845   cloudy wind NE and Blows a gale & very slippery
   nothing to be done   the Cellar full of water
   —Joseph Weare, York, Maine

Each autumn, rural New England households laid up their winter supplies of food in cellars and garrets. Root cellars offered particular advantages in food storage. Situated below the frost line, cellars were less affected by the variable temperatures that made most storerooms above ground unsuitable for long-term storage. The relatively stable, cool temperature usually prevented vegetables and fruits from freezing during the winter and kept food stores cool in the heat of the summer. While cellars ensured cold storage, they offered no guarantee against extreme cold, dampness, or vermin. Year after year, families coped with the vagaries of cellar and garret storage. Frost "crept into the cellar" and threatened or destroyed vegetables. "Torrential" rainstorms flooded cellars and drenched the stores of food. And rats made "sad havoc" and "great waste" on vegetables in the cellar and "great Devastation" in the corn in garrets.[8]

Working within the traditional technology of cold-cellar storage, early nineteenth-century advisers focused much of their efforts on improving knowledge about food spoilage. They cautioned readers against the problems that resulted from improperly aired or damp cellars and recommended various methods of coping with these problems. In 1796, *Thomas's Almanack* recommended, "When people take in vegetables, and close their cellars for the winter, it may not be untimely to recommend, that before cellars are closed, they be cleaned and aired—That vegetables (particularly cabbages) be put in such places as to prevent putrefaction; that they be carefully attended through the winter, and such as are defective removed." The advice for storing fruit resembled that for vegetables: initial care, and then a continued, careful watch through the winter. In 1850, the *New England Farmer* reprinted a comment from the *Albany Cultivator* on "Fruit in Cellars," which cautioned, "A great deal of win-

8. On frost: William Chamberlain, Diary, Mar. 5, 1833, Chamberlain Family Papers, 1830–1929, Maine Historical Society, Portland. On Apr. 11, 1840, William Chamberlain of Bristol, Maine, noted, "Opened the Cellar door and window—there has been no frost in the Cellar this winter"; on Dec. 7 of that year, he remarked, "Frost has made its appearance in the Cellar in several places." Thomas Smith of Portland, Maine, noted in April 1774, "It has not frozen in the house since the beginning of February," and on Dec. 8, 1784, "It has not frozen in the house yet" (Wm. Willis, ed., *Journals of the Rev. Thomas Smith, and the Rev. Samuel Deane* . . . [Portland, Maine, 1849], 278, 283). Similarly, Joseph Weare of York, Maine, noted for Jan. 31, 1838, "The frost has not yet got into our cellar," suggesting that he was awaiting the possibility of that prospect (Joseph Weare, Jr., Diary, 1803–1856, Maine Hist. Soc.).

On rain: Ann Bryant Smith (Portland, Maine), Diary, 1806–1807, Feb. 15, 1807, Maine Hist. Soc.; see also Chamberlain, Diary, May 21, 1832; Weare, Diary, Dec. 16, 1845.

On rats: Louisa Talbot (South Freeport, Maine), Dec. 11, 1868, Louisa Talbot and Susan Talbot Diaries, typescript, 1867–1868, B. H. Bartol Library, Freeport, Maine. William Chamberlain noted on May 25, 1836, "picked & sorted all the potatoes we have   there is abt 50 bushels of all sorts   the rats have the winter past made great waste of them" (Diary). On Jan. 18, 1791, Benjamin Gerrish of South Berwick, Maine, noted that "the Rats made Great Devastation among the corn" which was stored in the garret (Capt. Benjamin Gerrish, Diary, 1791, Maine Hist. Soc.).

ter fruit suffers early decay, in consequence of a deficiency of ventilation, especially during autumn, and after the fruit is deposited." Without proper care and tending, grains stored in garrets were also susceptible to vermin, fermentation, and mustiness. Advisers reminded their readers that regular care would ensure that their stores would remain good as long as they lasted.[9]

By the late 1820s, advisers were addressing the health consequences of poorly attended cellars. In the notes for the month of May, Robert Thomas's *Farmer's Almanac* for 1828 advised: "Attend to your cellars and be careful to have them cleansed, if you wish to preserve your health. There is too much neglect about this business." In 1846, Catharine Beecher explained: "A cellar should often be whitewashed, to keep it sweet. It should have a drain, to keep it perfectly dry, as standing water, in a cellar, is a sure cause of disease in a family. It is very dangerous to leave decayed vegetables in a cellar. Many a fever has been caused, by the poisonous miasm thus generated." At the same time, advisers continued to refine the technology of cold-cellar storage. In 1850, the *New England Farmer* offered a design for cellar bins: "Another cause of decay is the improper location of the shelves or bins, which are placed against or around the walls. By this inconvenient arrangement, the assorting of decayed specimens must be done all from one side, and the shelves must hence be very narrow, or the operator must stretch himself in a most irksome horizontal position."[10]

Many households followed these practices, at least in a general way. In the early summer, families routinely cleaned their cellars. In the late summer and early fall, Maine farmers made the necessary repairs to the cellar

9. *Thomas's Almanack*, 1796; S. W. Cole, ed., *New England Farmer* (Boston), II (Dec. 7, 1850), 395. In *The Maine Farmers' Almanac for the Year 1854*, Daniel Robinson recommended, "Gather fruit intended to be kept through the winter carefully, by hand, in the middle of a dry day; you may then put it down in dry sand as soon as picked; it should not lie in heaps to sweat." See also *Farmer's Almanac*, 1796; *Thomas's Almanack*, 1800.

On regular care, see *Thomas's Almanack*, 1800; Lydia Maria Child, *The American Frugal Housewife: Dedicated to Those Who Are Not Ashamed of Economy*, 22d ed., rev. (Boston, 1838), 9; *Maine Farmers' Almanac*, 1857, "Ventilating Grains." Among the recipes that William Chamberlain recorded in his daybook were "Recipe To Clean Musty Corn or Grain" and "Recipe to prevent musty wheat—from the Lion's Advocate" (Diary, Nov. 8, 1832, Apr. 20, 1835).

10. *Farmer's Almanac*, 1828; Catharine Beecher, *Miss Beecher's Domestic Receipt-Book: Designed as a Supplement to Her Treatise on Domestic Economy* (New York, 1846), 322; *New England Farmer*, II (Dec. 7, 1850), 394.

floor and prepared new bins for the root and vegetable crops that were carried into the cellar after the harvest. Although the bulk crops of vegetables varied, most farmers produced and stored a considerable supply of winter vegetables. In November 1836, William Chamberlain of Bristol, Maine, enumerated his vegetable harvest for the year: "potatoes . . . peas, beans beets F[rench] and E[nglish] turnips, cabbage, carrots, & parsnips Onions & good squashes." [11]

In October or November, Maine farmers routinely "banked up the house to keep the Cellar from freezing." They discovered that, when they did not bank their houses, they were more likely to lose their food stores. On December 24, 1811, William Wentworth of Bridgton, Maine, noted: "SNOWSTORM. My cellar not being sufficiently secured my potatoes are some frozen." Yet, after they had properly cared for their cellars, the only real safeguard against spoilage was constant care and a continued watch over their stores of food. In 1796, *Thomas's Almanack* concluded its recommendations for storing vegetables in root cellars with a qualified appraisal of the methods: "An individual cannot do much, but if what is above suggested, is attended to, benefits may follow." Households regularly experienced the limits of the old technologies of food preservation. Throughout the winter, men and women divided the tasks of "picking

11. Cellar cleaning was not a gender-specific task; however, within households, it may have fallen routinely to one member of the family. Elizabeth Wildes of Arundel, Maine, noted on May 22 and July 17, 1790, that her sister Abiel cleaned the "sellar" (Elizabeth Perkins Wildes, Diary, 1789–1793, Maine Hist. Soc.). In 1834, William Chamberlain "cleaned out the cellar" on June 2, and on August 14 he noted, "cleared out the cellar" (Diary).

On fixing cellars: On Sept. 14, 1842, William Chamberlain noted, "Employ'd Flooring over the cellar—hauled a load of sand from J Fountain's" (Diary). On June 5, 1850, Joseph Weare cleaned out his cellar; on Sept. 7, 1829, he noted, "cleaning out the Cellar and making New Potatoe pens" (Diary).

On Chamberlain's harvest: Chamberlain Diary, Nov. 25, 1836. On Oct. 31, 1833, Chamberlain "dug beets, parsnips french turnips carrots &c—& put into the cellar," adding these to the crop of potatoes already secured there. In other years, he made a list of his produce for the season. On Nov. 11, 1835, he noted, "Our crop this season consists of: [wheat, barley, oats, beans, peas, potatoes, corn] cabbages & turnips beets &c sufficient for the family nearly" (Diary). Joseph Weare usually recorded the storing of potatoes, beans, cabbages, turnips, apples, and pumpkins; occasionally he also listed other vegetables such as onions (Diary).

through" and "sorting" potatoes, apples, pumpkins, and beans to remove those that were rotten or going bad.[12]

Households had limited control over their food storage areas, and they could not prevent the inevitable waste of a portion of their food supply so long as they relied on the old technologies of food preservation. They sought improvements in those methods that would offer them more control and better safeguards. Advisers and householders alike realized that the best conditions for housing the different vegetables and fruits varied. Responding to—and encouraging—efforts to increase the longevity of food stores, the advice literature offered various and often contradictory "best" methods of preserving different vegetables and fruits. For example, *Thomas's Almanack* recommended, "Roots of beets, parsley, carrots, potatoes, parsnips, turnips, etc. take up in a dry day early [in November] to preserve them for the use of winter, clean them from dirt, and deposit them in dry sand" in the cellar. By contrast, Mary Eliza Rundell's cookbook advised, "The earth should not be cleaned from carrots, beet roots, parsnips and potatoes; these should be stored on a stone floor." Lydia Maria Child continued to advocate the old wisdom about parsnips in her cookbook: "Parsnips are good only in the spring." In 1825, Robert

12. On banking: Tobias Walker (Kennebunk, Maine), Diary, Dec. 9, 1844, Walker Family Papers, 1828–1893, Maine Hist. Soc. On Nov. 7, 1805, Joseph Weare noted, "banking up the house   Shutting up the cellar doors" (Diary). On Nov. 8, 1830, William Chamberlain recorded, "employed banking cellar" (Diary). In addition to adding protection against frost, this practice of banking the house also served to keep the first floor warm.

On Wentworth: William Wentworth, Diary, Dec. 24, 1811, Wentworth-Merrill Papers, Maine Hist. Soc. Wentworth's trials with potato storage continued. His potatoes also froze in 1818. In 1819, he finally built a proper cellar with rocks and stones. But not until 1833 (and again in 1838) did he note in his daybook, "November 1   banked the house." In spite of those efforts, in 1840, 1843, and 1844 frost still got into his cellar.

On root cellars: *Thomas's Almanack*, 1796.

On picking through: On May 25, 1836, William Chamberlain noted, "picked and sorted all the potatoes we have   there is abt 50 bushels of all sorts   the rats have made great waste of them" (Diary). On Apr. 29, 1856, Joseph Weare noted, "halling rotten potatoes out of Cellar" (Diary). On Nov. 7, 1806, Ann Bryant Smith noted, "Afternoon at home Separating the Sound apples from them that were not so—paring and filling some for pies" (Diary). Both Susan and Louisa Talbot noted that the "folks" were "picking over beans" on Apr. 5, 1867. Apparently this was a yearly procedure. On Feb. 12, 1868, Louisa noted "Mother picking over beans" (Susan Talbot, Diary, Louisa Talbot, Diary).

Thomas challenged that wisdom. He staged a scene in his February notes: One farmer, surprised to see his neighbor eating parsnips in February, asked where he got them. "Get them? Why I brought them up from my cellar, where I put them in the sand last fall. Now, . . . you will insist . . . that parsnips should not be dug until spring; and you can give no other reason for it than that your father and grandfather and great grandfather used to practice it. . . . Act and think for yourself. . . . By digging them in the fall, you see *we* have the use of them all winter, and certainly in the greatest perfection, possessing all their richness and nutritious qualities." [13]

The contradictions in the advice suggest that sure methods remained elusive. While some advisers and individuals continued to experiment with different methods of food storage, other individuals were more cautious in their embracing of that culture of innovation. For example, William Chamberlain was a risk-taker. In November 1830, he noted in his daybook, "Pulled part of the cabbages & sett them out for winter to be covered with sea weed." An asterisk directed attention to the bottom of the page, where he later noted, "This mode did not answer, lost them nearly all." Chamberlain may have been adapting a method of storage that had been recommended for potatoes as early as 1797: "Preservation of Potatoes against Frost. . . . This valuable root is very apt to be destroyed by Frost. . . . In a very dry spot, [dig] trenches . . . pile the potatoes into the shape of the roof of a house. . . . The drier they are when thus picked up the safer they will be." In 1806 and each autumn thereafter, William Wentworth of Bethel, Maine, dug holes sufficient to hold ten to fifteen bushels of potatoes. When he opened the holes in the late winter and early spring,

13. *Thomas's Almanack*, 1798; Mary Eliza Rundell, *A New System of Domestic Cookery, Formed upon Principles of Economy, and Adapted to the Use of Private Families* (Boston, 1807), vi. Thus, Lydia Maria Child, *American Frugal Housewife*, 35, recommended that celery roots should be "covered with tar to keep them moist," but Daniel Robinson, *Maine Farmers' Almanac* (1854), advised his readers to "[leave] the tops and leaves [of celery] open to the air."

On parsnips, see Child, *American Frugal Housewife*, 33; *Farmer's Almanac*, 1825. While Robert Thomas suggested that his advice was new, in fact, fall gathering of parsnips had already been advocated by *Thomas's Almanack*, 1798, and in 1807 by Rundell, *A New System*, vi. Some farmers continued to harvest parsnips by the traditional method. On Mar. 30, 1803, Joseph Weare "dug some parsnips" and again on Mar. 23, 1811, he noted, "dug the parsnips in the afternoon 2 bushels" (Diary). By the 1830s, William Chamberlain was harvesting his parsnips in October (Diary).

he often "[found] them considerably rotted" or "found some frozen. The weather has been so cold & the ground so bare that potatoes are considerably frozen." Often the potatoes that froze in the winter were rotten by April or May. In March 1824, Wentworth noted in his daybook, "Potatoes are rotting caused by frostbitten ones being mixed with them." [14]

While William Wentworth understood both the risks and their causes, he continued to practice this problematic method of outdoor storage until he was able to secure his cellar against frost. By contrast, in spite of the failure of some of his experiments, Chamberlain continued to seek new ways of securing his food stores. In January 1835, Chamberlain noted, "This day eat the last Connecticut sweet apples   we have never considered them as a winter apple & have wasted them in the fall   they may probably be kept sound until the first of March if gathered before they are fully ripe." [15]

The search for better methods may well have followed the growing impulse, in rural New England households, to adopt new methods and technologies that were being developed in agriculture and cookery in the first half of the nineteenth century. During the 1830s, most of the diarists noted either the borrowing or purchase of a threshing machine. Technological changes altered cooking as well; the most notable invention during that era was the cooking stove, which gradually replaced open-hearth cooking.[16] Even in commercial food preservation, mechanical innovations

14. Chamberlain, Diary, Nov. 1, 1830; Isaac Bickerstaff [Benjamin West], *The Town and Country Almanack, for . . . 1797* (Norwich, Conn.); Wentworth, Diary, Apr. 25, 1808, Feb. 21, 1810, Mar. 12, 1824.

15. On Dec. 24, 1811, Wentworth noted, "My cellar not being sufficiently secured my potatoes are some frozen" (Diary). Chamberlain, Diary, Jan. 30, 1835. Chamberlain continued to experiment. In 1832, he planted corn seed that had been soaked in copperas water for 10 days; in 1835, he noted, "I am led to think that in seting cabbages the plant ought not to be set in very wet earth" (May 30, 1832, July 16, 1835).

16. On threshing machines: In September 1835, William Chamberlain rented a threshing machine for his wheat and barley; because of the high cost of renting the machine, in 1839 he "Effected the purchase of a threshing machine" (Diary, Sept. 22–26, 1835, Sept. 21, 1839). In 1836, Joseph Weare "thrashed" out his wheat in Samuel Stone's threshing machine, and in 1838 he was "thrashing with the hand machine." In 1846, his son made a cultivator that they continued to use on the farm, and in 1855 his sons were winnowing out the barley in a winnowing machine (Diary, Oct. 17, 1836, Aug. 30, 1838, May 29, 1846, Nov. 15, 1855). On Jan. 11, 1841, Persis Sibley of Freedom, Maine, noted: "Nancy and I (for she hates tailoring as bad as I) have been out helping the threshers. . . . For the last three years all grain has been

had improved the technologies of canning and refrigeration by the 1850s, although the advantages of these methods for home preservation of bulk stores of food were still limited. However, Chamberlain's willingness in the 1830s to take risks as he sought new methods of securing his food stores may well have been the exception; Wentworth's caution was more likely the rule among New England farmers. Alan Taylor has argued that settlers in the Maine backcountry preferred "a mixed agriculture of summer wheat, winter rye, Indian corn, and potatoes [which] helped ensure family subsistence. They were wary of gambling their daily bread on the market and the vagaries of Maine's climate."[17] Similarly, most farmers

---

threshed by machines worked by horses" (Persis Sibley Andrews [Black], Diary, 1841–1853, Maine Hist. Soc.).

On cooking stoves: Mary Palmer Tyler of Brattleboro, Vt., remarked on Nov. 5, 1824, "This day by the advice of my husband I went to Mr. Hall and traded away our chaise for a cooking stove," in George Floyd Newbrough, "Mary Tyler's Journal," *Vermont Quarterly*, XX (1952), 24. Between 1837 and 1851, Maine diarists recorded the purchase of cookstoves. In 1837, William Chamberlain "put up a cooking stove   cost $40" (Diary, Dec. 7, 1837), and William Wentworth noted on Dec. 31, 1851, "We have sold 20 bushels of corn at 5/ pr. b. toward paying for a cooking stove" (Diary). In addition, a new variety of implements was available for cooking. In 1867, Louisa Talbot noted, "Mother and Marcia went to Yarmouth and bought a Doughnut Kettle" (Diary, Jan. 4, 1867).

17. On canning and refrigeration: According to Susan Strasser, *Never Done: A History of American Housework* (New York, 1982), 20, 22, commercial efforts to can vegetables began in the 1840s; innovations in commercial canning continued in the decades after the Civil War. By the 1850s, "self-sealing jars were sold in American country stores, but acid fruits and brined vegetables stored in them often spoiled because of poor seals." In addition, "domestic refrigerators were widely advertised and ice deliveries in urban areas were regular and reliable by mid-century, but per capita ice consumption remained low. Home iceboxes were rare; cookbooks and domestic manuals, which sometimes recommended use of refrigerators, suggested cold-storage methods for families who had none."

Those families who had refrigerators tended to use them as a short-term method of keeping fresh foods from going bad rather than for long-term storage of foods. An 1852 ad for Perkins's Upright Refrigerator recommended the appliance as a "handsome piece of furniture for the dining room," explaining that "you can have access to your ice at all times." In addition, "it will pay for itself in a short time by preserving your butter, milk, eggs, &c., &c." (*The Portland [Maine] Directory and Reference Book for 1852–53*, 298).

On mixed agriculture: Alan Taylor, *Liberty Men and Great Proprietors: The Revolutionary Settlement on the Maine Frontier, 1760–1820* (Chapel Hill, N.C., 1990), 75–76. Taylor focused on the ways that cold versus hot summers adversely affected the fruition and

were unwilling to risk their families' subsistence with storage methods that diverged too far from their traditional practices. Thus, improvements in the preservation of bulk stores of food remained largely within the old technology of cold-cellar storage.

Various foods required some amount of preparation or conversion before they could be stored. While these preservation methods called for greater initial efforts, they may have provided more control and perhaps greater safeguards. When farmer's almanacs and cookbooks offered "receipts" for salting, smoking or curing, and pickling meats, for preserving and keeping butter, for pickling cucumbers, and for making fruit preserves, they both promised a certain quality of product and claimed that the stores would remain good for months—or years.[18] Advice manuals also offered corrective recipes for restoring foods that had gone bad and for further extending the use of previously preserved foods; in both cases, the recipes promised to increase the longevity of the stores, and occasionally they promised better flavor. "Best" methods were offered as early as 1774 and continued to be offered in the nineteenth century. "Approved" and "new" methods were offered in 1793 and thereafter, and "sure" and "effectual" methods were offered in the early nineteenth century. Although authors made these claims about the information they

---

harvest of various crops. Yet farmers' wariness about the vagaries of the climate continued even after the crops had been stored.

Daniel Vickers suggests that New England farmers in general were cautious about their agriculture. In "Working the Fields in a Developing Economy: Essex County, Massachusetts, 1630–1675," in Stephen Innes, ed., *Work and Labor in Early America* (Chapel Hill, N.C., 1988), 69, Vickers argues that the marginal productivity of New England clearly affected the labor decisions that farmers made. In "Competency and Competition: Economic Culture in Early America," *WMQ*, 3d Ser., XLVII (1990), 4, Vickers argues that "the obsession with competency troubled early Americans far more than worries about the legitimacy of commerce."

18. In 1778, Nathan Daboll offered a recipe for pickling beef, promising, "Thus beef may be kept from fall until spring without being too salt" (*New-England Almanack*, 1778). Daniel Robinson included a recipe for preserving butter, claiming that the butter "will remain good a long time" (*Maine Farmers' Almanac*, 1841). *Thomas's Almanack* in 1789 printed a recipe for a "family cider" that would "keep sound for a twelve month." In 1808, he included a "Receipt for Pickling . . . cucumbers, or whatever your pickles consist of. . . . If rightly performed, this method will preserve pickles the year round, and forms a very agreeable sauce."

offered, significant differences in the various methods were rare. Often, the recipes simply recommended that the preservation process be repeated or that more of the same ingredients be added. In 1778, Nathan Daboll advised, "When the weather grows warm, the [meat] pickle may be boiled over again and scum'd; and by adding a little more salt, beef may be kept through the summer." In 1838, Lydia Maria Child recommended: "If you find your pickles soft and insipid, it is owing to the weakness of the vinegar. Throw away the vinegar, . . . then cover your pickles with a strong scalding vinegar, into which a little allspice, ginger, horseradish and alum have been thrown. . . . Pickles attended to in this way, will keep for years, and be better and better every year." [19] In most cases, the methods only extended the "taking care" strategy to converted foods in order to ensure their longevity.

When viewed as a whole, the changes in the methods and technology of food preservation between 1750 and 1850 primarily were adjustments or refinements in the old technologies, accompanied by an increased emphasis on taking care and taking control where it was possible.[20] In many ways, care was a first and last resort, as rural households came up against the limits of the old technologies of food preservation. The continual recourse, in the advice literature, to taking care suggests that writers were responding to new or increased expectations about the quality and longevity of foods that were preserved and stored, rather than trying to create that interest. Yet, they were unable to offer anything more than incremental elaborations of the old technologies of food preservation.

Those limits also suggest that, as long as families relied either on cellars for cold storage or on pickling methods of preserving for their bulk stores of food, food preservation would remain an uncertain link in the food process. Rural New England families had come to expect a winter fare that was varied by the addition of an array of roots, vegetables, fruits, and

19. *New-England Almanack*, 1778; Child, *American Frugal Housewife*, 85.

20. For example, Joseph Weare noted with increasing detail the process of food storage on his farm. Between 1803 and 1815, he would note that he killed a hog (or cow) and then cut up the hog (or beef). By 1816, his record changed; the second entry always read "cut up the pork [or beef] and salted it" (Diary, 1803–1815 passim). Similarly, his daybook notation for storing winter vegetables became increasingly specific. On Nov. 2, 1811, he noted, "Halled in the cabbage," on Nov. 15, 1825, he wrote, "put the cabbage in the cellar," and on Nov. 6, 1826, he recorded, "Pulled the cabbage and put them in the cellar" (Diary). While his notation changed, there is no indication that the methods that he employed had changed.

dairy products. Yet, while they continued to search for new techniques (taking better care or finding new recipes), they had not found household preservation methods that would guarantee that they could achieve those standards of variety.

The Division and Sharing of Responsibilities

October 18, 1850   finished diging potatoes for this year, raised about 100
    bushels if they do not rot in the cellar
April 29, 1856   halling rotten potatoes out of Cellar   red ones
    —Joseph Weare, York, Maine

January 10, 1867   Mother cutting and peeling rotten pumpkins. . . . January 12,
    1867   Mother stringing pumpkin.
December 11, 1868   Marcia not so well. . . . She is at work on the squashes, the
    rats are making sad havoc on the vegetables.
    —Louisa and Susan Talbot, Freeport, Maine

The methods in the old technology of food preservation ranged from bulk storage of crops to conversion of generally smaller quantities of food. For the most part, those general techniques followed the division of labor and responsibility between men (agriculture) and women (cookery) in rural New England. Yet, while some tasks and methods were clearly gender-specific, others were less easily designated as masculine or feminine.

In their farm daybooks, men detailed their responsibilities in the yearly cycle of food production and preservation. In turn, they recorded plowing, dunging, planting and sowing, hoeing and weeding, reaping and harvesting, threshing, husking, and storing their crops in the cellar and garret. Since many of their food-preservation responsibilities dealt primarily with storing their grains and vegetable crops as these were harvested, rather than with processing the foodstuffs, food preservation appears in their accounts as the final stage in their agricultural work.[21] In the same way, the

---

21. Every November, Joseph Weare noted, as the final chore of his agricultural activities for the year, "Shut up the Cellar Doors" (Diary). In the same way, Chamberlain noted, "Closed and banked the cellar" (Diary, Nov. 8, 1833). In both cases, this final task marked the completion of their agricultural routines for the year.

careful watch that they kept over their food stores during the winter was an extension of the routine of care and concern they had followed while their crops were growing. From start to finish, their task was to secure sufficient supplies of the staple foods that their families would need for the year.

Although men detailed their own work, they rarely, if ever, noted the work that women were doing, even when that work was connected to their own efforts in food production and preservation.[22] Indeed, men's accounts conveyed more of a sense of separate domains than women's accounts. For some men, the omission of women's work in their accounts may have been a consequence of men's perception of the purpose of their daybooks. They recorded the work they did, the amount of time that it took them, the use of hired help, and the results of those efforts. Such a record would help them determine in future years how to allocate their time. Given those concerns, there would be little need to detail the work done by women. In other cases, the gender-specific focus was a result of their preoccupation with the obstacles that they faced in carrying out their responsibilities. For farmers in northern New England, the shortened growing season gave little leeway if their first planting failed because of a late frost or extensive spring rains. Year after year, men in northern New England watched the vagaries of the weather, awaiting potential disasters. Their accounts became a record of their anxiety about the weather and its uncertain effect on the fruition of their crops and the security of their food stores. Throughout the growing season, William Chamberlain and his sons assessed the current state of the crops. Although some of the reports were optimistic, as crops appeared "forward," during "backward" times the prospect was quite gloomy. In June 1832, Chamberlain noted: "The season [is] such as to baffle the most experienced farmers— some are planting corn a 2d time—some the third, others are plowing the ground, planted, & sowing barley—others are planting potatoes beans &c in the corn hills & some determined to trust the corn first planted—it

22. Osterud, *Bonds of Community*, 160, found in upstate New York: "Women's diaries show a somewhat greater awareness of men's daily work than most men's diaries show of women's, but some men had sufficient contact with women's work, and regarded it as sufficiently important, that they recorded the labor of all the members of their household." The implication, of course, is that not all men highly valued women's labor: "Work might be devalued simply because women customarily performed it" (142).

is perfectly useless to ask advice or insult each other."[23] Since their crops, once harvested, were vulnerable to the limitations of cold-cellar storage, uncertainty continued through the winter and early spring.

Thus, farmers' daybooks focused on the long-range prospects of their responsibilities rather than the day-to-day food processing and domestic work that women did. By contrast, women often noted the seasonal progress of men's agricultural work on the farm as well as their own efforts in food preservation and meal preparation. In part, women's more detailed records grew out of their concerns and responsibilities in the final stages of the food cycle.[24] Women processed and converted portions of the crops that men produced, such as apples, pumpkins, and cucumbers. Since their work followed (and was therefore dependent upon) men's efforts, women had a clear stake in the successful completion of men's work. Women also prepared three meals a day, every day, for their families. The food stores that were secured by the combined efforts of men and women in rural households directly determined either the variety or the monotony of women's cookery and meal preparations. When Ann Bryant Smith of Portland, Maine, listed the array of foods stored in her home in March 1807—an assortment of foods that symbolized to her the well-being of her household—she included foods that had been harvested and

23. Chamberlain, Diary, June 14, 1832. All of the men's daybooks indicate uncertainty and anxiety in varying degrees about the risks that threatened the success of their crops each year. Each year, Thomas Smith categorized the present state of the crops: he might note "a growing season," often between July and September, or an "exceeding" dry or "melancholy" dry time, between May and mid-July. From this he attempted to predict the rest of the season: "never a greater prospect as to all the fruits of the earth" (1786); "melancholy dry time. All are now looking for an absolute famine" (1762); "drought . . . there is no prospect of any potatoes nor turnips nor any sauce at all" (July 1778). As was often the case, his predictions were off. Rains came in August 1778, and in September he noted, "Potatoes have grown to the wonder of all" (Willis, ed., *Journals of Rev. Thomas Smith*, 272–284; see also Samuel G. Webber, ed., "Diary of Jeremiah Weare, Jr., of York, Me.," *New England Historical and Genealogical Register*, LV, LXIII, LXIV, LXVI [1901, 1909, 1910, 1912]; Benjamin Gerrish, Diary; Joseph Weare, Diary).

24. Osterud, *Bonds of Community*, 214, argues, "In the realm of subsistence production the produce of men's labor passed through women's hands more often than the produce of women's labor passed through men's hands, for women did the final preparation of most farm produce for family consumption."

stored in bulk, foods that had been preserved, and groceries that had been purchased.[25]

On many farms women undoubtedly helped with the bulk storage of vegetables and fruits. But much of their labor and responsibility focused on converting and preserving foods that were processed before they were stored. Although cellar storage was sufficient for roots, tubers, and certain hardy vegetables, other vegetables had to be pickled before they could be stored. According to Harold McGee: "Pickling is the preservation of foods by impregnating them with acid, which discourages the growth of most microbes. . . . The vegetable is cooked to a soft consistency or put in a brine to draw out moisture that would dilute the vinegar. Then it is immersed in the vinegar, often with several spices." Yet, even after they had been processed, pickled foods still required tending and care through the year. In 1844, *The Improved Housewife* advised: "It is essential to the beauty and excellence of the pickles that they be always completely covered with vinegar. All kinds of pickles should be stirred up occasionally; the soft ones, if any, should be taken out, the vinegar scalded, and turned back scalding hot. If very weak, throw it away, and take new vinegar." Other conversion methods, such as drying, or making preserves and sauces, extended the usefulness of foods that had been stored in the cellar. As apples and pumpkins began to rot, the bad parts were cut out, and the rest was used in a variety of ways. In November 1806, Ann Bryant Smith noted, "Separating the Sound apples from them that were not so—paring and filling some for pies." In January 1867, Susan Talbot noted, "Mother cutting and peeling rotten pumpkins." Two days later, her sister Louisa recorded, "Mother stringing pumpkin."[26]

25. According to Smith, the wife of a Portland shopkeeper, their house was "well stored with bread, meat, pies, cake, milk, cream, tea sugar coffee, chocolate, bacon, chickens, mutton, meal, flour, corn, honey, chees, apples, potoes wine, brandy, gin, beer, cyder, preserves of various kinds—a good cow" (Diary, Mar. 12, 1807).

26. In their cookbooks, both Rundell, *A New System*, vi, and Child, *American Frugal Housewife*, 8, 33–35, included a range of detailed instructions for storing fruits and vegetables.

On pickling: Harold McGee, *On Food and Cooking: The Science and Lore of the Kitchen* (New York, 1984), 172–173; Mrs. A. L. Webster, *The Improved Housewife* . . . (Hartford, Conn., 1844), 153; see also Child, *American Frugal Housewife*, 85. In August 1795, Martha Ballard of Hallowell, Maine, noted, "Took care of my pickles; put them into vinegar" ("The Diary of Mrs. Martha Moore Ballard [1785–1812]," in Charles Elventon Nash, *The History of Augusta: First Settlements and Early Days as a Town* . . . [Augusta,

Thus women's food-preservation responsibilities ranged from shorter-term preparations to long-term methods of storage. Pie making was both a method of using fruit that was going bad and a means of short-term preservation. When women made pies, especially in the winter, they often made them in quantity. In January 1822, Mary Tyler noted, "Baking my mince pies all day—made 26—and six apple pies." In December 1844, Persis Sibley Andrews noted: "This is Thanksgiving-week. . . . I made 25 Thanksgiving pies, yesterday." In January 1867, Susan Talbot noted, "Mother baked 13 pies which I filled and sweetened and cut the apples for." Women also dried various fruits and vegetables, made preserves, jellies, and sauces from fruit and berries that they gathered, and stored the eggs that they gathered and the butter and cheese that they made. After mid-century, some women used home-canning methods for preserving small quantities of foods. In her diary, Louisa Talbot of Freeport, Maine, noted in July 1867, "Baked blueberry pies & putting some in cans and bottles."[27]

Maine, 1904], 346). While other accounts did not specify pickle making, they did note the planting and gathering of cucumbers. Most cookbooks offered recipes for pickling cucumbers and a variety of other vegetables.

On separating good fruit from bad: Ann Bryant Smith, Diary, Nov. 7, 1806; Susan Talbot, Diary, Jan. 10, 1867; Louisa Talbot, Diary, Jan. 12, 1867.

27. Newbrough, "Mary Tyler's Journal," *Vt. Qtly.*, XX (1952), 21; Persis Sibley Andrews (Black), Diary, Dec. 1, 1844; Susan Talbot, Diary, Jan. 8, 1867; Louisa Talbot, Diary, July 27, 1867.

Ann Bryant Smith's list of foods stored in the house included pies, suggesting that she used the method for short-term storage (Diary, Mar. 12, 1807). Even when women's pie-making efforts did not match the tremendous quantities listed here, their records suggest that they usually made more than one. In November of both 1790 and 1792, Elizabeth Wildes noted, "I baked some [apple] pyes" (Diary, Nov. 5, 1790, Nov. 3, 20, 28, 1792). See also "Selections from the Plymouth Diary of Abigail Baldwin, 1853–4," *Vermont History*, XL (1972), 218–223.

In October 1805, Martha Ballard cut both apples and pumpkins to dry ("The Diary of Martha Ballard," in Nash, *History of Augusta*, 426, 427). On July 20, 1839, Hannah Buxton's family finished the last of their dried pumpkin (Diary).

In the 1760s, Mary Holyoke of Salem, Mass., preserved damsons and quinces ("The Diary of Mrs. Mary [Vial] Holyoke, 1760–1800," in George Francis Dow, ed., *The Holyoke Diaries, 1709–1856* [Salem, Mass., 1911], 56, 59). Ann Bryant Smith preserved cranberries and made raspberry jam (Diary, July 29, 1806, Nov. 29, 1806). On July 8, 1843, Persis Andrews noted: "Strawberries are very abundant. . . . I have preserved as many as I can afford to" (Andrews [Black], Diary).

Mary Holyoke, Persis Sibley Andrews, Hannah Buxton, and Louisa and Susan Talbot

Within the household routine of food preservation, women's time-consuming efforts generally involved relatively small quantities of foods. Cutting, peeling, and paring each fruit or vegetable would be impractical for the crops that men stored in bulk. Women's work in food preservation was in many ways an extension of the time-consuming work that they did in meal preparation; in other words, women were always converting foods, whether for storage or for meals. However, given their cooking responsibilities, women benefited most from their extra efforts with pickles, sauces, and preserves, which offered variety for the daily fare that they prepared. Since women generally were converting smaller quantities of foodstuffs, they may have taken risks with their food-conversion techniques more readily than men took with their methods of bulk storage. There was room for experimenting with the combinations and quantities of ingredients and with the methods of processing foods, and the variety of recipes for pickles and preserves that cookbooks and almanacs offered encouraged innovation and experimentation.

Much of the work of food preservation and storage could be clearly assigned by a gender division of labor. In meat preservation, however, various tasks did not fit the conventional standards for allocating responsibilities. In their daybooks, men routinely noted a two-day procedure for processing meat. On the first day, they killed or slaughtered the hog or cow. Then, a day or two later, they "cut up the [meat] and salted it."[28] The actual range of tasks involved was more complex than that, and the usual gender orientation of responsibilities—bulk storage (men) versus conversion of foods (women)—offered ambiguous guidelines for dividing the work. The process of salting and pickling pork and beef most closely resembled women's conversion work in food preservation. Yet, perhaps since men had tended the livestock, they chose to continue their respon-

---

all made and stored butter or cheese or both; Ann Bryant Smith included eggs in her list of food stores (Smith, Diary, Mar. 12, 1807). On Aug. 27, 1844, Hannah Buxton noted, "10 doz of eggs laid down in lime water & lime" (Diary).

28. William Chamberlain, Diary; Joseph Weare, Diary. Occasionally, Chamberlain noted on the first day, "killing and drying two cows"; two days later, he would note, "employ'd salting Beef" (Diary, Dec. 7, 9, 1830). Chamberlain also included a series of recipes in his daybook, with the notation: "Recipes which may be of use from various authors 1841." Among these were a recipe for "Curing Hams" and a "Receipt for curing beef" (Chamberlain, Diary, Oct. 5, 1841).

sibility by preserving the meat. Men who lived near the coast were also in charge of another salting process: they "dressed" the ocean fish that they caught—often in large quantities. In general, men took responsibility both for securing the staple foods stored in quantity for the household and for handling the heavier tasks on the farm and homestead.[29]

Although men usually took responsibility for slaughtering, salting, and barreling most of the pork and beef, in some households wives "assisted" in that work, so that their husbands could complete the process in good time. That cooperative effort did not necessarily mean that women had an equal share in the decision making. In most households, the responsibility for the salting and barreling of pork and beef was gender-marked, not by the preservation technique, but by the quantities involved. On occasion, some women took responsibility for overseeing the slaughter and salting of meat. For the most part, those women had husbands who were absent, infirm, deceased, or involved in nonagricultural occupations. Although their efforts may have granted those women the opportunity to determine the quantity and variety of meats to be preserved, at least some women complained about that added responsibility, which diverted their attention from their own responsibilities. Hannah Buxton viewed meat preservation as a particularly burdensome task that she would have preferred to relinquish. In 1844, she noted, "Butchered pigs today for us   wt 426 lbs   8 months and 3 weeks old, business which I dislike." Similarly, Persis Sibley Andrews offered a detailed account of the division of labor in her household one year when her husband was away: "On Monday we had a swine & a beef slaughtered. . . . My girl is very young, only 15, & I was obliged to lead, tho' I knew not the way. We took care of the fat of the entrails & cleaned the tripe that day. Tuesday I stood by the man who

---

29. On May 30, 1831, Chamberlain noted his procedure for 800 alewives: "I put about ½ Bushl Liverpool salt   let them lye 48 hours then put in salt petre—let them remain in pickle 42 hours   then turned the pickle from them   they appear well seasond & saved for Smokg" (Diary). In his usual shorthand, Joseph Weare briefly noted his procedure for salting mackerel and other fish: "out a fishing last night   Dressing my fish in the forenoon" (Diary, Aug. 18, 1811).

For example, cider making was a heavier task than beer brewing, and men tended to be responsible either for making cider if they had their own press or for taking the cider elsewhere to be pressed. By contrast, on Mar. 2, 1807, Ann Bryant Smith noted, "Did all my housework   brued a kegg beer" (Diary).

cut & salted the meat—& saved such as we wanted fresh—then we cared for the sause—& the cows feet (I like the dish call'd 'cow-heels')—then we tried and strained all the tallow and lard. The hog weighed 280." [30]

Although most women noted the progress of their husbands' slaughtering routines in their diaries, the details of their accounts were reserved for their own preservation tasks, which generally involved smaller quantities of meat. In most rural households, women were entirely responsible for sausage making. However, the responsibility for smoking pork for bacon varied from household to household. In February or March of each year, William Chamberlain would put up bacon to smoke. He applied his usual

30. Buxton, Diary, Nov. 26, 1844; Andrews (Black), Diary, Dec. 7, 1844. On Jan. 5, 1768, Mary Holyoke recorded, "salted pork. Put bacon in pickle" ("The Diary of Mrs. Mary Holyoke," in Dow, ed., *The Holyoke Diaries*, 68). When her husband was alive, Elizabeth Wildes recorded, "We kild our hog." After her husband's death, her notation changed: "I had my hog put up. . . . My hog was killed" (Diary, Nov. 23, 1790, Oct. 22, Nov. 26, 1792). On Dec. 23, 1867, Louisa Talbot similarly noted, "father killing beef; Hannah helped clean the tripe" (Diary). It is worth noting that Hannah Buxton's husband was much older than she and quite ill; Mary Holyoke, whose husband was a doctor, was managing an "urban" household in Salem, Mass.; and Elizabeth Wildes's husband was dead. All of these women were more fully responsible for storing and preserving foods in their households than were women whose husbands were actively engaged in farming.

Although some women diarists did not mention the salting process, both published and manuscript cookbooks offered recipes for salting, pickling, and preserving meats. See Rundell, *A New System*, 47; Child, *American Frugal Housewife*, 40–43.

On responsibility and cooperation: Osterud, *Bonds of Community*, 150, has shown that, in rural New York households, "responsibility for particular operations was divided more clearly and consistently than was the actual performance of tasks." Laurel Thatcher Ulrich, *Good Wives: Image and Reality in the Lives of Women in Northern New England, 1650–1750* (New York, 1982), 50, argues: "Under ideal conditions day-to-day experience in assisting with a husband's work might prepare [a woman] to function competently in a male world. . . . But in the immediate world such activities could have a far different meaning. The chores assigned might be menial, even onerous, and, whatever their nature, they competed for attention with the specialized housekeeping responsibilities which every woman shared."

By contrast, when their wives were absent or dead, New England men turned "women's work" over either to hired young women or to single or widowed women relatives. Men made little record of women's activities even when women were absent, except to note the presence of a "girl" to do the work (Weare, Diary, Dec. 30, 1804). When William Chamberlain recorded, "We are alone   no girl" in 1849, he was noting the inconvenience this caused him (Diary, June 14, 1849).

interest in experimenting to that process. In January 1841, he noted in his daybook: "I find by observation from year to year that large legs intended for bacon ought to be in strong pickle at least two months then taken out and hung without smoke a week or 10 days to dry." In other households, women took responsibility for smoking bacon and hams.[31] As with salting meat, smoking bacon didn't have a gender orientation. Although the responsibility for that task shifted between men and women, in most households the responsibility, once assigned, was clearly allocated.

Thus, the strategy for dealing with a potentially ambiguous domain was to separate roles according to gender rather than to share the responsibilities and the decision making. There were occasions when labor was shared, but that cooperation did not pose a significant threat to the assumption that tasks and responsibilities had an appropriate gender orientation. Men may well have advocated a gender division of labor that gave them, to their way of thinking, responsibility for the important tasks that provided their families' subsistence. Women may have had different, although equally compelling, reasons for supporting a clear division of labor. Such an allocation of responsibilities ensured women's control over their domain and protected women's labor from constant diversion from their own work.[32]

---

31. Chamberlain, Diary, Jan. 29, 1841.

On sausage making: In December 1803, Martha Ballard noted, "I have finisht trying hogg's lard; cut sauches meet [sausage meat], and done other matters" ("The Diary of Martha Ballard," in Nash, *The History of Augusta,* 417; see also 397). In most years, within a week of the hog slaughter, Elizabeth Phelps of Hadley, Mass., "made sausages," "tryed the suet," and "cut the sausage meat" (Thomas Eliot Andrews, ed., "The Diary of Elizabeth [Porter] Phelps" [1763–1805], *NEHGR,* CXVIII–CXXII [1964–1968]). On Jan. 2, 1867, Susan Talbot noted: "Mother taking care of meat. She and Lil making sausages and hogshead cheese" (Diary).

On smoking bacon and ham: Ann Bryant Smith noted on Mar. 2, 1807, "Did all my housework . . . hung our bacon up and began smoking it" (Diary). In February 1768, six weeks after she put her bacon in pickle, Mary Holyoke noted, "put bacon up Chimney" ("The Diary of Mrs. Mary Holyoke," 68); in March 1827, her daughter noted, "put up chimney 8 legs bacon" ("Diary of Mrs. Susanna [Holyoke] Ward, 1793–1856," in Dow, ed., *The Holyoke Diaries,* 178). On Feb. 28, 1839, Hannah Buxton briefly noted, "finished smoking hams" (Diary).

32. Osterud, *Bonds of Community,* 139, 184. Osterud argues, "In practice, women and men stepped out of their conventional work roles when the good of the farm seemed to require them to do so" (139). Yet the exchange of labor between husbands and wives was

The claims of advice manuals, as they addressed their separate audiences, attempted to perpetuate that gender division of labor. *Thomas's Almanack* encouraged agricultural improvements for market and commercial purposes. In 1791, the almanac included a "Best Method to Salt and Preserve Beef" that was "a sure way of putting up beef to remain good and fit for any market for the space of years." The "scientific" explanations that accompanied some of the methods and recipes implied that a better understanding of the process might improve the result. In 1804, the almanac offered a "New Method to Preserve Cider," explaining that "fermentation in cider proceeds from small particles of apples remaining." In 1809, it included a recipe for preserving potatoes, advising that they be "plunged into boiling water for the space of a minute," then taken out and dried. "The effect . . . is to destroy the power of vegetation which, from that saccharine matter produced in that process, occasions the sweet taste which potatoes possess in the spring and beginning of summer."[33] Although farmers could and did use the same preservation techniques for both market produce and household food stores, the almanac's emphasis on producing the greatest quantity at the least risk was oriented toward a market advantage. Cookbooks advised women in their household work. Their proclaimed concern was with quality—preparing good, healthful foods, in an economical and "systematic" way.[34] Although the advisers

---

not always equal; women were more likely to assist men than vice versa. "This asymmetry suggests that [some families] saw field work as more crucial to their welfare than most household tasks" (166).

33. *Thomas's Almanack*, 1791, 1804, 1809. In the 1803 "Agricultural," the almanac noted, "for the benefit of our country readers we insert the following profitable method of making butter in winter, by which means we hope they will be able to reduce the price of that article." Similarly Daniel Robinson offered a recipe "To keep apples and pears for market" (*Maine Farmers' Almanac*, 1836).

David Jaffee, "The Village Enlightenment in New England, 1760–1820," *WMQ*, 3d Ser., XLVII (1990), 329, 332, argues that Robert Thomas's *Farmer's Almanac* "introduced significant additions to the genre of almanacs and advocated reading as a source of entertainment, edification, and useful knowledge for rural folk. For enterprising farmers' sons the agrarian economy offered an increasing range of commercial possibilities." "Thomas anticipated change in the countryside only if newer theories of agricultural 'improvement' were consonant with farmers' beliefs."

34. On market advantage: Thus Daniel Robinson spoke to farmers' household needs, but in market improvement language. In 1854, he noted, "If you pack in casks, or boxes, such as you design for the table, they will *richly repay* your trouble in their extra relish and

actually offered many of the same techniques and recipes, they disguised the similarities by presenting them from different gender orientations—a market economy versus a household economy.

The tendency of the advice literature to associate production for markets with men's farm work and household consumption with women's domestic work did not reflect the gender orientation of some responsibilities in many families. Most men in rural New England were burdened by the responsibility of providing their families' yearly subsistence. At the same time, women had long been market producers; their butter, cheese, and eggs traditionally were an important source of income for their families.[35] Clearly, some components of food preservation suggest that, from the perspective of rural families, the gendered orientation of the marketplace in early industrial America was not as absolute as it appeared in urban households. Yet, as men became increasingly involved in commercial agricultural production, they might have begun to perceive their wives' efforts

---

flavor, particularly your turnips, which may thus be kept sweet over until spring" (*Maine Farmers' Almanac*, 1854).

On cookbooks: Child, *American Frugal Housewife*, on the title page, claimed that her cookbook was "dedicated to those who are not ashamed of economy," and she began the 1838 edition with a chapter entitled "Odd Scraps for the Economical." Rundell, *A New System*, title page, claimed to be offering a "Regular system" that was "Formed upon Principles of Economy."

35. In some areas, women participated in an informal system of trade and exchange. See Laurel Thatcher Ulrich, "Housewife and Gadder: Themes of Self-sufficiency and Community in Eighteenth-Century New England," in Carol Groneman and Mary Beth Norton, eds., *"To Toil the Livelong Day": America's Women at Work, 1780–1980* (Ithaca, N.Y., 1987), 21–34. Other women participated directly in market exchanges. While Jemima Weare made the butter on her family's homestead, her father-in-law noted in his daybook, "Octa & Jemima went to Portsmouth   Carried Butter got 21 cents pr lb" (Weare, Diary, Aug. 25, 1853). Eliza Buckminster Lee, *Memoirs of Rev. Joseph Buckminster, D.D., and of His Son, Rev. Joseph Stevens Buckminster* (Boston, 1849), 32, described the work of women in Piscataqua, N.H.: "The wives of fishermen were the market women of Portsmouth. . . . There were families that had been furnished by the selfsame women long years, from blooming youth to wrinkled age, with eggs, berries, chickens, spun yarn, knitted stockings, &c, coming as regularly as the Saturday came." Joan M. Jensen, *Loosening the Bonds: Mid-Atlantic Farm Women, 1750–1850* (New Haven, Conn., 1986), 81, argues that, in the Philadelphia area, "an extensive butter trade developed quickly after 1750. Farm families [and in particular farm women] produced increasing amounts of butter for market from that time [on]."

simply as supplementing their own greater efforts in market production. Although households differed in the way that they established their gender division of responsibilities, still the general assumption of men's final authority in production for household and market meant that men would make the agricultural and husbandry decisions that determined the composition of the staple food supply and ultimately shaped women's efforts to prepare their families' daily fare.

Dietary Decisions

Tuesday January 3 1832   the frost has examined the contents of a great many
   cellars   blasted much of the saus with his chilling breath
Thursday December 27 1832   the amount of snow fallen this season about 12
   or 14 inches   no severe cold days—saus in the cellar uncovered has not been
   touched by frost
Thursday March 5 1833   Clear & cold   the frost has crept into our cellar &
   threatens to ruin all our saus
Thursday January 9 1834   there has been no very cold weather or blocking
   snow this winter   no frost in our cellar
January 16 1840   a severe cold day—frost made its appearance in the cellar but
   has done no injury as yet
   —William Chamberlain, Bristol, Maine

In their farm daybooks, most men omitted any mention of (and concern about) the further processing of foods once their responsibilities ended. That omission suggests that they made the decisions about the kinds and quantities of crops they would plant and the livestock they would raise based on their perceptions of their responsibilities and goals. They raised foodstuffs for a variety of purposes (since their goal was to produce sufficient crops for their families' subsistence, for animal fodder, and for the market) rather than simply for the eventual use of those foods in their families' daily fare. At the same time, the quantities of vegetables, fruits, meat, and grains that they stored for their households were most at risk from the weather and vermin. Since men took the responsibility for providing a sufficient diet for their households, they might have felt necessarily entitled to make the final decisions about the best means of securing and preserving those stores of food for their households.

Allan Kulikoff has argued: "From the mid-eighteenth century to the end of the nineteenth, [yeomen] controlled the operation of their farms. *They* decided what crops to produce, how to divide family labor among farm tasks, whether and when to grow crops for distant commodity markets." Kulikoff suggests that a particular system of decision making and power emerged from that yeoman farm role in rural families: " 'Yeoman' is gender-specific and implies dominance of both the household and the polity by domestic patriarchs." Carolyn Merchant proposes certain implications of this division of authority for women's control over their "domain": "[In horticultural communities] women's access to resources to fulfill basic needs may come into direct conflict with male roles in the market economy. . . . In the nineteenth century . . . as agriculture became more specialized and oriented toward market production, men took over dairying, poultry-raising, and truck farming, resulting in a decline in women's outdoor production. Although the traditional contributions of women to the farm economy continued in many rural areas and some women assisted in farm as well as home management, the general trend toward capitalist agribusiness increasingly turned chickens, cows, and vegetables into efficient components of factories within fields managed for profits by male farmers." [36]

Whether men's proprietary control over agricultural production and bulk food preservation simply was assumed by men, was granted by women, or was a continuation of traditional decision making, by implication it placed a greater value and a higher priority on men's agricultural responsibilities and goals than on women's often more ephemeral domestic and culinary efforts. And men's final authority over the composition of the food supply had a significant impact on women's work in food preservation and meal preparation. Perhaps it was understood by both men and women that, if men significantly altered their production routines (either for subsistence or economic reasons), they were responsible for ensuring that the household could purchase foods that were no longer produced on

36. Kulikoff, "The Transition to Capitalism," *WMQ*, 3d Ser., XLVI (1989), 141–144; Carolyn Merchant, "Gender and Environmental History," *JAH*, LXXVI (1989–1990), 1118. Kulikoff argues further that, "properly used, the term ['yeoman'] points to power relations that denied female individualism and insisted that women were part of a male-dominated household. The wives of yeomen (or 'yeoman women'?) were precisely that—*wives*, without separate political or social identities" (144).

the farm so that women could continue to carry out their culinary responsibilities.[37] But while men made the major decisions about the bulk staples of the diet, women did have some control over the variety of their families' daily fare. They produced, gathered, and stored other foods that added flavoring and variety and increased the nutritional value of the meals they prepared.[38]

Perhaps women's efforts in preserving a wide range of "supplementary" foods (where supplementary was relative to men's standards of quantity rather than standards either of variety or nutrition) provided sufficient compensation for their lack of control in determining the staples of the household food supply. And, in another way, women might have accepted that production orientation in dietary decision making. The often limited variety of staple foods that men produced undoubtedly frustrated many women's meal preparation efforts. Indeed, men's goals of production both for their families' yearly subsistence and for the market were not necessarily compatible with women's goal of a varied daily fare. But when women's preservation efforts failed, for the most part they lost some of the variety that enhanced their cooking efforts. At least for Maine farmers, the responsibility and the risk involved in their production and preservation efforts affected their families' basic subsistence.[39]

Farmers in northern New England may well have felt that the goals of achieving a competency had to be granted more importance than the goals of taste. As long as they had to rely on the old technologies of food preservation, then ultimately they could not guarantee a sufficient subsistence for their families. And as long as households produced much of their own food supply, women's domestic domain would remain dependent on their husbands' agricultural and preservation efforts—and decisions. For rural New England households, the concept of a "corporate family economy" described the combination of separate efforts performed "for the good of the farm." But, since the good of the farm usually was mea-

37. Joseph Weare planted barley and wheat, both separately and in combination, between 1803 and 1848. However, he always planted less wheat than barley, and finally in 1849 he stopped planting wheat altogether. But by 1853 (and perhaps sooner), he was purchasing one barrel of flour a year for the use of his household (Diary).

38. McMahon, "A Comfortable Subsistence," *WMQ*, 3d Ser., XLII (1985), 50. See also Carole Shammas, "The Domestic Environment in Early Modern England and America," *Journal of Social History*, XIV (1980–1981), 14, 17.

39. Taylor, *Liberty Men and Great Proprietors*, 75–76.

sured in terms of men's agricultural and economic responsibilities, men held most of the authority for decision making. Nonetheless, the preservation of home-produced foods provided an overlapping link between men's and women's domains and responsibilities. The ambiguity of that "shared" domain enabled the perpetuation of the fiction of a "corporate family economy." It also restricted "domesticity" and the "separation" of the household economy from the market economy. Thus, it might have prevented the full granting to rural women of certain authority even in their domestic efforts.

T H E assumption of an inherently gendered orientation and division of labor offered ambiguous guidelines when applied to the realities of daily life in rural New England. The ambiguities were especially pronounced in the tasks and responsibilities of food preservation. The preservation, conversion, and storing of food was a transitional phase between food production and meal preparation. Men did the initial work of producing most of the foods on the farm: they harvested their crops, salted their meat, and continued their watch over their bulk stores of food. While women also had important roles as food producers, tending kitchen gardens and managing the dairy, much of their effort was oriented toward converting and preserving a range of supplementary foods and then cooking meals for their households from the supplies of food that had been laid by. For men, food preservation was the final stage of their work in the food cycle; for much of women's work, it was the beginning.

Food preservation was the least-gendered phase in the foodways process, even though most families sought and found ways to allocate the various tasks and responsibilities by gender. In this transitional stage, the processing of foods shifted from men's hands to women's hands. Yet, while men's and women's coordinated efforts in food preservation created at best a shared domain rather than shared responsibilities, preservation did allow for a smooth transition of tasks. As Osterud has suggested for rural New York families, the "gender system was riddled with contradictions."[40] Nonetheless, "sphere" does work as a metaphor for these rural households. Most men saw those spheres as a reality. Given their focus on their own responsibilities, they did not concern themselves with women's tasks; this unconcern effectively limited their opportunities for perceiving conflicts in the gender division of roles and responsibilities. By contrast,

40. Osterud, *Bonds of Community*, 226.

women saw those gendered spheres as metaphors for the reality of daily tasks, since their work could not be separated from the work that men did. That difference in perception might have helped prevent conflicts over tasks and responsibility where the gender orientation was ambiguous. Women tended to adjust their work and responsibilities depending on the success of men's work, and they might have relinquished the authority for certain decisions, where there were potential conflicts, in exchange for preserving their authority in other areas.

The strategy for dealing with ambiguity in gender relations was to avoid overlap and to divide tasks and responsibilities. The necessity for creating that strategy might have come not just from the transitional nature of food preservation in the foodways process; it might also have been related to the uncertainties inherent in the technologies of food preservation.[41] While families attempted to perpetuate a gender division of roles and responsibilities, they sought to overcome the uncertainty (and ambiguity) of food preservation as a link in the foodways process. Through their search for new methods of food preservation, they attempted to lessen the traditional disparity in their diet caused by the fluctuations between seasons of abundance (fresh foods) and seasons of monotony (stored provisions). But their efforts to control seasonal fluctuations in their yearly fare were limited by the old technologies of food preservation. Beyond "taking care," they had not found methods of preserving an array of foods over an extended period that would ensure the variety that they sought for their winter and spring fare.

Thus, through the middle of the nineteenth century, food preservation remained an ambiguous realm (and a measure of the ambiguities) in the experience of rural New England households. Their assumptions about the gender orientation of work led them to separate rather than share overlapping tasks and responsibilities. And the old technologies of food preservation prevented them from fully escaping the disparities created by the traditional seasonal orientation of their diet.

41. To return to Lévi-Strauss's raw-cooked dichotomy, food preservation techniques fall in between the "raw" and the "cooked"; and while the products of those techniques might initially be categorized as "converted," by the end of the winter, without continued care, they could switch to the opposite pole and become "rotten."

Donald C. Jackson

Roads Most Traveled

Turnpikes in Southeastern Pennsylvania

in the Early Republic

> Roads are the veins and arteries of the body politic, for through them flow the
> agricultural productions and the commercial supplies which are the lifeblood
> of the state. . . . But roads belong to that unappreciated class of blessings, of
> which the value and importance are not fully felt because of the very great-
> ness of their advantages, which are so manifold and indispensable, as to have
> rendered their extent almost universal and their origin forgotten.
> —W. M. Gillespie, 1849

When historians' thoughts turn to the Transportation Revolution of the nineteenth century, most attention centers on the canals and
railroads built across North America. The importance of these methods
of personal and material conveyance is recognized in special museums,
journals, and professional societies. In addition, every year thousands of
modern-day enthusiasts book nostalgic excursions on canal boats and an-
tique trains, journeys that avowedly offer direct participation in America's
travel heritage. In contrast, old roads and turnpikes, the other compo-

The author thanks both Darwin Stapleton of the Rockefeller Archive Center and Charles
Rumsey of the University of Wisconsin–Stevens Point for providing material on early road
history from their personal research files.

nent of the nineteenth-century transportation troika, receive scant public attention and are the subject of no popular appreciation societies. Ironically, many of the roads and turnpikes built in the late eighteenth and early nineteenth centuries still function as part of America's contemporary highway system. Their right-of-way may be realigned and their width increased, but many roads laid out in the precanal era (pre-1825) still carry significant amounts of traffic along routes paralleling or complementing more modern highways. In fact, the continued popularity of old turnpikes as transportation corridors, with their traffic jams, rush hour congestion, and idiosyncratic right-of-way alignments, may be the very reason that the public resists considering them a component of America's nostaligic past.

The importance of early road construction may be unappreciated by today's citizenry, but two hundred years ago the issue attracted great public attention. Federalists recognized the importance of improved travel, and the desire to rationalize the nation's transportation system continued during Thomas Jefferson's presidency.[1] Interest in early highway expansion found most prominent expression in the "Report on Roads and Canals" prepared by Jefferson's Treasury secretary Albert Gallatin. Gallatin observed in 1808 that "the general utility of artificial roads and canals is at this time so universally admitted as hardly to require any additional proofs," and his voluminous report incorporated technical and economic data on scores of existing turnpike companies. Roads were not considered simply to be secondary adjuncts of a system focused on canals (and later railroads); instead, as William Gillespie later noted, they constituted the "veins and arteries of the body politic," and efforts to improve them comprised a significant movement during the early years of the Republic.[2]

1. W. M. Gillespie, *A Manual of the Principles and Practice of Road-making . . .*, 3d ed. (New York, 1849), 15. The extensive literature on Federalism, Hamilton, and Jefferson does not warrant listing here, but it is worth noting that Jacob Ernest Cooke, *Alexander Hamilton* (New York, 1982), 109–120, describes the quarrel between Hamilton and Jefferson as "a skirmish, not a war." Although the two leaders did not share a common interpretation of the U.S. Constitution, many of their views concerning economic development were relatively compatible. As E. James Ferguson has pointed out, "In his second inaugural, he [Jefferson] suggested that Treasury surpluses beyond the amounts necessary to discharge the public debt on schedule might be 'applied in time of peace to rivers, canals, roads, arts, manufacturers, education, and other great objects . . .'" (Ferguson, ed., *Selected Writings of Albert Gallatin* [New York, 1967], 228).

2. Albert Gallatin, "Report on Roads and Canals [Apr. 6, 1808]," in Thomas Cochran,

Pennsylvania's road history is particularly noteworthy because of the proliferation of turnpikes in the commonwealth during the precanal era. America's first major toll road, the Philadelphia and Lancaster Turnpike, was completed in late 1794 at a cost of $465,000. By 1821 a total of eighty-four separate turnpike companies in the state operated over eighteen hundred miles of roads that required almost $6,000,000 to build.[3] In contrast to the federally financed National Road built to connect the Potomac River watershed with the Ohio Valley, the early turnpikes of Pennsylvania were primarily financed by private investors who supplied approximately $4,100,000 to cover construction costs. This compared with the state government's contribution of about $1,850,000 to the commonwealth's pre-1821 road building effort (usually expressed by purchasing stock in privately chartered turnpikes). Pennsylvania's road-building initiative represented a dramatic commitment to improving overland transportation, a commitment that both reflected and energized nationwide efforts in the field.

Among scholarly historians, the import of early road improvement in America has been documented by George Rogers Taylor (in *The Transportation Revolution: 1815–1860*), Joseph Durrenburger (*Turnpikes: A Study of the Toll Road Movement in the Middle Atlantic States and Maryland*), and Frederic J. Wood (*The Turnpikes of New England and Evolution of the Same through England, Virginia, and Maryland*), while several

---

ed., *The New American State Papers: Transportation* (Wilmington, Del., 1972), I, 17–275; Gillespie, *Principles and Practice of Road-making*, 15.

3. Commercial interests sought to improve Philadelphia's ties with Lancaster and the interior of the state because of concern that trade in the Susquehanna River Valley would otherwise gravitate to Baltimore. See James Weston Livingood, *The Philadelphia-Baltimore Trade Rivalry, 1780–1860* (Harrisburg, Pa., 1947). Charles I. Landis, "Philadelphia and Lancaster Turnpike: The First Long Turnpike in the United States," *Pennsylvania Magazine of History and Biography*, XLII (1918), 1, XLIII (1919), 84, describes the social and legislative origins of this pioneering turnpike company.

"Report from the Committee on Roads, Bridges and Inland Navigation [Mar. 23, 1822]," *Journal of the Senate of the Commonwealth of Pennsylvania* (Philadelphia, Lancaster, Harrisburg, 1790–), XXXII (1821–1822), 654–670 (hereafter cited as *Pennsylvania Senate Journal*), is the source for statistics on the state's early 19th century turnpikes. More references on early turnpike and toll bridge companies are listed in Adelaide R. Hasse, ed., *Index of Economic Material in Documents of the States of the United States: Pennsylvania, 1790–1904* (Washington, D.C., 1919–1922).

British books have explored developments from a European perspective.[4] In addition, a plethora of descriptive books and articles discuss road history in a local or regional context, including Philip Jordan's *National Road*, John Faris's *Old Roads Out of Philadelphia*, and Charles Landis's 1918 article, "History of the Philadelphia and Lancaster Turnpike: The First Long Turnpike in the United States."[5] Many treatments of America's

4. George Rogers Taylor, *The Transportation Revolution: 1815–1860* (New York, 1951); Joseph Austin Durrenberger, *Turnpikes: A Study of the Toll Road Movement in the Middle Atlantic States and Maryland* (Valdosta, Ga., 1931); Frederic Wood, *The Turnpikes of New England and Evolution of the Same through England, Virginia, and Maryland* (Boston, 1919); William Albert, *The Turnpike Road System in England: 1663–1840* (Cambridge, 1972); and Christopher Taylor, *Roads and Tracks of Britain* (London, 1979). For other British books, see Geoffrey Hindley, *A History of Roads* (Secaucus, N.J., 1972); Albert C. Leighton, *Transport and Communication in Early Medieval Europe, AD 500–1100* (Newton Abbot, 1972). Other important scholarly works include Philip Elbert Taylor, "The Turnpike Era in New England" (Ph.D. diss., Yale University, 1934); Charles Alexander Williams, "The History and Operations of the Pennsylvania Turnpike System" (Ph.D. diss., University of Pittsburgh, 1954); Christopher Colles, *A Survey of the Roads of the United States of America, 1789,* ed. Walter W. Ristow (Cambridge, Mass., 1959); Robert F. Hunter, "Turnpike Construction in Antebellum Virginia," *Technology and Culture,* IV (1963), 177–200. Useful overviews of transportation history include Balthasar Henry Meyer and Caroline E. MacGill, *History of Transportation in the United States before 1860* (Washington, D.C., 1948); J. L. Ringwalt, *Development of Transportation Systems in the United States . . .* (Philadelphia, 1888); and Nathan Rockwood, *One Hundred Fifty Years of Road Building in America* (New York, 1914).

5. Philip D. Jordan, *The National Road* (New York, 1948); John T. Faris, *Old Roads out of Philadelphia* (Philadelphia, 1917); Landis, "Philadelphia and Lancaster Turnpike," *PMHB,* XLII (1918), 1, XLIII (1919), 84. Also see S. F. Hotchkin, *The York Road, Old and New* (Philadelphia, 1892); Horace R. Barnes, "Organization and Early History of the Conestoga Navigation Company," Lancaster County Historical Society, *Historical Papers and Addresses,* XXXIX (1935), 49–61; Charles I. Landis, "The Beginnings of Artificial Roads in Pennsylvania," Lanc. Co. Hist. Soc., *Historical Papers and Addresses,* XXIII (1919), 99–107; Chester D. Clark, "The Old Centre Turnpike," Northumberland County Historical Society, *Proceedings,* VIII (1936), 113–126; H. Frank Eshleman, "History of Lancaster County's Highway System (from 1714 to 1760)," Lanc. Co. Hist. Soc., *Historical Papers and Addresses,* XXVI (1922), 37–80; S. F. Hotchkin, *The Bristol Pike* (Philadelphia, 1893); J. Orin Oliphant and Merrill W. Linn, "The Lewisburg and Mifflinburg Turnpike Company," *Pennsylvania History,* XV (1948), 86–119; and D. K. Turner, "The Turnpike Roads," in *A Collection of Papers Read before the Bucks County Historical Society,* II (Riegelsville, Pa., 1909), 565–575. Also see Norman B. Wilkinson, comp., *Bibliography of Pennsylva-*

early roads appeared during the first part of the twentieth century, when increased automobile travel prompted interest in highway history. This literature stressed general economic data concerning the length of roads, their cost, and the ultimate financial fate of the organizations responsible for their operation. Locally oriented articles often featured anecdotal travel narratives that highlighted the seemingly poor conditions of early highways or the history of colorful or significant people or establishments associated with a particular road. In the 1950s George Taylor Rogers succinctly incorporated much of the existing scholarship into *The Transportation Revolution,* and since that time interest in the field has been largely dormant.[6]

This essay examines roadbuilding in early nineteenth-century America by focusing on developments in southeastern Pennsylvania and, after describing basic technological issues associated with the construction of road systems, centers its attention on the work of engineer and surveyor Robert Brooke in laying out the Germantown and Perkiomen Turnpike in 1802. By studying the particular issues addressed by Brooke in planning this toll road we can better comprehend the technical skills and political accommodations required to build turnpikes. Economic issues related to early turnpikes illuminate the significance of road improvement within broader social and commercial contexts, and, although this essay considers specific activities associated with only a few of Pennsylvania's early turnpikes, it elucidates issues relevant to the mid-Atlantic economic community as a whole.

---

nia History (Harrisburg, 1957), 102–106, 255–261; and Carol Wall, ed., *Bibliography of Pennsylvania History: A Supplement* (Harrisburg, 1976), 20–21, 62–63.

6. Since Taylor's book some good studies have appeared that discuss road development, most notably John Flexer Walzer, "Transportation in the Philadelphia Trading Area, 1740–1775" (Ph.D. diss., University of Wisconsin, 1968). Walzer discusses road layout and construction in the 18th century (146–214) and notes the role of stream crossing on transportation development (215–253). Toll road development in northwestern Rhode Island in the early 19th century has also been documented in Daniel P. Jones, *The Economic and Social Transformation of Rural Rhode Island, 1780–1850* (Boston, 1992). In contrast, James T. Lemon, *The Best Poor Man's Country: A Geographical Study of Early Southeastern Pennsylvania* (Baltimore, 1972), is exemplary in many respects but pays scant attention to road development. Similarly, Diane Lindstrom, *Economic Development in the Philadelphia Region, 1810–1850* (New York, 1978), presents an excellent discussion of canal and railroad development but devotes little attention to turnpikes.

Turnpikes and Early Road Construction

In twentieth century America, the term "turnpike" usually refers to a heavily traveled thoroughfare that exacts a special fee, or "toll," from motorists. The tolls are then used by a state-sponsored authority (like the Pennsylvania Turnpike Commission) to pay off construction debt and fund maintenance costs. In contrast, motorists using America's extensive Interstate highways (originally promoted as a National Defense Highway system) are usually not charged a toll. Instead, the federal government, in concert with state highway commissions, finances construction costs by placing a special tax on gasoline. In the late eighteenth and nineteenth centuries, American citizens and their elected representatives did not consider it prudent to finance widespread road construction directly through the use of federal or state tax revenues. Beginning with the state-chartered Philadelphia and Lancaster Turnpike in the 1790s, private turnpike companies were often authorized to take control of an important road or right-of-way, pay for the cost of upgrading and maintaining the roadbed, and then charge travelers a toll for using the improved highway. The idea of privatizing the highway system by relinquishing control of major routes to nongovernmental commercial groups originated in England in the 1660s and was later transplanted to the United States. Better-quality roads served the commercial interests of many Americans, and in this context the chartering of turnpike companies was widely, but not universally, heralded. Toll roads often attracted the ire of the traveling public, who resented having to meet the demands of tollkeepers every five miles, and, by the beginning of the twentieth century most of the early turnpikes had been turned into "free roads," often by legislative action.[7] Today, the concept of toll roads survives in the form of a few state-sponsored turnpikes,

7. The origins of turnpikes in Great Britain are documented in Albert, *Turnpike Road System*, 6–56; and in William Albert and P.D.A. Harvey, eds., *Portsmouth and Sheet Turnpike Commissioners' Minute Book, 1711–1754* (Portsmouth, 1973), xi–xxviii. In medieval Europe, traders were at times required to pay a toll for the privilege of traveling along specific roads or rivers. See Leighton, *Transport and Communication in Early Medieval Europe*, 41–42.

Durrenberger, *Toll Road Movement*, 153–165, describes the conversion of early turnpikes to nontoll roads. Pennsylvania's 1911 Sproul Highway Act specifically authorized the conversion of existing toll roads into state-owned highways.

although the vast majority of America's modern highways are funded by gasoline tax revenues.

Specifically, the term *turnpike* derives from a large, balanced log (or "pike") that extended across the roadway at each tollhouse. When lowered, the pike prevented horse and wagon traffic from passing; after payment of the required toll, the pike would be lifted (or turned) by the toll collector and the roadway cleared for passage. Although the technique of using a turning pike to control traffic past tollhouses was often altered or abandoned altogether, use of the word *turnpike* to signify a toll road, and *not* a free road, became widely recognized. Here, *turnpike* primarily refers to a prominent road for which tolls are paid to finance construction and maintenance costs. In a few cases, there were roads that did not charge tolls but were similar in size and form to those operated by chartered turnpike companies.[8] *Turnpike* here will also denote the most important, and expensive, highways built in the early years of the American republic.

In building a simple road, the first task is to clear away any impediments that might obstruct a traveler, the most obvious being trees, stumps, brush, and stone boulders. After removing such barriers (which is not always easy), it becomes possible for the roadbuilder to begin preparing a smooth, level surface free of disrupting bumps or depressions. The use of hand-held rakes and hoes or horse-drawn scrapers to level a cleared roadway represents a basic act that distinguishes an improved road from a path worn clear by years of foot travel or animal migration. Carried out by farmers and rural laborers (perhaps working under the direction of county or township supervisors), this leveling did not need to utilize sophisticated equipment in order to improve noticeably a right-of-way's ability to carry sustained traffic.[9] The techniques of clearing and leveling roadways in the

8. In the last decade of the 18th century the Pennsylvania state government authorized monetary grants (or subsidies) to pay for the improvement or repair of specific roads in the commonwealth. Grants were made both to local authorities and to contractors and ranged in size from a few hundred to a few thousand dollars. About 50 grants (mostly predating 1795) are listed in an appendix to the *Journal of the Senate of the Commonwealth of Pennsylvania* (Lancaster, Pa., 1807–1808), 347, 464.

9. In 1555 an act of Parliament delegated responsibility for British road repair to local parish authorities. Albert, *Turnpike Road System,* 14–15, reports that "parishioners were obliged to spend four days [later increased to six] a year working on the roads under the supervision of surveyors appointed by the parish churchwardens. . . . Most surveyors served

colonial era might appear primitive to a twentieth-century observer, but the effects of such activity were appreciated by travelers with firsthand knowledge of the alternative.

Along with providing a clear, relatively level surface, a roadbuilder usually sought to select a right-of-way representing the shortest, most effective route between two points. In this context, the roadbuilder relied upon the skills of a surveyor in laying out a route across the landscape. The procedures of land measurement had been well established in Europe before Britain's settlement of North America, and colonial surveyors were able to call upon long-standing techniques and traditions in their efforts to lay out a road system.[10] In basic terms, the process of surveying breaks down into a few tasks:

(1) The measurement of distance. In the colonial era this was facilitated by use of an iron chain 66 feet long known as Gunter's chain (after Edmund Gunter, an Englishman who invented it in 1620). Conveniently, eighty chains equaled one mile, and ten square chains equaled one acre. One chain also equaled four poles (or rods), each 16.5 feet long. To calculate the distance between two points, a Gunter's chain was merely stretched out as many times as necessary in order to connect the two points. For example, if it required 4.25 lengths of Gunter's chain to connect two points, then they were 280.5 feet apart.

(2) The measurement of direction. This involved the use of magnetic compasses to orient the survey line to a north-south axis (as determined by a region's magnetic declination). Sightings made between two points

---

for only one year, and this lack of continuity militated against any technical advance." Albert does not consider this system to represent a radical innovation, but rather a formalization of an ancient common law obligation. Also see Sidney Webb and Beatrice Webb, *English Local Government: The Story of the King's Highway* (London, 1913), 14–61.

10. Early surveying in England is documented in A. W. Richeson, *English Land Measuring to 1800: Instruments and Practices* (Cambridge, Mass., 1966). Pre-19th century surveying techniques and equipment in Pennsylvania are described in John Barry Love, "The Colonial Surveyor in Pennsylvania" (Ph.D. diss., University of Pennsylvania, 1970). Love refers to numerous early surveying books, including William Leybourn, *The Compleat Surveyor: Containing the Whole Art of Surveying of Land . . .* (London, 1653); John Carter, *The Young Surveyor's Instructor* (Philadelphia, 1774); and Robert Gibson, *A Treatise of Practical Surveying . . .* , 4th ed. (Philadelphia, 1785). Documentation for material on surveying included within this article is drawn from Love unless otherwise noted.

could provide directional readings by use of a vernier, an instrument that allowed the measurement of horizontal angles. A vernier divided the circular base of the tripod-mounted sight into 360 degrees and provided a means of determining how much a sight line diverged from magnetic north-south.

(3) The measurement of elevation. The measurement of relative heights was not something that the typical colonial land surveyor needed to worry about in determining the location and size of various tracts of land. However, for surveyors involved in laying out roads and canals, the measurement of elevations became important. The key to the process lay in using a spirit level that contained an enclosed glass cylinder of water with a small air bubble. By attaching this level to either an open or a telescopic sight, elevation changes could be determined by placing a vertical measuring rod some distance away and then reading through the sight the relative height of the second location. By moving across the terrain and making a series of "backsighting" and "foresighting" readings between evenly spaced points, it was possible for a surveyor to calculate changes in elevation that did not need to be corrected relative to the curvature of the earth. Differences in elevation could also be determined using readings from a "theodolite," an instrument that enabled a surveyor to measure vertical angles; by taking distance and angular measurements between two points, trigonometry could be used to calculate vertical distances. However, in the eighteenth and early nineteenth centuries theodolites were expensive, and they did not find much use among general surveyors.

With the expanded use of wagons and other wheeled vehicles in colonial America in the later eighteenth century, the importance of determining changes in elevation became much more important for roadbuilders. Whereas a horse and rider might easily travel up and down steep grades, a horse-pulled wagon might not. Consequently, in planning roadways that could more efficiently handle wagon traffic, surveyors needed to determine the grades associated with various routes. By the early nineteenth century, turnpike proponents recognized that a grade of five degrees (a vertical rise of 462 feet for every mile traveled) represented the maximum that wagons could easily traverse.[11]

11. Gallatin, "Report on Roads and Canals," in Cochran, ed., *American State Papers: Transportation*, I, 35–36, recommends a maximum grade of five degrees.

The increase in wagon traffic also prompted a desire to protect roadway surfaces from erosion and rutting caused by wheel rims. The deleterious effects of wheeled traffic were heightened by rain and moisture that could soften the roadbed and create a quagmire of mud. The construction of drainage ditches alongside highways and the grading of rights-of-way with central crowns that forced standing water off to the sides eventually became standard methods of countering this problem. For low-lying and swampy areas, the solidity of the roadway surface was frequently enhanced by placing wooden logs lengthwise across the highway. Known as "corduroy roads" because of their ridged surface, these designs represented an early attempt to improve the quality of earthen roads.[12]

Stone Roads and Stone Arch Bridges

In terms of technology, high-quality road construction in the late eighteenth and early nineteenth century usually depended on the use of crushed stone for the surface of the roadbed, especially for the mid-Atlantic region. In New England, many of the early nineteenth-century turnpikes used packed earth instead of crushed stone for their roadway surfaces, a

12. Gallatin, *ibid.*, 36, notes the importance of road drainage. C. Taylor, *Roads and Tracks of Britain*, 13, notes the use of timber to stabilize road surfaces as far back as 3000 B.C. Françoise Audouze and Olivier Büchsenschütz, *Towns, Villages and Countrysides of Celtic Europe*, trans. Henry Cleere (Bloomington, Ind., 1992), 145–147, includes a photograph of a wooden trackway built circa 1000 B.C. Llewellyn Nathaniel Edwards, *A Record of History and Evolution of Early American Bridges* (Orono, Maine, 1959), 20, reports that "the term 'corduroy' as applied to the ridged surfaces of early timbered bridge floors and roadways gained acceptance from a supposed resemblance to the king's corduroy cloth." Readers should distinguish early corduroy roads from wooden "plank roads" built during the late 1840s and early 1850s. Plank roads were built with smoothly cut timber designed to provide a level, hard surface for the roadway. For a few years plank roads appeared to be viable, but maintenance problems with rotting wood proved intolerable, and the Plank Road Movement quickly lost its allure. The proliferation of plank roads depended upon a highly mechanized saw-milling industry, and, as a consequence, the technology did not become popular until the 1840s (because of their late date and limited utility, plank roads are not discussed in this article). See William Kingsford, *History, Structure, and Statistics of Plank Roads, in the United States and Canada* (Philadelphia, 1851), for a favorable view of this type of construction. G. Taylor, *Transportation Revolution*, 29–31, provides a more realistic assessment of plank roads as a transportation technology.

design feature made feasible because sleighs in winter months could ride directly on a solid coating of snow and ice. In more temperate climates, the existence of a permanent winter snowpack was less certain and could not be relied upon. Thus, in the Philadelphia hinterlands stone pavement became necessary to ensure year-round travel.[13] Stone surface offered means of eliminating, or at least severely reducing, the erosive action of rain on the right-of-way and helped prevent wagon wheels from gouging out deep ruts.

Quarrying, breaking, placing, and replacing rocks along a roadway represented a time-consuming and expensive process that required most of the capital raised by many turnpike companies. For example, in 1804 the final ten-mile section of the Germantown and Perkiomen Turnpike cost a total of almost $107,000, with more than $76,900 of this expended on "leveling, forming, shouldering and stoning." Given that the right-of-way had carried traffic for several decades before the chartering of the turnpike, practically all of this latter sum would have been spent for quarrying, breaking, and placing the stone pavement. Significantly, this $76,900 did not include more than $20,000 expended for masonry bridge construction or any funds for "ploughing the road, rolling the road, making side drains [or] . . . setting mile-stones."[14]

The right-of-way for turnpikes was usually stipulated to be fifty to sixty feet wide. However, only a portion of this width was intended to be covered with stone pavement. For example, the Pennsylvania legislature

13. The use of stone for road surfaces dates to Roman times. C. Taylor, *Roads and Tracks of Britain*, 66, describes the Roman road between Salisbury and Dorchester as including sections in which "the agger was made by spreading a layer of large flints on the ground surface and covering these with a large bank of rammed chalk. The road surface itself, a layer of gravel obtained from Pentridge Hill . . . had been laid on top of this chalk bank." The relative cheapness of New England's early earthen-surfaced turnpikes is documented in Gallatin, "Report on Roads and Canals," in Cochran, ed., *American State Papers: Transportation*, I, 33–36, 204–209. In this 1808 report, Gallatin specifically refers to how Pennsylvania's roads require stone surfacing if they are to be suitable for wagon traffic during the winter months.

14. Construction costs for the Germantown and Perkiomen Turnpike are detailed in William Dary and Sam. W. Fisher to Gen. Wm. Macpherson, Oct. 28, 1807, reprinted in Gallatin, "Report on Roads and Canals," in Cochran, ed., *American State Papers: Transportation*, I, 228–230. The technology of leveling and draining roads dates to Roman times while the use of milestones to demarcate highway routes was characteristic of early British turnpikes. See C. Taylor, *Roads and Tracks of Britain*, 66, 158.

authorized the Germantown and Perkiomen Turnpike to build a toll road
with a total width of fifty to sixty feet that included an "artificial road" at
least twenty-eight feet wide. This artificial road was to be

> bedded with wood, stone, gravel, or any other hard substance well com-
> pacted together, and of sufficient depth to secure a solid foundation to
> the same, and the said road shall be faced with gravel or stone pounded,
> or other small hard substance, in such manner as to secure a firm . . .
> [and] even surface.[15]

The rest of the road's right-of-way was to be left with an earthen surface
for use by travelers during dry weather. Often known as "summer roads,"
these portions of a turnpike could also be a convenient way to divert traffic
around a section of stone pavement undergoing repair.

The techniques employed in building Pennsylvania's early stone roads
are not well documented, but we are fortunate in having a short "Directions for Making Roads" that appeared in a 1799 issue of the *Philadelphia Magazine and Review*. Written by an anonymous roadbuilder, who
claimed "many years . . . [of having] had opportunities of trying experiments upon this subject," this article stressed the importance of leveling
a firm, solid earthen foundation for a road. Over this foundation, he recommended:

> The stones should be spread equally over the surface, and settled with
> a light sledge [hammer]; in this operation, such stones as are too large,
> must either be broken or carried away; over this a layer of small stones,
> not larger than eggs, should be scattered, and settled with hammers be-
> tween the interstices of the largest. Over this a small quantity of any
> hard clay, just sufficient to cover the stones, should be spread; if mixed
> with gravel it will be better—but if gravel alone were used, it would
> fall through the stones and be wasted. . . . In a month or two [of cattle
> and carriage traffic], the clay and gravel will be worn away, and the
> corners of the large stones will appear—men should now be employed
> to break the stone with hammers, weighing about two pounds and an
> half. . . . After another months, or six weeks, the road must be broken,
> with care, in the same manner; and, with proper intervals, it should be

15. Gallatin, "Report on Roads and Canals," in Cochran, ed. *American State Papers: Transportation*, I, 232.

broken from time to time, as often as may be necessary—four times is, in general, sufficient.[16]

From the above description, the construction of permanent stone pavements clearly constituted a laborious and time-consuming process. The technology was not complicated, but it represented a substantial and extended commitment to the breaking of rocks. The author counsels employing strong men capable of handling large hammers (with handles "four to five feet long") who are to carry out the work of breaking rocks directly on top of the road being paved. However, such techniques were not always recommended. For example, in the 1830s an English author urged that the breaking of stone in small pieces be carried out on the side of the road and only those stones of proper size be transferred for placement in the pavement. This author also noted:

> The proper mode of breaking stones, both for effect and economy, is in a sitting posture. This work can be done by women, boys, and old men past hard labour.[17]

During the late eighteenth and early nineteenth centuries two British engineers, Thomas Telford and John McAdam, devised special construction methods for stone pavements that eventually became internationally recognized standards for roadway design. Telford's method, which involved the construction of a solid rock foundation that supported a surface of smaller crushed stone, was originally used for military road construction in Scotland. Although usually quite durable, Telford's method

---

16. "Directions for Making Roads," *Philadelphia Magazine and Review*, I (1799), 123–125 (this reference courtesy of Darwin Stapleton). In Great Britain, numerous tracts on road building and repair were published in the 17th and 18th centuries, including Thomas Proctor, *A Profitable Worke to This Whole Kingdome, Concerning the Mending of All Highwayes, as Also for Waters and Iron Workes* (London, 1610); William Mather, *Of Repairing and Mending the Highways* (London, 1696); John Hawkins, *Observations on the State of the Highways, and on the Laws for Amending and Keeping Them in Repair . . .* (London, 1763); Henry Homer, *Enquiry into the Means of Preserving and Improving the Publick Roads . . .* (Oxford, 1765); Alexander Cumming, *Observations on the Effects Which Carriage Wheels, with Rims of Different Shapes, Have on the Roads* ([London], 1797). The above references, and many more, are listed in Harwood Frost, *The Art of Roadmaking . . .* (New York, 1910), 505–508.

17. John W. Parker, *The Roads and Railroads, Vehicles, and Modes of Travelling, of Ancient and Modern Countries . . .* (London, 1839), 78.

required prodigious amounts of labor and was so expensive that it could not be employed in America except using inferior, bastardized techniques. As an alternative to Telford's method, McAdam advocated using a single layer of crushed stone placed directly upon earthen foundations. Counseling the construction of roadway surfaces with a slight crown along the center line to prevent rainwater from ponding on the roadway, McAdam promoted a design in which a conglomerate of small, hard, and irregularly shaped stones could bond together under the pressure of traffic and form a solid, yet resilient, road surface.[18]

In terms of early American highway construction, it appears relatively meaningless to ask which method, Telford's or McAdam's, was the most influential. In 1822 the Committee on Roads, Bridges, and Inland Communication within the Pennsylvania Senate reported upon the status of road building within the commonwealth and took note of a new pamphlet on McAdam's system, which apparently had received little or no attention in America before that time. The committee's report observed:

> The construction of stone and other certified roads is a science which few men understand, and yet which few men hesitate to undertake; and it is no doubt from a want of ordinary skill in preparing and applying the materials of which our roads are composed, and in shaping their surface, and of ordinary judgement in the application of labour, that most of our roads have been constructed so inexpensively, and some of them so badly.[19]

The committee report stressed the small size of stones used in the McAdam method ("The stones are broken so fine as that none of them exceeds six *ounces* in weight, in order that a more speedy consolidation may be produced") and indicated that the average depth of Pennsylvania's existing stone roads was about twelve inches, compared with about ten inches for McAdam pavements. In retrospect, explicit use of the McAdam system seems to have been practically unknown in Pennsylvania (and the rest of the United States) before 1820. In place of McAdam's method, which required a great amount of labor to break the stone into extremely small

---

18. For more on methods of road construction promoted by Telford and McAdam, see Hindley, *A History of Roads*, 63–67; and Gillespie, *Principles and Practice of Road-making*, 195–210.

19. *Pennsylvania Senate Journal*, XXXII (1821–1822), 658.

pieces, early Pennsylvania roadbuilders were content to use larger rocks to form a thicker, rockier, yet easier-to-build pavement. Unfortunately, the use of larger stones did not facilitate the compaction of a road surface into a hard, erosion-resistant mass. But compared with earthen roads, even loose, rutted stone surfaces offered substantial advantages to travelers, especially those using wheeled vehicles in inclement weather.

Stone not only created a relatively durable road surface; it also served as a key material in the construction of permanent highway bridges. When thoughts turn to early American bridges, the image of the rustic wooden-covered bridge is usually brought to mind; or, for devotees of structural engineering, the most significant early nineteenth-century bridges are represented by Timothy Palmer's Permanent bridge over the Schuylkill River in Philadelphia (wood, completed 1805), Louis Wernwag's Colossus over the Schuylkill in Philadelphia (wood and iron, completed 1813), or the early iron-chain suspension bridges of James Finley and his associates.[20] Early stone-arch bridges are often dismissed as insignificant (one modern bridge historian even claimed, "It is probably not incorrect to say that psychologically Americans were as temperamentally unsuited to build with stone as it was economically unfeasible for them to do so"). In truth, they functioned as invaluable components of many early roads and turnpikes. This was especially true in southeastern Pennsylvania, a region that prompted Treasury secretary Gallatin to remark in 1808 that, "in the lower counties of Pennsylvania, stone bridges are generally found across all the small streams."[21]

The importance of permanent bridges in promoting year-round road transportation can be better understood after considering a 1797 description of a traveler's encounter with a rain-swollen creek. Since this quotation is taken from a treatise advocating bridge construction, the propagandistic orientation of the text should be acknowledged. But anyone who has stood near a stream on a cold winter's day and contemplated a crossing can appreciate the insight underlying the following scenario:

20. See Lee H. Nelson, *The Colossus of 1812: An American Engineering Superlative* (New York, 1990), for discussion of large-scale bridge building in America in the early 19th century. An important contemporary source is Thomas Pope, *A Treatise on Bridge Architecture* . . . (New York, 1811).

21. David Plowden, *Bridges: The Spans of North America* (New York, 1974), 9; Gallatin, "Report on Roads and Canals," in Cochran, ed., *American State Papers: Transportation,* I, 35.

[The] traveller stops, and surveys the turbulent torrent that hides an unknown bottom,—he hesitates—doubts whether to risk a passage or not; at last, by delay grown impatient, he with fear and trembling cautiously moves forward and perhaps arrives in safety on the opposite bank; but alas! too frequently the rash, or fool-hardy driver, is carried down the stream, and all is lost![22]

In light of this quotation, it is not surprising that in 1805 the Falls Turnpike embossed its corporate seal with the slogan "Difficulties Made Easy" featuring an etching of a large wagon crossing a stone-arch bridge.[23] Stone-arch bridges represented to the traveling public a technology that guaranteed their safety in crossing streams no matter what the season. Fords (locations where rivers could be crossed by wading through a relatively shallow section of the stream) represented uncertainty to the traveler who wished to avoid the terrors of a turbulent current. For example, Elizabeth Drinker described a harrowing trip to Lancaster in 1778:

This day we forded three large rivers, the Conestoga the last, which came into the carriage, and wet our Feet, and frightened more than one of us.[24]

Wooden bridges represented an improvement over fords, but they were subject to rot and decay, especially if they were not "covered" and thus protected from the elements. Bridge historian Llewellyn Edwards has described many colonial-era wooden bridges as "pile trestle-type structures . . . built across mud flats, bogs and shallow water." Designed in many ways to mimic simple maritime wharves, structures of this type were also "built outward from stream bank and shore locations to provide means of transfer to water-borne rafts." As an alternative to wooden trestles, stone arches promised safety uncompromised by structural uncertainties associated with exposed wood.[25] For this reason, and the fact

22. Charles W. Peale, *An Essay on Building Wooden Bridges* (Philadelphia, 1797), iii.

23. Stock Certificate for the Falls Turnpike Road Company (ca. 1805), Anthony Kennedy Papers, Historical Society of Pennsylvania, Philadelphia. The Falls Turnpike connected the lower Susquehanna River Valley with the city of Baltimore.

24. April 1778 diary entry by Elizabeth Drinker, quoted in Faris, *Old Roads out of Philadelphia*, 121.

25. Edwards, *Early American Bridges*, 21. The uncertainties inherent in exposed wooden bridges were experienced by George Washington upon his return from the Constitutional

that the mid-Atlantic region did not lack readily available building stone, masonry bridges became a key part of many early turnpike systems.

Although the ancient art of stonemasonry was among the basic artisanal skills brought across the Atlantic by settlers from Great Britain and northern Europe, the subject of stone quarrying and masonry construction techniques has heretofore attracted little historical attention among colonial historians.[26] The skills required to build small-scale stone-arch bridges certainly existed in Pennsylvania during the early eighteenth century, and considerable data on the construction of a 1730s masonry bridge in Philadelphia's dock district still survive. The technology of building simple, small-scale stone bridges was widely practiced in Britain during the eighteenth century, and during this era basic design treatises for stone-arch construction became available in Europe. As these books migrated to the New World, builders could have undertaken bridge projects with increased confidence in the theoretical considerations underlying their work.[27] However, small-scale stone-arch bridge construction constitutes a

---

Convention to Mount Vernon. His diary entry of Sept. 19, 1787, reported: "Lodged at the head of Elk [Elkton, Md.]—at the bridge near to which my horses (two of them) and Carriage had a very narrow escape. For the rain which had fallen the preceeding evening having swelled the Water considerably there was no fording it safely. I was reduced to the necessity therefore of remaining on the other side or of attempting to cross on an old, rotten and long disused bridge. Being anxious to get on I preferred the latter and in the attempt one of my horses fell 15 feet at least the other very near following which (had it happened) would have taken the Carriage with baggage along with him and destroyed the whole effectually [*sic*]. However, by prompt assistance of some people at a Mill just by and great exertion, the first horse was disengaged from his harness, the 2d. prevented from going quite through and drawn off and the Carriage rescued from hurt." Donald Jackson and Dorothy Twohig, eds., *The Diaries of George Washington*, V, *July 1786–December 1789* (Charlottesville, Va., 1979), 186.

26. A significant exception is Harley J. McKee, *Introduction to Early American Masonry: Stone, Bricks, Mortar, and Plaster* (Washington, D.C., 1973).

27. Harrold E. Gillingham, "The Bridge over the Dock in Walnut Street," *PMHB*, LVII (1934), 260–269. The stone-arch bridge over Pennypack Creek in the Frankford section of Philadelphia is reputed, in part, to date to 1697. An early 20th-century view of this structure appears in Faris, *Old Roads out of Philadelphia*, 290. Ted Ruddock, *Arch Bridges and Their Builders, 1735–1835* (Cambridge, 1979), provides an excellent illustrated history of stone-arch bridges in Great Britain but does not cover North American bridges. Examples of stone-arch design books include Stephen Riou, *Short Principles for the Architecture of*

conservative technology that also can be successfully implemented using traditional, empirically based methods of building. The typical compressive strength of any stone used in construction is so great that the stresses induced in arches with spans of less than one hundred feet never approach critical limits. In this context, early American stone-arch bridges did not involve any daring feats of engineering skill, and probably for that reason histories of American bridge construction often ignore them.

Although the arch spans of the stone bridges built to service Pennsylvania's early roads and turnpikes were not particularly noteworthy (at least by contemporary European standards), several structures were quite substantial in their overall size. For example, in the 1790s, Abraham Witmer constructed a nine-span stone-arch bridge over Conestoga Creek in Lancaster with an aggregate length exceeding three hundred feet. Similarly, in the late 1790s local authorities in Montgomery County built a six-span stone-arch structure over Perkiomen Creek near Collegeville with a total length approaching the size of Witmers Bridge. Although less massive than the Perkiomen and Witmer bridges, numerous other stone-arch spans were erected in the late eighteenth and early nineteenth centuries as key components of the region's highway system. By the mid-nineteenth century, covered wooden bridges would develop into a truss technology widely used for many small river crossings, but at the beginning of the century this proliferation had yet to occur.[28]

Stone-Bridges (London, 1760); and Charles Hutton, *The Principles of Bridges: Containing the Mathematical Demonstrations of the Properties of the Arches, the Thickness of the Piers, the Force of the Water against Them, etc. . . .* (London, 1801).

28. Charles I. Landis, "Abraham Witmer's Bridge," Lanc. Co. Hist. Soc., *Historical Papers and Addresses*, XX (1916), 155–174. See Faris, *Old Roads out of Philadelphia*, 79, 270, 300, for photographs of the three-span arch bridge over Darby Creek on Baltimore Pike a few miles south of Philadelphia, the seven-span arch bridge over Nesheminy Creek on Old York Road, and the three-span arch bridge over Poquessing Creek on Bristol Pike in Torresdale.

The covered wooden-truss bridge commonly associated with popular conceptions of a Currier and Ives Americana did not begin to proliferate until the 1820s, largely because extensive amounts of saw milling were required to produce the cut timber necessary for each structure. Covered wooden bridges, in which the main structural members were protected by roofs and clapboarding, were built as early as the first decade of the 19th century, but it took many more years before the technology flourished. In the 1950s, Richard Sanders Allen inventoried more than 500 existing wooden covered bridges in New England and the mid-Atlantic region and reported that the oldest surviving structure dated to 1825; fewer

Taken together, the use of stone for pavement surfaces and for permanent arch bridges constitutes one of the most distinctive features of early turnpike construction in southeastern Pennsylvania. Neither of the uses represents a particularly complicated form of technology, but both offered a way to dramatically improve the material conditions of overland travel. The utilization of stone and masonry is easy to overlook when studying the history of modern industrialization and transportation, because it seems so ancient and primitive in its origins. However, technology need not be complicated to be effective. Without being too fanciful, road construction in the region can aptly be characterized as representing an aspect of America's Stone Age of technological development that complemented the more widely recognized Wooden Age documented by Brooke Hindle and others.

Pennsylvania's Early Road System

When Europeans first settled in the Delaware Valley in the seventeenth century, they made use of existing Indian paths and trails for their early overland travels.[29] After William Penn set out to promote settlement of his colony in the 1680s, such trails continued to form the basis of a primitive highway system. With most colonists residing near the Delaware River, waterborne traffic played a significant role in the commercial life of early Pennsylvania, and it reduced any immediate need for extensive effort to be placed on improving roadways. However, as farmers moved inland, farther away from large waterways, the colonial government directed more attention to developing a transportation system capable of serving all colonists, not just those located close to navigable rivers.[30]

---

than 10% predated 1840. See Richard Sanders Allen, *Covered Bridges of the Northeast* (Brattleboro, Vt., 1957), 108–113; and Allen, *Covered Bridges of the Middle Atlantic States* (Brattleboro, Vt., 1959), 107–114.

29. The early history of Pennsylvania trails is described in Paul A. W. Wallace, "Historic Indian Paths of Pennsylvania," *PMHB*, LXXVI (1952), 411–439.

30. Several early actions by the Provincial Council of Pennsylvania to authorize road construction were approved at a "Councill held att philadelphia die Jovis, 29th October, 1696." These included an "order for the Laying outt a sufficient Road, the nearest and best that may be had from the Lowermost ferry upon Skuilkill, Comonlie called Benjamin Chambers' ferry, Into the Town of Philadelphia," and an "ordr for Laying outt a Road from

During the colonial era two basic types of roadways were authorized by government authorities. For a few heavily traveled roads designed to tie Philadelphia into a nascent intracolonial road network, the Provincial Council authorized the construction of Kings Highways. This type of thoroughfare included the Great Southern Road leading south from Philadelphia through Chester to Wilmington, Delaware, and its northern counterpart, the Bristol Road, which extended up the Delaware River toward Trenton and points north. Other routes authorized by the Provincial Council included the Great Wagon Road west from Philadelphia to Lancaster and the York Road, which extended north from Philadelphia through Doylestown and crossed the Delaware River at New Hope. In adhering to English traditions, most local roads were authorized by the county-based General Court of Quarter Session, which entertained petitions from landowners desiring roads, appointed "viewers" to examine the locations of proposed roadways, and established legally binding highway rights-of-way based on the reports and surveys submitted by these viewers. For both Kings Highways and roads authorized by the Court of Quarter Sessions, the responsibility for construction and maintenance fell to local authorities under the jurisdiction of the county courts. Before 1762 all male inhabitants within Pennsylvania were technically required to spend at least a few days per year laboring to clear and repair roads. After 1762 this road-work requirement could be fulfilled by the payment of a property tax to township officers, who in turn used the tax proceeds to hire laborers.[31]

Up to the time of the American Revolution practically all transportation responsibilities were carried out by locally based officials. A major exception concerned the financing and construction of the Military Road, or Forbes Road, which led west from Carlisle to the Ohio Valley (that is, Fort Pitt). This road facilitated the transport of army troops during the Seven Years' War (1756–1763) and was built directly by the provincial government without recourse to local authorities. In addition, some ferrying operations, such as Wright's Ferry across the Susquehanna River at the present-day town of Wrightsville, were administered by nongov-

New worke, in Newcastle Countie . . . to the king's road in Chester Countie" along a route "the nearest and most convenient that may be had, and Least prejudicial to the Lands and Improvements of the nighbourhood." *Minutes of the Provincial Council of Pennsylvania*, I [Mar. 10, 1683–Nov. 27, 1700] (Philadelphia, 1852), 499.

31. Wilbur C. Plummer, *The Road Policy of Pennsylvania* . . . (Philadelphia, 1925), 7–25.

ernmental authorities in furtherance of the colony's overall transportation development.[32]

After the Revolution the basic nature of Pennsylvania's transportation systems remained unchanged, with local township and county officials retaining control over most roadways. Beginning in the 1780s some direct state subsidies were authorized for the construction of highways by both the commonwealth itself and individual counties. At the same time, county and township authorities acting through the auspices of local courts continued to lay out, approve, and construct roadways. In addition to other means of raising revenue, in the 1780s the state legislature experimented with lotteries as a way to finance highway construction. During the eighteenth century, lotteries were a popular method of obtaining money from citizens who would otherwise resist attempts to increase taxation. However, lotteries apparently played a minor role in financing of highway construction, especially when compared with the movement to privatize road building through the formation of state-chartered turnpike companies. As stated earlier, interest in shifting financial responsibility for road construction and maintenance away from the public sector toward private investors began in seventeenth-century England. However, privately financed turnpike companies were not authorized in the American colonies before the Revolution. In the 1780s some proprietors, such as Abraham Witmer in Lancaster, were authorized by the state legislature to build and operate toll bridges for their personal benefit (and at their own risk). By the early 1790s the desire by merchants and the traveling public to improve the commonwealth's transportation system, while also allowing investors to make money, led to the chartering of the first privately financed turnpike in the United States.[33]

Officially chartered in 1792, the Philadelphia and Lancaster Turnpike Road Company quickly raised $465,000 from investors in the two termi-

---

32. A good synopsis of the history of Wright's Ferry illustrated with early 19th-century engravings appears in Gerald S. Lestz, *Artists' Album / Lancaster County* . . . (Ephrata, Pa., 1988), 64–71.

33. On state-chartered turnpike companies, see Plummer, *Road Policy of Pennsylvania,* 26–46. The financial and legal status of Witmer's toll bridge is discussed in Landis, "Witmer's Bridge," Lanc. Co. Hist. Soc., *Historical Papers and Addresses,* XX (1916), 155-174. The original charter for the Philadelphia and Lancaster Turnpike Company is reprinted in Gallatin, "Report on Roads and Canals," in Cochran, ed., *American State Papers: Transportation,* I, 230–235.

nal cities to pay for the new facility. Running sixty-two miles from the west bank of the Schuylkill River in Philadelphia to the stone-arch Witmers Bridge over Conestoga Creek in Lancaster, the new turnpike closely followed the old Great Wagon Road originally laid out between the two communities in the 1730s. Rather than assume responsibility for clearing a new right-of-way, the turnpike company poured its money into the placement of a stone pavement along the length of the roadway and financed construction of a new, three-span stone-arch bridge across Brandywine Creek at Downingtown. By late 1795 the turnpike was open to all travelers willing to face the open palms of fifteen tollkeepers stationed roughly every five miles along the right-of-way.[34] The commercial success of the undertaking soon became evident, and within a few years other communities and other investors were clamoring for turnpike charters of their own.

Despite the size and prominence of the original Philadelphia and Lancaster Turnpike, very little is known about the technical skill and engineering effort that went into the highway's construction. Conceivably, the renowned British engineer William Weston played a role in planning and implementing the turnpike project. Certainly by the time work on the road started, Weston had become a fixture of the Philadelphia region's public works improvement effort, particularly in regard to the development of canals along the Schuylkill River. However, there is no surviving historical documentation related to the engineering of the Lancaster Turnpike or to Weston's possible association with this prominent project. As an alternative, this essay turns away from the Philadelphia and Lancaster Turnpike and focuses on the Germantown and Perkiomen Turnpike Road Company, an organization that received a charter from the state in February 1801.[35]

---

34. Charles I. Landis, "Philadelphia and Lancaster Turnpike," *PMHB,* XLII (1918). Additional data are also presented in Landis, "The First Long Turnpike in the United States," Lanc. Co. Hist. Soc., *Historical Papers and Addresses,* XX (1916), 205–340.

35. James T. Mitchell and Henry Flanders, comps., *The Statutes at Large of Pennsylvania, from 1682 to 1801* (Harrisburg, Pa., 1896–1911), XVI, 525–540, contains a copy of the chartering legislation passed on Feb. 12, 1801. Weston's involvement with projects such as the Delaware and Schuylkill [Canal] Company and the Schuylkill and Susquehanna [Canal] Company is documented in Darwin H. Stapleton, *The Transfer of Early Industrial Technologies to America* (Philadelphia, 1987), 52–68. Based on the proximity of these projects to the Lancaster Turnpike and the status that Weston held as a professionally trained civil engineer from Britain, it is reasonable to believe that his expertise would have been called upon in the construction of a prominent, and costly, toll road. However, no direct evidence

The planning that preceded construction of this latter turnpike is well documented, thus it is possible to use this road to help assess the nature of turnpike engineering in southeastern Pennsylvania at the beginning of the nineteenth century.

Planning the Germantown and Perkiomen Turnpike

First settled by German immigrants beginning in the 1680s, Germantown is a community about ten miles northeast of the colonial center of Philadelphia on the northern side of the Schuylkill River.[36] The road from Philadelphia to Germantown constituted one of the colony's most traveled thoroughfares because it carried both intracommunity traffic as well as regional traffic directed toward the upper reaches of the Schuylkill River valley. The road between Germantown and Philadelphia quickly developed into a heavily traveled transportation corridor while the route beyond Germantown (which developed out of an old Indian trail and was usually known as Manatawny Road or Reading Road) served to connect the agricultural hinterland of the Schuylkill Valley with the commonwealth's major urban center.

The charter for the Germantown and Perkiomen Turnpike Road Company authorized construction and operation of a twenty-six-mile-long toll road extending from Philadelphia to Germantown (a distance of about ten miles) and thence another sixteen miles through rural Montgomery County to the newly completed stone-arch bridge over the Perkiomen Creek at Collegeville. This toll bridge had been financed by state and county authorities in the 1790s, and, after opening for service in November 1799, the masonry structure provided a major impetus for upgrading the roadway between Perkiomen Creek and Philadelphia. Thus, just as the Philadelphia and Lancaster Turnpike was designed to bond agricultural

---

of Weston's association with the Philadelphia and Lancaster Turnpike Company is known to exist.

36. For more on the history of 17th- and 18th-century Germantown, see Stephanie Grauman Wolf, *Urban Village: Population, Community, and Family Structure in Germantown, Pennsylvania, 1683–1800* (Princeton, N.J., 1976); the early road history of Germantown is discussed on 23–26. Wolf notes that the original route for the Germantown Road (which developed into the turnpike) evolved from an old Indian trail.

interests in the Lancaster area with Philadelphia, the Germantown and Perkiomen Turnpike was intended to facilitate commercial interchange between the northern part of the Schuylkill Valley and the Germantown-Philadelphia economic community. To finance construction of the improved roadway, the company raised $285,000 in capital from scores of private investors.[37] Stephanie Grauman Wolf notes that "summer" residents from Philadelphia who built homes in Germantown in the late eighteenth century (in a migration encouraged by yellow fever epidemics) were the primary backers of the new turnpike company, and "in 1798, this wealthy group of 'newcomers' engineered the passage of a turnpike bill in the state legislature, which turned the whole community [of Germantown] irrevocably into a suburb of its large, metropolitan neighbor."[38]

Benjamin Chew, who served as president of the Germantown and Perkiomen Turnpike Road Company, was most prominent of these English Quaker newcomers and lived in the community's "most elaborate house." The new turnpike was opposed by some local tradesmen who anticipated that the improved road could attract regional commerce into Philadelphia (only five miles away) that previously centered on Germantown. However, Chew and other summer residents who invested in the turnpike welcomed the transfer of commercial activity to Philadelphia, presumably because it would increase toll revenue but also because they sought to keep Germantown "as quiet and rural as possible" and preserve an environment suitable for a wealthy country gentry.[39]

For engineering expertise in planning the new turnpike, the company turned to Robert Brooke, a thirty-one-year-old surveyor who, according to Benjamin Latrobe, had worked with William Weston in building the aborted Delaware and Schuylkill Canal in the mid-1790s. Born in 1770 in The Trappe, a village near Collegeville in the Schuylkill Valley, Brooke

---

37. F. G. Hobson, "Perkiomen Bridge—When and by Whom Built—Lottery—History up to Date," Historical Society of Montgomery County, Pennsylvania, *Bulletin*, X (1956), 121–130. Gallatin, "Report on Roads and Canals," in Cochran, ed., *American State Papers: Transportation*, I, 226–235, describes the Germantown and Perkiomen Turnpike and lists the capital expended on construction at $285,000. Cost breakdowns for each five-mile stretch of the road further document these expenditures. A complete listing of the more than 200 original stockholders in the Germantown and Perkiomen Turnpike Road Company is provided in Norman Keyser, "Early Transportation in Germantown," in *Germantown History* (Germantown, Pa., 1915), 42–43.

38. Wolf, *Urban Village*, 14.

39. Wolf, *Urban Village*, 36, 54–55.

descended from an English family that immigrated to Pennsylvania in 1699.[40] A fourth-generation American, Robert Brooke attended the College of Pennsylvania (subsequently the University of Pennsylvania) during 1792, and, although he never formally graduated, he is listed in university records as being a member of the class of 1793. During his year in college he apparently concentrated his efforts on surveying and mathematical courses because of their relevance to his later professional work as an engineer and surveyor.[41] In contrast with many land surveyors of the colonial

40. See Daniel Hovey Calhoun, *The American Civil Engineer: Origins and Conflict* (Cambridge, Mass., 1960), 20, for evidence of Latrobe's stating in 1816 that Brooke "undoubtably acquired under him [William Weston], much of the valuable knowledge he possesses." The Brooke Family Biographical File in the Montgomery County Historical Society, Norristown, Pa., contains data on the genealogy of the Brooke family tree. Also see Wilfred Jordan, *Colonial and Revolutionary Families of Pennsylvania: Genealogical and Personal Memoirs* (New York, 1933–), IV, 176–177, VI, 704–706, VIII, 406–407, XII, 608–610. Succinctly put, Robert Brooke's great-great-grandfather John Brooke immigrated to Pennsylvania from Yorkshire in 1699, bringing his son Matthew with him. Matthew fathered William Brooke, who in turn fathered John Brooke in Limerick, Pa., in 1740. John Brooke married Elizabeth May in 1762, and their union resulted in the birth of Robert Brooke in 1770. During the Revolutionary war John Brooke served as captain in the Sixth Battalion of the Pennsylvania Militia; after the war he returned to his farm in the Limerick area, where he lived until his death in 1815. See "Nineteenth Century Real Estate Offerings," Hist. Soc. of Montgomery Co., *Bulletin*, X (1956), 249, for notice of Robert Brooke's participation in the sale of his late father's estate.

41. See the *University of Pennsylvania Biographical Catalogue of the Matriculates of the College, 1749–1893* (Philadelphia, 1894), 33; and W. J. Maxwell, comp., *General Alumni Catalogue of the University of Pennsylvania* (Philadelphia, 1917), 24, for evidence of Robert Brooke's matriculation to the University of Pennsylvania. During this period it was not unusual for students to enter the school at the age of 15 or 16. Brooke would have been 22 when he started, making him older than many of his classmates.

For more on the school's curriculum offerings in the 18th century, see Francis James Dallett, comp., *Guide to the Archives of the University of Pennsylvania, from 1740 to 1820* (Philadelphia, 1978). In particular see references to the Jasper Yeates student notebook (file #1654) described on p. 131; also see Edward Potts Cheyney, *History of the University of Pennsylvania: 1740–1940* (Philadelphia, 1940), 82–86, for evidence that the school's offerings included courses on surveying, conic sections, and "fluxions" during the period of Brooke's attendance. A notebook attributed to Robert Brooke from his days as a student at the College of Pennsylvania in the Rawle Papers, HSP, indicates that he obtained considerable training in spherical trigonometry and other sophisticated mathematical techniques of use to a skilled surveyor.

era, Brooke gained proficiency in measuring ascents and descents, a skill necessary for laying out gradients associated with all types of overland transportation systems, including highways, canals, and railroads.[42]

By 1800 Brooke was operating as a professional land surveyor based out of his home in the Northern Liberties section of Philadelphia. In January 1802 the officers of the Germantown and Perkiomen Turnpike Road Company requested him to "take ascents and descents necessary to report thereon with Draft [map] and Profile" for the section of the turnpike between milestone No. 5 and "the Fork of the Road on Chestnut Hill." This survey work was to cover the second five-mile section of the turnpike measured out from Philadelphia. Brooke may have previously surveyed the first five miles of the turnpike, but documents relating to such work do not survive. From Philadelphia to the Chestnut Hill section of Germantown the route of the new turnpike was, at least in the collective mind of the turnpike company, already determined to be the existing right-of-way of the old Germantown Road. Consequently, Brooke was asked to report only upon the gradients associated with this route. In April 1802 Brooke accompanied the officers of the company "for the purpose of inspecting the 2nd section of 5 miles of the Road," and upon the completion of this field trip he received a new, more broadly conceived mission, focusing on the remaining distance between Chestnut Hill and the Perkiomen Bridge.[43]

On April 15, the turnpike company authorized Brooke to survey the remaining distance to be traversed by the toll road. In particular, he was to examine two routes, "the one called the Reading Road, the other passing the house of John Huston, Esq.," for the purpose of determining the distance associated with each route. Along with preparing detailed survey data and sketch maps, Brooke was directed to "take a general view of the intermediate country and report his opinion" on whether a new right-of-

42. Love, "The Colonial Surveyor," 15–155, describes the techniques and training involved in surveying work during the 18th century. Knowledge of the actual techniques employed by Brooke in carrying out his surveying work is speculative because there is no record of the instruments he possessed.

43. J. M. Saul to Mr. Brooke, Jan. 28, 1802, Germantown and Perkiomen Turnpike Road Company to Mr. Brooke, Apr. 8, 1802, Germantown and Perkiomen Turnpike Road Company Papers, HSP. All material on Brooke's association with the company is drawn from this source. J. M. Saul served as secretary for the turnpike company. Brooke's public status as a surveyor is documented in *The New Trade Directory For Philadelphia, Anno 1800* (Philadelphia, 1799), 171. Northern Liberties is north of Philadelphia's old center city.

way could be laid "in part on a new line [or] partly on either of the present roads so as to gain [that is, reduce] distance" for the new turnpike. In addition, Brooke was formally encouraged "to report other observations as may occur to him, so as to promote the interests of the Company and the Publick."[44] Clearly, the nature of Brooke's new assignment differed significantly from his earlier task of surveying the second five-mile stretch of road. Now he had the responsibility for examining at least two distinct routes and recommending what the best course of action would be for the turnpike company. Brooke immediately set out to study the terrain between Chestnut Hill and the Perkiomen Bridge, a region of southeastern Pennsylvania that was not far removed from his birthplace and one that he probably already knew quite well. Working quickly, he submitted a formal report to the company on May 12, less than a month after receiving the assignment.

Brooke's May 12 report is several pages long and distills down a large amount of data related to surveys of various routes.[45] Along with precise measurements of distance, the report discusses numerous other factors relevant to deciding on an ideal route for the turnpike. Brooke focused attention on four potential rights-of-way for the road; two of them represent the Reading Road and the John Huston route noted in the company's directive of April 15 while the others concerned a cutoff of the Reading Road route and a completely new route of Brooke's own devising. Aside from issues of distance, Brooke reported on the nature and availability of stone. ("From Wissahickon Creek to Barren Hill, a light slaty gravel, some appearance of soft sandstone quarries. From thence to the cross lane below Hickory Town, marble and limestone quarries. And from thence to Hickory Town hard sandstone.") He addressed the suitability of foundations necessary for potential new stone-arch bridges. ("The Wissahickon Creek crosses the proposed [Huston] Rout[e] very obliquely. Bad foundation for a bridge—Valley wide—Banks low and falling off into a springy swamp on the west side—Building stone inconvenient.") However, the basis for Brooke's ultimate recommendations extended beyond strictly technical issues and encompassed a broader view of his role as the company's engineer.

44. Resolution in the Minutes of the Germantown and Perkiomen Turnpike Road Company, Apr. 15, 1802, Germantown and Perkiomen Turnpike Papers.

45. Robert Brooke to President and Managers of the Germantown and Perkiomen Turnpike Road Company, May 12, 1802, *ibid.*

Within the context of deciding between the Reading Road and Huston routes, Brooke advocated laying out a hybrid of the former road that would cut out a needlessly circuitous section and reduce the turnpike's length by eighty poles (approximately 1,300 feet, a pole being 16.5 feet long). This adjusted right-of-way would necessitate bypassing Hickorytown, a village about halfway between Germantown and the Perkiomen Bridge, and Brooke recognized that this might have political ramifications. However, he did not feel constrained to recommend only what he thought would prove politically viable. Brooke actively embraced the company's directive to make "observations . . . so as to promote the interests of the Company and the Publick" and took the initiative in proposing a completely new, "more advantageous" route.

During the colonial era two primary roads extended out of Philadelphia on the north side of the Schuylkill Valley. One of these passed through Germantown and constituted the basis of the first two sections of the new turnpike. The second, known as the Ridge Road, ran closer to the river and passed near the area that would later become the community of Manayunk. The old Germantown Road and the Ridge Road came within a few hundred yards of one another a short distance past Chestnut Hill, and at this point the Reading Road could easily draw traffic from the Ridge Road. Beyond this de facto confluence of the old Germantown Road and the Ridge Road, the latter route continued several miles further to the community of Norristown located on the banks of the Schuylkill River. What Brooke proposed was to bypass practically the entire length of the old Reading Road above Chestnut Hill by rerouting the turnpike along the upper section of the Ridge Road leading into Norristown. From Norristown he proposed constructing a new road running along a straight line to the Perkiomen Bridge:

> The ground along this route is generally firmer, less liable to wet weather springs, and not intersected by so many small streams of water. It is more regular in its surface and course and affords better materials than the Hickorytown route . . . the only real objection . . . to this [new] route is the expense of building a bridge over the Skippack [Creek].

Obviously the technical advantages associated with the terrain traversed by the Ridge Route appealed to Brooke. But, just as important, he stressed the economic advantages of the new right-of-way:

I am also of the opinion that it would be in the interest of the company to take the Norristown Road . . . on account of the increased toll, for if you took the Hickory Town Road you leave the lime kilns, marble quarries and the Norristown and New Lancaster Road [a newly improved road connecting Lancaster with Norristown] so far to the left as nearly to exclude all the travelling from that quarter.

Noting that the route through Hickorytown would primarily provide transportation access for farmers, he contrasted this with the advantages of running the turnpike through "the marble and lime stone quarry country" and asserted that the "Ridge Route" through Norristown would be "an acquisition to Germantown and Chestnut Hill by increasing the trade [carried by the turnpike]."

After receiving Brooke's May 12 report advocating adoption of a right-of-way through Norristown, the turnpike company declined to make any immediate decisions concerning his proposals. By August, the company decided to authorize a more complete survey of the Ridge Route, and on September 22 Brooke submitted a detailed study of the right-of-way.[46] This survey indicated that the new route was slightly shorter (by 26.45 poles, or about 400 feet) than the Reading Road through Hickorytown and presented no difficult grades that might prove troublesome to wagon traffic. Brooke also reported that "tolerable good building stone" for erecting a new bridge over Skippack Creek was readily available, and, in fact, quarrying of this stone could help reduce the height of the road as it passed over Skippack Hill. Bolstering the arguments first presented in his May 12 report, Brooke took care to highlight offers of free stone made by several landowners who held property near the Ridge Route. For example:

Mr. Moore, a stone cutter . . . says if the Board should conclude to take the Norristown Route . . . they may raise and take as much marble and limestone from his quarries (which are about 20 or 30 poles from the road) as they may want for the road, gratis . . . [and] Mr. Alexander Crawford also says that the Board may raise and take away as much limestone as they may want for the road, gratis . . . [Mr. John Markley] also says that the Board may have as much gravel and stone out of his land as they want for the road without any charge.

46. Robert Brooke to President and Managers of the Germantown and Perkiomen Turnpike Road Company, Sept. 22, 1802, *ibid.*

Brooke's forceful advocacy of a turnpike right-of-way that would avoid the old Reading Road might have elicited support from landowners along the Ridge Road to Norristown, but it did little to comfort commercial interests in the community of Hickorytown. Hickorytown stood to become an isolated, backwater community if the turnpike opted for the Ridge Road right-of-way, or even if the Reading Road was realigned to decrease the turnpike's length by thirteen hundred feet. Neither prospect prompted enthusiasm among the village's merchants and residents. In late November 1802 the company requested Brooke to attend a meeting of the turnpike managers, apparently for the purpose of responding to criticisms raised at a November 10 meeting of Hickorytown residents. After meeting with the managers on December 2, Brooke provided a written defense of his previous recommendations and specifically referred to a report prepared by Andrew Norney, a resident of Hickorytown, which denigrated the accuracy of Brooke's survey work. Brooke did not back down from his belief that routing the turnpike through Norristown was in the company's best interest, but the managers clearly were under pressure to drop the idea of bypassing Hickorytown. On December 16 the turnpike company held another meeting with Brooke to discuss the situation, and, although records related to this conference do not survive, the outcome of criticisms to Brooke's earlier reports soon became clear.[47] Simply put, the right-of-way for the Germantown Turnpike as built in 1803–1804 closely followed the right-of-way of the Reading Road, which, in turn, followed an old Indian trail.

Despite the reluctance of the Germantown and Perkiomen Turnpike Road Company to adopt the Ridge Route, or even straighten out sections of the old Reading Road right-of-way, Brooke did not break off relations with the company. Although he apparently played no role in actually constructing the turnpike's stone pavement or masonry bridges (these tasks were handled by several independent contractors and stone masons), Brooke prepared another, more detailed, survey of the route through Hickorytown in April 1803. In early 1803 he also assumed responsibility for surveying two major feeder roads that connected into,

47. John Johnson to Robert Brooke, Nov. 25, 1802; Nov. 10, 1802, meeting in Hickorytown, noted in Robert Brooke to President and Managers of the Germantown and Perkiomen Turnpike Road Company, Dec. 6, 1802, *ibid*. A brief notice announcing the Dec. 16, 1802, meeting survives within the Germantown and Perkiomen Turnpike Road Company Papers. Records related to the meeting itself do not survive.

and served the economic interests of, the Germantown Turnpike.[48] The Cheltenham and Willow Grove Turnpike ran along the right-of-way of the old York Road (modern-day State Route 611) while the Chestnut Hill and Springhouse Turnpike extended along the·established route running from Germantown to Bethlehem (old State Route 309). These two toll roads were in operation by 1805, helping to supplement traffic on the Germantown Turnpike. In 1806 Brooke received a commission to resurvey the entire length of the existing Philadelphia and Lancaster Turnpike. Beyond his turnpike work, Brooke helped lay out the Chesapeake and Delaware Canal, which connected the Susquehanna River watershed with Delaware Bay, and he remained involved in water improvement projects along the length of the Schuylkill River. In 1821, at the age of fifty-one, Brooke's engineering career came to a premature end when he passed away in Philadelphia after suffering a "lingering illness."[49]

Robert Brooke is not remembered as a great contributor to America's

48. Robert Brooke's involvement in later surveys of the Germantown Turnpike and his work surveying the two new "feeder" turnpikes is documented in the Germantown and Perkiomen Turnpike Papers and in the Cheltenham and Willow Grove Turnpike Road Company Record Book Papers, HSP. On independent contractors: *Aurora, General Advertiser* (Philadelphia), Mar. 24, 1804, includes an advertisement entitled "Contractors Wanted" that solicits "all persons desirous of entering into contract for forming and making the road between the 15 mile stone and the bridge over Perkiomen creek, or any part thereof . . . no contractor to undertake more than half a mile at one time—the contractors to find their own tools—the road to be completed early in the month of October next."

49. A description of the survey Robert Brooke commenced on Nov. 3, 1806, is available in the Philadelphia and Lancaster Turnpike Road Papers, HSP. Brooke's involvement with the Chesapeake and Delaware Canal is documented in the Robert Brooke Collection, Hagley Library and Museum, Wilmington, Del. This file contains copies of survey records made by Brooke that are retained by the New York Public Library. Brooke's work along the Schuylkill River in 1816 is noted in Charles Peterson, "The Spider Bridge: A Curious Work at the Falls of the Schuylkill, 1816," *Canal History and Technology Proceedings*, V (1986), 243–259. Peterson's article contains a circa 1816 drawing made by Brooke showing the layout of Josiah White's wire mill at the Falls of the Schuylkill. Brooke's work in surveying the Union Canal between the Schuylkill River and the Susquehanna River is noted in the *Report of the President and Managers of the Union Canal Company of Pennsylvania to the Stockholders* (Philadelphia, 1818), 9–11.

Brooke's death on Nov. 4, 1821, is noted in *Poulson's American Daily Advertiser*, Nov. 6, 1821. The newspaper indicates that he "died on Sunday evening the 4th inst. after a lingering illness."

early technological development, and the purpose of discussing his work in planning the Germantown and Perkiomen Turnpike is not to help elevate him to such status. Rather, the point is to highlight the care and insight that one early nineteenth-century turnpike surveyor brought to his work and the engineering sophistication he exhibited in conceiving what he considered to be the best route to cover the distance between Chestnut Hill and the Perkiomen Bridge. In analyzing this problem, Brooke followed a holistic approach that considered the influences of terrain, geology, material supply, and market demands. As such, he was not content merely to survey a predetermined route and hand over his data to the company. Instead, Brooke considered his assignment within a broader economic context that went beyond completing a mere technical assignment and involved assessments akin to those we frequently associate with engineers of the twentieth century.

Ultimately, Brooke's work to develop what he considered a more economically rational and technically superior right-of-way was rendered moot by local political maneuverings. However, Hickorytown residents who resisted his proposals were not acting irrationally or in bad faith; they sought only to protect economic interests derived from the manner in which the Reading Road had been laid out during the colonial era. We can never know whether the plans to meld the upper Ridge Road into the lower Germantown Turnpike would have borne out Brooke's optimistic prognostications. But Brooke's efforts to eliminate some of the inefficient twists along the upper Germantown Turnpike that were obviated by lobbying from Hickorytown residents are easy to appreciate from the perspective of twentieth-century motorists who still must drive along roadways with rights-of-way first blazed as Indian trails.

If nothing else, Brooke's experience can serve as evidence that as early as 1802 the politics of highway development were already a prime determinant of road design. Today, the seemingly idiosyncratic layout of old highways can be readily accepted as convincing proof that early builders had little understanding of how to plan or build good roadways. This reinforces the idea that early roads were always poorly designed and that the Turnpike Movement constituted a transportation failure that necessarily required the development of canals and railroads to rectify. In contrast, Brooke's work with the Germantown and Perkiomen Turnpike, although it did not prove politically feasible for the company to implement fully, indicates that early nineteenth-century road planning in southeastern Pennsylvania had reached a highly developed level of skill and understanding on the part of at least one engineer-surveyor. The approach advanced by Brooke also demonstrates that efforts to rationalize transportation development in the Federalist and Jeffersonian era were not confined to planning undertaken in a broad-based, national context (perhaps best exemplified by Gallatin's 1808 "Report on Roads and Canals"). In this sense, Brooke's work should not be seen as unique or distinctive, but rather as symptomatic of more general trends in early nineteenth-century American society.[50]

Turnpike Economics and Cultural Significance

A key issue underlying any consideration of early turnpikes concerns the economic viability and profitability of their operation. It is easy to portray the evolution of turnpikes-to-canals-to-railroads as resulting from the economic deficiencies of the transportation method being replaced. Although turnpikes and roads could not realistically compete with canals or railroads in the transport of heavy bulk shipments (like coal), highway traffic hardly came to a halt in the 1830s. No matter how widespread canal and railroad construction became, active trunk lines and branches of these systems ultimately served only a fraction of the United States. True, the regions and markets directly served by canals and railroad were critically important to the United States economy, and their significance should not be denigrated in an attempt to inflate the value of turnpikes. However, operations of canals and railroads expanded upon and supplemented, rather than completely replaced, highway systems. The former technologies might have served as the nation's transportation arteries, but roads were the capillaries that pumped life into many regional economic systems.

In assessing the value of turnpikes and roads to economic development, historians are limited by the paucity of surviving statistical data. Although in-depth financial records for early nineteenth-century turnpikes cannot be located for the great majority of companies, many did submit reports to state legislatures on a semiregular basis. From such statements (and other

---

50. Jefferson himself, of course, is remembered for advocating a variety of initiatives to rationalize and improve the character of American economic life, including coinage and a uniform system of weights and measures. See Silvio E. Bedini, *Thomas Jefferson: Statesman of Science* (New York, 1990), 203–206.

scattered documents) it is possible to assess the basic profitability of some
Pennsylvania turnpikes in both the pre- and postcanal era. The tolls authorized by the state legislature for the Philadelphia and Lancaster Turnpike
Company in 1794 established a standard for later turnpikes and indicate
the manner in which travel was taxed by Pennsylvania toll roads. Someone walking along the Lancaster Turnpike would not have been subject to
tolls, but practically every other use of the road entailed cash payment. The
rate schedule lists several of the toll categories and the payment required
for each ten miles of travel.[51]

Horse and Rider:	6.25 cents
	(¹⁄₁₆th of a dollar)
Score of Sheep:	12.5 cents
	(⅛th of a dollar)
Score of Hogs:	12.5 cents
Score of Cattle:	25 cents
Wagon, charge for each horse (2 oxen = 1 horse):	
Wheels over 12 inches in width:	2 cents
Wheels between 10 and 12 inches:	3 cents
Wheels between 7 and 10 inches:	5 cents
Wheels between 4 and 7 inches:	6.25 cents
Wheels less than 4 inches:	12.5 cents

The toll rate correlated closely with the damage that a particular form of
travel might inflict upon the roadway and is reflected in the differentiated
toll rates for wagons depending upon the width of their wheels. Narrower
wheels were much more likely to gouge out ruts; thus they engendered a
higher toll than a similar wagon with wider wheels that actually might
work to smooth out and compact the road surface.

51. Legislation authorizing the Philadelphia and Lancaster Turnpike Company, including a listing of toll rates, appears in Cochran, ed., *American State Papers: Transportation*, I,
240–245. These tolls were to be received from "any person riding, leading or driving any
horses, cattle, hogs, sheep, sulkey, chair, chair, chaise, phaeton, cart, waggon [*sic*], wain,
sleigh, sled, or other carriage of burden or pleasure."

Although some turnpike projects proved to be financially weak, many provided steady, if not spectacular, dividends for dozens of years. For example, both the Philadelphia and Lancaster Turnpike and Germantown and Perkiomen Turnpike were reliable sources of income for their investors. The Lancaster Pike cost approximately $465,000, and for the first years of its operation (1796–1802) its annual tolls increased from approximately $11,000 to more than $20,000.[52] However, repair expenses came to $49,361 for this same period, which, in concert with wages for the tollkeepers, limited dividends to about 1.85 percent per year. For the next five-year period (1802–1807) tolls accumulated reached $120,000, and dividends averaged more than $10,000 per year, a figure corresponding with a return on investment of slightly more than 2 percent. Between 1813 and 1815 the tolls averaged about $45,000 per year, with dividends running about $20,000 (approximately 4.5 percent); for the 1823–1825 period toll revenue averaged about $28,000 annually, with dividends near $12,000 (about 2.5 percent). From 1833 to 1835 the average annual toll dropped from almost $35,000 to $14,000 while dividends plunged from $14,548 (about 3.1 percent) to $4,311 (about .9 percent). This decline in profitability clearly reflected competition from canals, but the old turnpike still continued to generate a small profit. In fact, even in the mid-1840s, fifty years after first opening, the Philadelphia and Lancaster Turnpike Road Company collected annual tolls averaging $7,000 while paying dividends of about $3,000 (.6 percent).[53]

The profitability of the Germantown and Perkiomen Turnpike Road Company was slightly more lucrative than its Lancaster-bound counterpart. Between 1805 and 1807 toll revenue averaged about $21,000 while dividends reached a little more than $10,000 (around 3.5 percent return on

---

52. Financial data taken from a company report published in the *Pennsylvania Senate Journal*, XIII (1802–1803), 245–248. This financial analysis is based upon a calculation of return on investment that does not consider depreciation. It also does not factor in operating profits used to supplement the officially reported amount of capital investment. The point is to discuss profitability in a basic sense and not to undertake sophisticated economic analysis using the limited data available.

53. 1802–1807: *Pennsylvania Senate Journal*, XVIII (1807–1808), 157–159. 1813–1815: *Journal of the Twenty Sixth House of Representatives of the Commonwealth of Pennsylvania . . .* (Harrisburg, 1815 [–1816]), 583–585. 1823–1825: *Pennsylvania Senate Journal*, XXXVI (1825–1826), 405–406. 1833–1835: *Pennsylvania Senate Journal*, XLVI (1835–1836), II, 361–363. 1840s: *Pennsylvania Senate Journal*, LVII (1847), II, 396–398.

investment). In 1811 the investment return leaped to more than 6 percent ($17,100 in dividends) while in 1815 dividends were 5 percent ($14,250). During the 1820s the profitability of the Germantown Turnpike remained steady, with annual dividends averaging more than $9,000 (more than 3 percent); even in the mid-1840s the company was paying out dividends of more than 3 percent on annual toll revenues of more than $18,000. As primary feeder roads to the Germantown Turnpike, the Cheltenham and Willow Grove Turnpike Road Company and the Chestnut Hill and Spring-house Turnpike Road Company were fortunate enough to share in the good fortune of the former road. In fact, the latter turnpikes were excellent long-term investments, with returns on investment averaging about 5 percent during their early years of operation (1805–1815). After operating for forty years these two companies remained profitable, with the Chestnut Hill and Springhouse Turnpike generating a 5 percent dividend in 1845 and the Cheltenham and Willow Grove Turnpike producing an amazing 8.9 percent return on its original investment of $80,000 in the same year.[54]

Of course, not all Pennsylvania turnpikes were money-makers. For example, in 1812 the Ridge Turnpike Road Company began operating over the right-of-way of the old Ridge Road running into Philadelphia from Norristown, but the toll road never proved to be profitable.[55] After almost fifteen years of service (in which it competed for traffic with the German-town and Perkiomen Turnpike), the Ridge Turnpike took in annual tolls

54. Germantown and Perkiomen: 1805–1807: *Pennsylvania Senate Journal*, XVI (1805–1806), 197, XVII (1806–1807), 9–10, XVIII (1807–1808), 132–133. 1811, 1815: *Pennsylvania Senate Journal*, XXII (1811–1812), 114–115; *Journal of the Twenty Sixth House of Representatives of Pennsylvania* (1815–1816), 97–98. 1820s, 1840s: *Pennsylvania Senate Journal*, XXXIII (1822–1823), 100, XXXIV (1823–1824), 189–190, XXXV (1824–1825), 504–505, XXXVI (1825–1826), 89, LVI (1846), II, 41–43, LVII (1847), II, 211.

Cheltenham and Willow Grove; Chestnut Hill and Springhouse: Financial data for these turnpikes appear in *Pennsylvania Senate Journal*, XVI (1805–1806), 188, XVIII (1807–1808), 280–281, 342–343, XXII (1811–1812), 109–110, LVI (1846), II, 250–254; *Journal of the Twenty Sixth House of Representives of Pennsylvania* (1815–1816), 293–294.

55. The Ridge Turnpike competed with the Germantown and Perkiomen Turnpike but was never able to challenge the economic superiority of the older toll road. The upper section of the Ridge Turnpike followed along the rather advantageous route advocated by Brooke in his 1802 reports. However, the section of the road that approached Philadelphia contained a lengthy, steep grade (to the south of what is now Manayunk), and this apparently did not constitute a great attraction to travelers in wagons. Traveling on the route might have been considered tolerable if free, but not if it entailed paying tolls.

of about $10,000 while annual expenses (including repairs) came to about $10,000, leaving no money for anything else. Thus, for 1825 the Ridge Turnpike had no money for dividends and operated under a reported debt of almost $140,000. Similarly, the Centre Turnpike Company attempted to manage a toll road from Reading in the upper Schuylkill Valley to Sunbury in the Susquehanna Valley. As reported by the son of the famous British scientist Joseph Priestley (whose father sought political refuge in Sunbury in the 1790s), by the early 1820s the seventy-three-mile-long Centre Turnpike was in serious financial difficulty after operating fitfully for more than a decade.[56] The company suffered from such things as subscribers who reneged on their promises to invest in stock, sketchy record keeping, poor construction work, and a heavy debt owed to both the state and to banks. Costing more than $200,000 and generating average annual tolls of less than $8,000, the Centre Turnpike Company represented a classic failure of a long-distance toll road that could not attract enough traffic to maintain itself as a worthwhile capitalistic investment. As exemplified by the Ridge Turnpike and the Centre Turnpike, not all Pennsylvania toll roads of the early nineteenth century proved to be good investments.

Moving beyond simple economic analyses of profitability, it is worth noting that the value of turnpikes is not reflected solely by the profit-and-loss statements of various operating companies. Improved roads and turnpikes could provide potential economic benefits to a region or community regardless of whether dividends were paid to investors. For example, a sizable proportion of toll revenues was returned to the local community through the salaries paid to tollkeepers (who were individually paid upwards of $250–$350 annually).[57] Similarly, the cost of keeping a road surface in good repair was quite large, especially for heavily traveled turnpikes, and expenditures for quarrying, breaking, and placing stone almost invariably went to pay for the services of locally based contractors. Toll revenues were not necessarily lost to a community, in the sense that ulti-

---

56. *Pennsylvania Senate Journal,* XXXVI (1825–1826), 161. See Joseph Priestley, Jr., Secretary, "To the Stockholders of the Centre Turnpike Company leading from Reading to Sunbury," circa May 1822, broadside no. 26, American Philosophical Society Library, Philadelphia.

57. Salaries for tollkeepers on the Lancaster Turnpike were reported as being "from two hundred and fifty to three hundred and fifty dollars per annum" at the beginning of the 19th century. See Elliston Perot to William MacPherson, Sept. 7, 1807, included in Gallatin, "Report on Roads and Canals," in Cochran, ed., *American State Papers: Transportation,* I, 240.

mately much of the money was returned to farmers and contractors active in the local economy.

Further evidence of the local economic significance of turnpikes can be found in a letter from a local businessman in the Cumberland Valley near Chambersburg, Pennsylvania, to the renowned railroad financier Jay Cooke. Written in 1859, a time when turnpikes are usually considered to be passé within the context of America's transportation development, this letter stated:

> Considerable sums of money has [sic] been expended in our county in the last few years in making turnpikes, which every person knows are unproductive, but they are always satisfactory to the farmers. . . . These turnpikes are seven in number which, run from various directions from our town and will therefore be of advantage to the railroad.[58]

In this instance, the efforts of railroad promoters to attract local investors to support railroad construction was being impeded by the diversion of local capital into turnpike projects. Cooke and his associate obviously considered turnpikes to represent inferior investment opportunities, at least compared with railroads, and they bemoaned the seeming lack of sophistication being demonstrated by farmers in the Cumberland Valley. But, at the same time, they acknowledged the "advantage to the railroad" inherent within transportation services provided by the local turnpikes. Within a context strictly limited to the profitability of an operation as defined by dividend payments paid to an outside investor, turnpikes might not have appeared as viable investments. However, people depending upon a turnpike for commercial interaction within a regional economic market might well consider a toll road to be successful even if it never turned a profit per se on their investment in its construction.

Work by Daniel Jones focusing on the political economy of northeastern Rhode Island in the early nineteenth century brings to light a different type of local reaction to turnpike projects, a reaction which was decidedly negative. In rural areas near the Connecticut–Rhode Island border, groups of farmers resisted efforts by Providence-based merchants to expand commercial ties to the city's hinterland, at times lobbying the state legislature to reject petitions requesting turnpike charters and at times threatening violence against turnpike officers and toll collectors.[59] Rhode

---

58. A. Armstrong to Jay Cooke, Jan. 15, 1859, Jay Cooke Papers, HSP.

59. Jones, *Transformation of Rhode Island*, 46–57, 68–81, characterizes opposition

Island, with its traditions of political autonomy dating back to the time of Roger Williams, may represent an extreme case in terms of turnpike opposition. Certainly such opposition was much less overt in Pennsylvania, and my research into the history of turnpikes that served Philadelphia has uncovered only one case in which farmers attempted to exert political force to block implementation of a toll road.

In the first decade of the nineteenth century, farmers northwest of Norristown opposed authorization of the Ridge Turnpike Road Company. Resistance to the toll road developed after the Germantown and Perkiomen Turnpike had been approved, and it appears that the farmers wanted to keep open one free route to Philadelphia for use during good weather. Within a decade this opposition dissipated, and state authorization of the Ridge Turnpike occurred in 1811.[60] In a similar vein, the previously noted opposition of local Germantown tradesmen to the new, improved road to Philadelphia indicated a concern over the economic changes that would be wrought by turnpike development. However, we should also remember that the reaction of commercial interests in Hickorytown was exactly the opposite, as they actively lobbied to ensure that the Germantown and Perkiomen Turnpike did *not* pass them by. Although hardly as dramatic or as politically significant as the campaigns waged by antiturnpike forces in Rhode Island, the effort to block turnpike charters by some farmers and tradesmen indicates that toll road expansion in Pennsylvania did not occur without a modicum of social debate over its merits.

As indicated by the toll rates listed above for the Lancaster Turnpike, travel on turnpikes could constitute an expensive proposition and thus prompt a hostile reaction by the traveling public. In its most basic form,

---

against turnpikes as a reaction by farmers who did not wish to be drawn into a capitalist market economy controlled by urban merchants and industrialists. Turnpike opposition is also discussed in Philip Taylor, "The Turnpike Era in New England," 115–118; and in Albert, *Turnpike Road System*, 25–29. On Connecticut–Rhode Island resistance, see Jones, *Transformation of Rhode Island*, 50–51. In 1799 the Glocester town meeting successfully opposed an effort to convert an important section of the Providence-Killingly road into a turnpike. In 1801 the tollkeeper for the Providence and Norwich Turnpike Society was "afraid that charging tolls might result in the destruction of his tollgate and house by disgruntled local travelers."

60. Various aspects of this controversy are documented in the Jonathon Roberts Papers, HSP; and "First Ridge Turnpike Plan Aroused Opposition," scrapbook A-8-15, Montgomery County Historical Society Archives, Norristown, Pa.

this hostility became manifest in efforts to evade payment of tolls; cut-offs that enabled travelers to bypass tollhouses became popularly known as "shunpikes." [61] Farmers and local travelers might have appreciated the benefits that turnpike developments brought to them, but those benefits did not mean that they relished the thought of helping to pay dividends to capitalist investors residing in distant cities. The popular discontent with having to pay tolls for local traffic is reflected within Pennsylvania's Act Regulating Turnpike and Plank Road Companies, enacted in 1849. This act specified:

> No toll shall be demanded from any person or persons passing and repassing from one part of his, her or their farm to any part of the same; and all persons with their vehicles or horse going to or from funerals, or places of public worship, or of military trainings or elections, shall be exempt from the payment of toll, when traveling on such turnpike road.[62]

If possible, travelers sought to avoid toll payments either by legislative fiat (as evidenced in the above law) or by using roads that avoided tollbooths. Of course, a huge number of local roads existed throughout the early nineteenth century that, at least during periods of good weather, were capable of carrying substantial quantities of traffic. However, these roads, which were usually devoid of substantial stone pavements, offered comparatively little advantage to travelers in Pennsylvania when the weather turned sour. It is in this context that the greater social and economic value of turnpike development comes into focus.

The cost of quarrying, breaking, and placing stone on road surfaces constituted the primary expense of turnpike companies, and this effort was geared toward increasing the highway's year-round serviceability. The importance of undertaking such work is specifically reflected in a seasonal breakdown of the toll revenues of the Germantown and Perkiomen Turnpike. During the 1820s this company reported its financial status semiannually, with statistics grouped into two six-month segments, one running from November through April and the other from May through October. An observer might assume that most traffic on the turnpike would take place during the summer and fall, when the weather was

---

61. See G. Taylor, *Transportation Revolution,* 28, for discussion of "shunpikes."

62. Approval of this law appears in the *Journal of the Fifty-Ninth House of Representatives of the Commonwealth of Pennsylvania* (Harrisburg, 1849), I, 134.

most suitable for travel and when harvests would be brought to market. However, the company's records reveal that between November 1821 and October 1825 a cumulative total of almost $47,000 in tolls was taken in during the winter and spring months (November–April) while summer and fall (May–October) tolls only slightly exceeded $31,000.[63] Thus, 60 percent of the company's revenues were obtained at times when it might logically be inferred that traffic would be at a minimum.

A conclusion to be drawn from the above observation is that stone-surfaced turnpikes played an important role in allowing the economic and social life of southeastern Pennsylvania to thrive during the inclement winter months. More personal testimony verifying this phenomenon is provided in two early nineteenth-century letters written by a female correspondent who had no apparent financial interest in promoting turnpike development but who simply commented to a family friend about how the higher quality of turnpike road surfaces related to her day-to-day life in the wintertime.

> Here the severe cold of three days last week has been followed by moist drizzling weather, a compleat thaw so that off the turnpike the roads are dreadful.
>
> The turnpike is finished and we can now go to town at all times and in all weather.[64]

Leaving aside other issues related to turnpike construction, these two observations, in concert with the data documenting fall and winter toll receipts on the Germantown and Perkiomen Turnpike, provide telling evidence why the onset of capital-intensive road building activity assumed great cultural significance. In a related context, William Cronon commented upon the importance of reliable, year-round travel between an urban commercial center and its hinterlands in describing the development of mid-nineteenth-century Chicago. In noting how railroads connected Chicago to the surrounding prairie, he observes:

> The railroads also alleviated many of the worst effects of winter. The period from November to April had always been the dullest season

---

63. *Pennsylvania Senate Journal*, XXXIII (1822–1823), 100, XXXIV (1823–1824), 189–190, XXXV (1824–1825), 504–505, XXXVI (1825–1826), 25–26.

64. Deborah Logan to Dr. George Logan, Dec. 23, 180[5?], Deborah Logan to Albanus Logan, Jan. 25, 180[2?], Maria Dickenson Logan Papers, HSP.

of the year, when trade ground to a virtual halt for merchants and farmers alike. With the railroad, rural farmers could travel to urban markets whenever they had the need and funds to do so, even in the deep cold of February. . . . As one railroad promoter wryly remarked, "It is against the policy of Americans to remain locked up by ice one half of the year." [65]

For early nineteenth-century Philadelphia, stone-paved turnpikes served a function analogous to the role later played by railroads in nurturing Chicago's development. The environmental character and scale of each city's hinterland differed; thus, new transportation technologies were required to meet commercial needs as the nineteenth century evolved. But the importance of travel and trade that could proceed despite the onslaught of winter transcends the myriad technological differences between Chicago's railroads and Philadelphia's turnpikes and highlights a cultural contribution shared by both systems. If indeed by the late 1840s it was "against the policy of Americans to remain locked up by ice one half of the year," then the roots of this antipathy in Pennsylvania clearly lay in the earlier proliferation of stone-paved turnpikes.

Historians studying America's early transportation systems may continue to consider roads and turnpikes of secondary importance because canals and railroads offered significant economic advantages in terms of bulk carriage and (for railroads) speed. Unquestionably, turnpikes were not always profitable as commercial investments, and at times they might have engendered hostility among travelers who disliked paying tolls. However, their value to the social and economic life of southeastern Pennsylvania extends beyond these limitations. A primary conclusion of this essay is that the history of early roads should be viewed as more than crude prelude to the expansion of canals and railroads; and, in this context, Robert Brooke's work in conceptualizing the Germantown and Perkiomen Turnpike can be considered to presage later systematic approaches to transportation design. Although the seemingly primitive nature of stone technology underlying turnpike development might impell belief that road building's effect on society or on the natural environment remained rela-

---

65. William Cronon, *Nature's Metropolis: Chicago and the Great West* (New York, 1991), 75–76. The internal quote originally appeared in John A. Wright, "Effects of Internal Improvements on Commercial Cities: With Reference to the Pennsylvania Central Railroad," *Hunt's Merchants' Magazine*, XVI (1847), 263–272.

tively insignificant, we should acknowledge that mere simplicity of form does not limit a technology's ability to shape patterns of human development. As Richard White noted in his environmental history of Island County in Washington State:

> Early roads conformed to natural boundaries, yet their impact on natural systems was considerable . . . no matter how primitive these tracks were, they became avenues for invading plants by stripping away native vegetation and thus brought changes in the natural landscape . . . [thus showing how] far-reaching effects can be obtained from relatively unsophisticated technology.[66]

By analogy, White's view of how simply built roads could dramatically affect the environment of Puget Sound can be extended to the more general impact of roads on cultural life in the early Republic. Nineteenth-century turnpikes may appear less striking and significant than the canals and railroads that eventually overshadowed them, but their import is no less real.

66. Richard White, *Land Use, Environment, and Social Change: The Shaping of Island County, Washington* (Seattle, Wash., 1982), 40.

R OBERT B. G ORDON

Custom and Consequence

Early Nineteenth-Century Origins of

the Environmental and Social Costs

of Mining Anthracite

When a new technology is first used, its practitioners must make design decisions that are difficult to alter at a later time. The proprietors of a canal might choose a narrow width to keep construction costs low and, later, after they have tunnels and bridges in place, find that increased traffic requires a larger waterway. In the meantime, factories and residences would have been built along the route, making the desirable engineering

The research on which this paper is based was supported by a fellowship at the Philadelphia Center for Early American Studies. Many of the ideas discussed here were developed in conversations with the fellows of the Transformation Project of the Philadelphia Center and on a field trip through the anthracite districts in company with Carolyn Cooper, Greg Galer, and Patrick Malone. I thank Judith McGaw for organizing the year devoted to early American technology at the Philadelphia Center and for valuable comments on earlier drafts of this paper, Michael Zuckerman for stimulating discussions of the role of technology in American history, Patrick Malone for organizing the field trip, David Salay for arranging entrance to several sites, and M. C. Korb for showing us the open pit anthracite mine operated by the Bethlehem Mines Corporation.

improvement difficult to execute.[1] Equally important, and often more intractable, are social customs and usages established in the early years of a new technology that subsequently cause inefficiencies, environmental damage, or accidents that need not be consequences of using the technology. These customs and usages persist because managers and workers make them part of their routine. Claiming that new and better procedures are too costly, managers might threaten a community with loss of jobs if forced to change their ways. Workers might resist change through formal work rules sanctified by past-practices clauses in contracts or, in some cases, by law.[2] Alternatively, resistance might appear in less formal but equally powerful, unspoken attitudes toward safety and the environment. Many social and economic problems today originated in decisions and practices put in place during the first years that new technologies were practiced. They have contributed to the demise of integrated steelworks in the United States, the failure of railroads and public transportation to compete effectively with motor vehicles, and the arrested development of nuclear power. Historical studies can give us insights on how these difficulties arose, their consequences, and how we might deal with them.

In 1825, as Americans began mining anthracite, they established a new industry in the United States. Anthony F. C. Wallace, in his anthropological study of St. Clair, Pennsylvania, a late-nineteenth-century anthracite town, shows how the pattern of anthracite mine ownership and management led to frequent accidents and widespread social ills.[3] Here I will explore how, within a decade of 1825, adventurers in the coalfields had established both the Pennsylvania technique of mining anthracite underground and the customs and usages among miners, managers, and owners. Although many of these choices soon proved deleterious to everyone, the mining community retained them into the twentieth century. They contributed to the high environmental and social costs of mining and to the

---

1. The full potential of rail transportation in many areas of eastern North America, where it is most needed today, cannot be realized because of this problem.

2. An example is the legal definition in some states that a day's work for a train crew is 100 miles traveled, probably realistic when established more than 100 years ago but an anachronism today that has helped shift transport of goods to the less energy-efficient highway system.

3. Anthony F. C. Wallace, *St. Clair: A Nineteenth-Century Coal Town's Experience with a Disaster-prone Industry* (New York, 1987).

premature replacement of coal by other sources of energy that were not encumbered with comparably debilitating technical and social practices.

In this exploration, we will find that, by about 1840, Americans achieved economical and serviceable solutions to the technical problems in combustion engineering and transportation systems they needed to make anthracite a useful fuel. However, their progress with mining technology lagged, and a technological unbalance in the coal-fuel energy system ensued. The safe and efficient operation of anthracite mines remained dependent on the craft skills of individual miners. The organization of mine operation created barriers to the full exercise of miners' skills. Even without these inhibitions, craft skills alone would not have made mining safe or free of unnecessary environmental damage. The failure of the social organization of the mining communities and mining companies to deal with the limitations of mining skills led to serious consequences for all connected with anthracite production.

The geographical separation of the mines from those who benefited from using anthracite exacerbated the high social cost of coal mining. While canals and railways distributed anthracite to industries that created much wealth throughout the northern and Middle Atlantic states, the social and environmental costs of this wealth remained concentrated in the mining districts, out of the purview of mineowners (who lived elsewhere) and the consumers, who benefited from abundant, cheap fuel.

History and Geography

Americans began making a fundamental change in their industries, described by some business historians as the "American industrial revolution," about 1820, and it was well under way within twenty years.[4] In place of the fall of water and burning of wood, they adopted coal as their principal energy source while simultaneously replacing wood in engineering structures and machinery with iron, much of it made with coal fuel.

During their Wooden Age, Americans used the abundant, renewable, but bounded natural resources of wood and water to produce primary materials (such as iron and glass) and to manufacture finished products.[5]

4. Alfred D. Chandler, Jr., "Anthracite Coal and the Beginning of the Industrial Revolution in the United States," *Business History Review*, XXVI (1972), 141–181.

5. Technology in this period is discussed in Brooke Hindle, ed., *America's Wooden Age: Aspects of Its Early Technology* (Tarrytown, N.Y., 1975).

These resources were widely distributed and accessible to almost anyone who wanted to use them for an industrial enterprise. They were renewable because forests could provide a sustained yield (and were sometimes so managed) and a water privilege could continue to deliver power for hundreds of years. They were bounded because of the physically determined upper limit on the power that could be obtained at any one water privilege and the biologically determined sustained yield possible from a forest. It was uneconomic to transport fuel wood great distances, and (before the advent of hydroelectricity) waterpower had to be used where it was generated. Industries based on wood and water resources were dispersed across the countryside. Because small- to medium-sized water privileges were abundant and access to woodlands was relatively easy along the East Coast of North America, entrepreneurs with quite modest means could acquire the resources they needed to initiate an industrial enterprise. This natural resource base that facilitated entry into industry also discouraged proprietors from concentrating their enterprises in large urban centers. The resulting dispersion contributed to industrial innovation by encouraging formation of many small enterprises sufficiently separated to develop independently while close enough for artisans and proprietors to exchange and compare ideas easily.[6] A happy coincidence of natural resources and cultural factors helped Americans gain the competence in manufacturing that made them formidable competitors in international markets for tools, machinery, and textiles by midcentury.

In examining the decisions that shaped the anthracite industry in Pennsylvania, we will look at mining as part of an energy system that included the production of the fuel, its transportation, and its use. After about 1825, Pennsylvanians opened coal mines, and capitalists constructed canals and railways that could deliver coal at low cost to distant places; and manufacturers thus could move away from their dependence on energy derived from wood and the fall of water. As they gradually shifted to energy from mineral coal, they changed the industrial landscape. The Pennsylvania coal resources were abundant—the amount miners could dig at a cost lower than that of alternative fuels was much larger than the demand ever became. Because of its relatively high energy density, mineowners could ship coal at a cost low enough to make it competitive with other energy sources wherever canals, inland waterways, or railroads reached. The importance

6. Robert B. Gordon, "Hydrological Science and the Development of Waterpower for Manufacturing," *Technology and Culture*, XXVI (1985), 204–235.

of the natural resource characteristics that had favored small, dispersed units of production diminished. With factories powered by coal-fired boilers and steam engines, proprietors could concentrate their works in cities. They could leave to city officials and taxpayers the problems of supplying the social services needed by workers that had to be undertaken by factory owners at rural industrial sites. They avoided the limits on growth and the need to schedule work around power sources that waxed and waned with changes in the weather over the seasons. Managers could concentrate on the commercial aspects of their businesses.

Cheap anthracite allowed city dwellers to change urban domestic life. By 1825, the increased cost of hauling cordwood greater distances as woodsmen depleted nearby forests made fuel dear in most American cities. By replacing wood fuel for domestic heating with anthracite, city residents reduced both the cost of heat and the air pollution caused by wood smoke. Architects adopted anthracite fuel to make effective heating systems for large buildings. Steamboats and, later, locomotives with coal-fired boilers delivered raw materials for manufacturing more cheaply and reliably than draft animals or sailing vessels could. All these factors helped entrepreneurs who placed industries in cities.

Whenever we substitute one natural resource for another, we have to overcome technological barriers.[7] Americans needed new techniques in three different areas if they were to make large-scale use of anthracite. First, they had to design new grates, fireboxes, and furnaces and work out new firing procedures, because anthracite did not burn as easily as wood or bituminous coal. Since anthracite was not a common fuel in Britain or Europe, Americans had to create the technique for burning it. Next, they had to construct new canal and railway systems to carry coal from the mining districts to customers. Finally, they had to develop the technique of mining anthracite on a large scale.

COMBUSTION TECHNOLOGY

Promoters of anthracite had to make a determined effort to convince Americans that they offered a useful fuel. At first, most people did not believe that anthracite would burn: "The coal found was so different from any previously known, that it was deemed utterly valueless—more

7. This is discussed in Robert B. Gordon, Tjalling C. Koopmans, William D. Nordhaus, and Brian J. Skinner, *Toward a New Iron Age? Quantitative Modeling of Resource Exhaustion* (Cambridge, Mass., 1987), chap. 3.

especially as no means could be found to ignite it."[8] Two properties of anthracite caused difficulties when people first tried to use it. First, since anthracite is nearly pure carbon and contains little volatile matter, it must be heated to a high temperature before it will ignite. Then, when burning, a cold draft of air will extinguish it. Second, anthracite burns with a very short flame, making it difficult to distribute the heat it generates over the surface of a boiler or furnace.

Although professional scientists participated in the development of anthracite combustion technology, experimentation by many individuals who gradually learned the necessary skills established its value as a fuel (see Table 1). American innovators developed three major applications of anthracite: heating homes and public buildings; firing boilers for stationary steam engines, steamboats, and locomotives; and smelting and refining iron. Jesse Fells began the development of the technology of heating with anthracite through his experiments with grates in 1808. Eliphalet Nott and others secured many patents for stoves between 1828 and 1835.[9] Experimenters found designing grates and fireboxes to burn anthracite under boilers in steamboats and locomotives a more difficult problem. However, by about 1840, they had largely overcome the engineering barriers to the substitution of coal for wood. Thereafter buyers chose their fuel by relative prices rather than technological capabilities.

American innovators had some of their most difficult technical problems and greatest successes applying anthracite to ferrous metallurgy. Pennsylvania smiths had used anthracite in their hearths in the late eighteenth century. After the often-cited demonstration by Josiah White and Erskine Hazard in 1812, others adopted it for heating iron preparatory to forging or rolling and for melting iron at foundries as rapidly as the supply expanded.[10] No one had to solve large technological problems in these applications.

8. Ele Bowen, ed., *The Coal Regions of Pennsylvania* . . . (Pottsville, Pa., 1848), 26. There are many accounts of efforts made to gain acceptance of anthracite for other fuels in the United States. Two comprehensive discussions are H. Benjamin Powell, *Philadelphia's First Fuel Crisis: Jacob Cist and the Developing Market for Pennsylvania Anthracite* (University Park, Pa., 1978); and Frederick Moore Binder, *Coal Age Empire: Pennsylvania Coal and Its Utilization to 1860* (Harrisburg, Pa., 1974).

9. Bowen, ed., *Coal Regions*, 18; Binder, *Empire*, chap. 1.

10. For example, William Wood of New York City demonstrated the use of anthracite for melting cast iron in a cupola furnace in 1820. See *Cursory Review of the Schuylkill Coal in Reference to Its Introduction into New York* (New York, 1823).

TABLE 1. *Development of Anthracite Combustion Technology*

1768.	Obadiah Gore successfully uses anthracite in blacksmithing; its use by smiths becomes common thereafter.
1775.	Anthracite is reported to have been taken to the United States Armory at Carlisle for the manufacture of firearms.
1805.	Professor James Woodhouse demonstrates that anthracite can be burned in an air furnace.
1806.	Oliver Evans at the Mars Iron Works uses anthracite to melt cast iron in a cupola furnace.
1807.	In a trial at the Philadelphia waterworks, anthracite is "deemed rather an extinguisher than an aliment of fire," and the rest is used as gravel on the neighboring garden walks.
1808.	Jesse Fell designs grate that is satisfactory for burning anthracite.
1812.	Josiah White and Erskine Hazard discover utility of anthracite in heating metal at their wire works (often described as the first industrial use of this coal). Anthracite used in distillation of liquor and at four rolling and slitting mills near Philadelphia.
1813.	Joshua Malin uses anthracite to melt several hundred tons of pig iron.
1818.	Coal stoves made by Texler in Bethlehem are being widely sold in the Wyoming Valley.
1825.	Anthracite is used to fire boilers of a stationary steam engine at the Phoenix Nail Works, French Creek, Pennsylvania.
1826.	Marcus Bull conducts experiments on the heating power of coal. Lehigh Coal and Navigation Company begins experiments with anthracite as fuel in towboats.
1827–1831.	Grates and coal stoves come into general use in towns of eastern United States.
1828.	The Baltimore and Ohio Railroad uses anthracite in steam locomotives, but has trouble with burned-out grate bars and fire boxes.

TABLE 1. Continued

1829.	The Collins Company, the celebrated axmaker in Collinsville, Connecticut, replaces charcoal with anthracite in forge fires.
1831.	The Delaware and Hudson Canal Company supplies anthracite to ferryboats on the Hudson River, but difficulties in its use continue until Eliphalet Nott designs a suitable boiler in 1835.
1838.	The Reading Railroad begins development of locomotives using anthracite fuel.

Sources: Frederick Moore Binder, *Coal Age Empire: Pennsylvania Coal and Its Utilization to 1860* (Harrisburg, Pa., 1974), 18, 49, 73, 90, 93, 112, 122; Ele Bowen, ed., *The Coal Regions of Pennsylvania* . . . (Pottsville, Pa., 1848), 18, 19, 22, 27; Howard N. Eavenson, *The First Century and a Quarter of American Coal Industry* (Pittsburgh, 1942), 141–143; Benjamin H. Powell, *Philadelphia's First Fuel Crisis: Jacob Cist and the Developing Market for Pennsylvania Anthracite* (University Park, Pa., 1978), 51, 52, 57, 91; James M. Swank, *History of the Manufacture of Iron in All Ages* . . . (Philadelphia, 1892), 363.

American ironmasters used large amounts of charcoal as they smelted iron in blast furnaces and then converted the resulting pig in finery forges to wrought iron.[11] A basic difficulty arose when they attempted to convert pig iron to wrought iron with mineral coal instead of charcoal: sulfur in the coal contaminated the iron. British ironmasters had solved this problem through a series of developments extending over fifty years that

11. The historical and archaeological record of the use of anthracite in ironmaking before 1850 is poor and uneven. A blast furnace plant is a prominent feature on the landscape and represents a large capital investment. A blast furnace making 50 tons of pig iron per week could supply 20 or so puddling furnaces, which were often dispersed in small ironworks instead of being concentrated in one large plant. For these reasons, the smelting branch of the 19th-century iron industry attracted most of the attention of contemporary writers; the archaeological record is similarly biased because blast furnace stacks survive a long time (usually without their associated plant) and many have been made into industrial monuments. No wrought iron–works has yet been studied by industrial archaeologists, and no puddling furnace is on public display (or even survives) in the United States. Much research on the anthracite iron industry will have to be done before the different contributions are sorted out, but it does appear that conversion of pig to wrought iron with coal was effected first and was followed by success in smelting pig iron with anthracite.

included Henry Cort's success in 1784 with what later became known as the puddling process. They avoided contaminating the iron with sulfur by using a reverberatory furnace to keep the fuel separated from the iron. Even though it was a key step in the transition from wood-and-water– to coal-and-iron–based industry in America, historians of technology have yet to study the introduction of the puddling process in the United States. Secondary sources show us that, by 1830, a number of American iron-works had puddling furnaces operating. English ironmasters always fired these furnaces with bituminous coal that produced a long flame and believed that anthracite was unsuitable fuel for puddling. Nevertheless, by redesigning the firebox and adding a forced-draft blower to the puddling furnace design then used in Britain, American ironmasters succeeded with anthracite. The technique was well established among them by 1841.[12] Thus, Americans made an important (and as yet undescribed) development in pyrotechnology. They achieved what was considered impractical in Britain and made possible large-scale production of wrought iron in eastern Pennsylvania, where the long-flame coal needed for conventional puddling furnaces was not available.

Pennsylvanians began to experiment with anthracite as a substitute for charcoal in blast furnace smelting in 1821.[13] The early experiments came to naught because an anthracite fire will not burn hot enough to melt iron without a preheated air blast, and only furnaces with cold-blast were then in use. James Neilson patented his hot-blast stove in Britain in 1828, and by 1831 Frederick Geissenhainer was experimenting with the use of a hot, high-pressure blast for smelting with anthracite in the United States. By 1840 Americans had anthracite smelting successfully under way at seven blast furnaces, and they blew in six more the following year (see Table 2). Again, we lack an adequate history of this new technique. However, its rapid development and importance are clear; by 1854, when the first reliable statistics were compiled, anthracite-smelted iron accounted for 45 percent of all the pig made in the United States.[14]

12. John Fritz, *The Autobiography of John Fritz* (New York, 1912), 147; Thomas Turner, *The Metallurgy of Iron* (London, 1900), 306; Frederick Overman, *The Manufacture of Iron* (Philadelphia, 1850), 271; Walter R. Johnson, *Notes on the Use of Anthracite in the Manufacture of Iron* (Boston, 1841), 11.

13. Bowen, ed., *Coal Regions*, 31.

14. Peter Temin, *Iron and Steel in Nineteenth-Century America: An Economic Inquiry* (Cambridge, Mass., 1964), 52.

TABLE 2. *Anthracite-Fired Blast Furnaces in 1841*

Location	Proprietor	Date of Blowing In	Output per Week
Mauch Chunk	Baughman, Guiteau & Co.	1838	8 tons
Pottsville	Marshall, Kellog & Co.	1839	28
Roaring Creek	Burd Paterson & Co.	1840	40
Phoenixville	Reeves	1840	28
Danville	Biddle, Chambers & Co.	1840	35
Catasauqua	Crane Iron Co.	1840	50
Danville	George Patterson	1840	32
Danville	Biddle, Chambers & Co.	1841	70 (each of 2)
Stanhope	Stanhope Iron Co.	1841	56 (each of 4)

Source: Walter R. Johnson, *Notes on the Use of Anthracite in the Manufacture of Iron* (Boston, 1841), table 1.

TRANSPORTATION TECHNOLOGY

The advent of stoves and grates suitable for burning anthracite created opportunities for the proprietors of the new canal and railway systems to increase traffic between the coalfields and the major eastern cities. Additional traffic was generated by established industries based on waterpower, such as the textile mills along the Merrimack River, as they adopted anthracite for auxiliary power to be used in case of drought or to expand the capacity of an existing plant beyond that which could be sustained by its water privilege. As the supply of cheap coal available in urban centers improved, proprietors of industries dispersed at rural waterpower sites were able to move so as to take advantage of urban settings for their works. Few manufacturers set up works in the coalfields, however, until the late nineteenth century, when many silk mills moved there to use the labor of miners' wives and daughters. Consequently, the transportation networks were an essential component of a system that made mineral fuel widely available for ironmaking and manufacturing.

In North America, anthracite occurs only in a small area of eastern Pennsylvania. The geological processes that formed the anthracite also raised surrounding mountains that greatly restricted the possible routes by which it could be moved to market. Anthracite was formed through conversion of plant remains to nearly pure carbon at the high temperature and pressure resulting from deep burial and the subsequent deformation associated with mountain building. Although hard coal, anthracite is not as resistant to erosion as most other rocks; and, since the landscape of eastern Pennsylvania has been subjected to continued erosion for about 100,000,000 years, coal is found only in basins (see Figure 1) where it has been protected by more enduring types of rock (see Figure 2). The upturned strata of grey, hard, massive conglomerate and sandstone make ridges that surround the coal-bearing rock within the basins; where there were no protective ridges, weathering processes carried away the coal.

Only a few large rivers, such as the Susquehanna and the Schuylkill, penetrate the ridges surrounding the coal basins. Canal and, later, railroad builders used the water gaps (Figure 1), the easiest routes into the basins. Whoever controlled the water gaps controlled the most economical coal trade until others opened alternative routes by either tunneling through the protective ridges or constructing incline planes.

Since there is an extensive secondary literature on the Pennsylvania anthracite canals and railways, we need only review the leading events. When White and Hazard convinced themselves of the value of anthracite fuel, they first leased the Lehigh Coal Mine Company, which others had started late in the eighteenth century, and then organized the Lehigh Coal and Navigation Company to improve navigation on the river and, later, construct a canal that allowed them to move coal to Philadelphia at low cost (Figure 1). The proprietors of the Schuylkill Navigation began bringing coal from the southern field to Philadelphia in 1825, allowing miners to rapidly increase production in this field. Others entrepreneurs opened the Delaware and Hudson canal in 1829 to bring the product of the northern coalfield to New York and Albany and began the North Branch canal along the Susquehanna in 1834 to open a route to Baltimore. Railroad proprietors later captured much of the coal traffic, beginning with the Philadelphia and Reading in 1844 and the Lehigh Valley in 1855. However, it was the completion of the water routes and the developments in combustion technology that set off a coal boom in eastern Pennsylvania in the years after 1825.[15]

15. John N. Hoffman, *Anthracite in the Lehigh Region of Pennsylvania, 1820–45,* Smithsonian Institution, United States National Museum Bulletin 252 (Washington, D.C., 1968); W. Julian Parton, *The Death of a Great Company: Reflections on the Decline and*

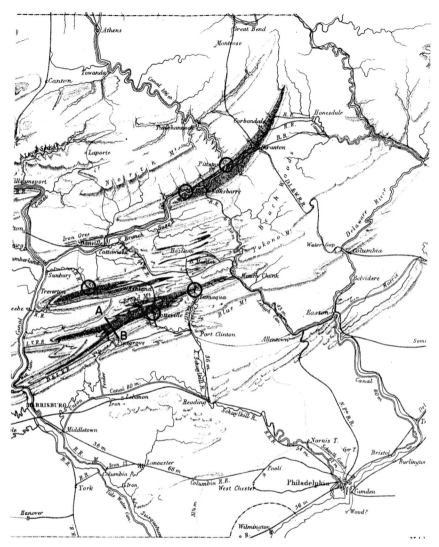

FIGURE 1. The Anthracite Coalfields of Pennsylvania. *From Henry D. Rogets,* The Geology of Pennsylvania . . . , 2 vols. *(Philadelphia, 1858), II, following 1018.* The principal water gaps are marked by circles. Coal was carried out of the mining regions by the railroads and canals shown. The northern coalfield is penetrated by the Susquehanna River at two water gaps, the middle field by the Shamokin Creek gap, and the southern field by the Schuykill and Little Schuykill rivers. Canals and some of the railroads penetrated the coal fields at the water gaps; other railroads used planes or tunnels; those that did not had difficult, circuitous routes into the coal basins. Detail of A–B is shown in Figure 2.

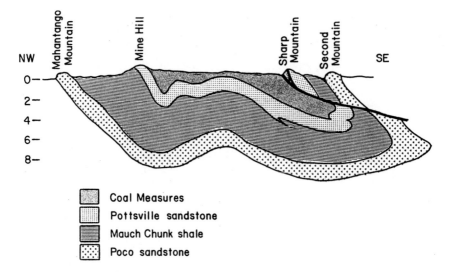

FIGURE 2. Cross Section of the Western End of the Southern Coalfield. *Based on the Geological Map of Pennsylvania (Pennsylvania Geological Survey, 1960), detail of A–B, Figure 1.* The erosion-resistant rocks that form the ridges surrounding the coal basins are the Pottsville and Pocono sandstone and conglomerate. Anthracite occurs as veins in the rocks designated "coal measures." At the southeast end of the section these formations have been offset by a major fault. The scale is in thousands of feet.

Mining Technology

GEOLOGY OF THE MINES

The processes of mountain formation that created the coal basins also folded the coal veins (see Figure 3). Multiple folds often brought one vein to the surface at several places. Anyone could draw coal from these outcrops without the aid of sophisticated mining technique or a large investment in equipment or mine development, so it was easy for the early adventurers to obtain coal with very modest means and no special skills. But, when these adventurers exhausted the surface outcrops and found they had to mine at depth, they encountered geological complexities for which they were quite unprepared. Geological uncertainties could be as perplexing for an individual mine, or even a single coal vein, as they were for a coal basin.

Twentieth-century strip mines allow us to see some of the geologi-

Fall of the Lehigh Coal and Navigation Company (Easton, Pa., 1986); Binder, *Empire*, 136–144, 146–147; Bowen, ed., *Coal Regions,* table facing 56.

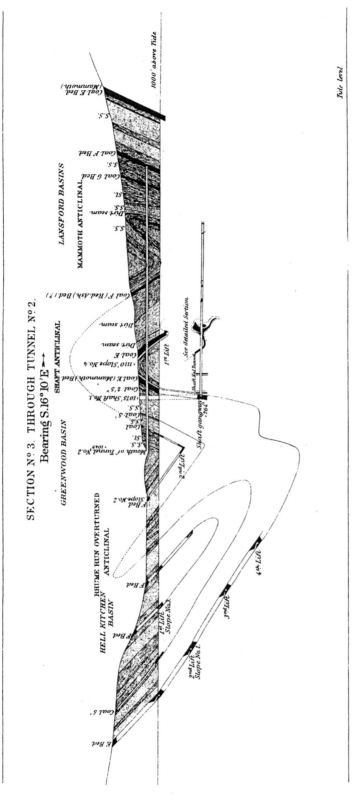

FIGURE 3. Cross Section of the Eastern End of the Southern Coalfield near Lansford. *Drawn by the Second Geological Survey of Pennsylvania in 1883,* Atlas, *Southern Coal Field.* The section shows the structure determined from observations at the surface, in tunnels, and in the mines. The course of the strata at depth, where there are no data, and the course where they formerly existed before being removed by erosion have been estimated. Note the limited data available as late as 1883 for constructing these extrapolations and that no faults are shown. Mining in this area was carried out by the Lehigh Coal and Navigation Company.

cal structures encountered by miners working underground. Coal veins worked by underground mining in the nineteenth century now stand exposed on the walls of some twentieth-century open-pit mines (see Figure 4). The steep dip of the veins, the small-scale folding, and the displacement of the folds by faults made mining this coal by underground methods a difficult task. Miners frequently encountered faults (places where the strata are displaced along a nearly plane surface) while following a vein. Then they could not mine more coal until they found the vein again. As a rule, neither the miner nor the mine operator had any reliable way of determining the direction or distance that the coal had been offset by a fault, and the temptation to continue digging through the intervening rock in the hope of recovering the vein led many mine operators into bankruptcy.[16]

Where there were no faults, miners still encountered large, unexpected changes in thickness of a vein or bulges of the roof into the coal. Often, the coal within a vein varied in quality and in the proportion of carbonaceous shale mixed in the coal. Miners might unexpectedly encounter underground watercourses or gas pockets, which could suddenly increase the volume of water or firedamp (methane, a gas than can form explosive mixtures with air) entering a mine, and roof rock that rapidly lost strength when exposed to air and increased the risk of rock falls within the mine.[17]

Geological advice. Because of growing interest in the state's mineral wealth, the Pennsylvania legislature appropriated funds for the First Geological Survey of Pennsylvania in 1836. Henry Darwin Rogers and a team of talented assistants did much meticulous fieldwork up to 1841, when the legislature curtailed support for the survey.[18] As the first detailed study of a mountain system outside the Alps, the Pennsylvania survey contributed much to the development of geological science. International recognition of Rogers's work led to his appointment as professor of natural history at the University of Glasgow in 1858.

Regardless of how it eventually contributed to geologists' understanding of the large-scale structure of mountain belts, the First Survey failed to

16. A dramatic account of the ruin of mine operators through this cause is given by Bowen, ed., *Coal Regions*, 41.

17. H. M. Chance, *Mining Methods and Appliances Used in the Anthracite Coal Fields,* Second Geological Survey of Pennsylvania, 1883 (Harrisburg, Pa., 1883), 10, 400.

18. The origins and the causes of the demise of the First Survey are described by Wallace, *St. Clair,* 201–215.

help the mine operator or miner in the anthracite districts. First, the initial reports, published in the 1840s, were "small, fragmentary, and hastily written."[19] Second, mine operators and owners were disinclined to accept the reports. Throughout the nineteenth century, geologists had no convincing theory to account for the phenomena they so carefully observed in the field. They advanced and vigorously defended sharply divergent interpretations. Rogers advocated the theory that the folds of the mountains in Pennsylvania had been formed by waves on the liquid interior of the earth caused by gigantic earthquakes. J. P. Lesley, one of Rogers's assistants, observed, "Unfortunately Mr. Rogers was led by overstrained poetic sentiment to imagine for these tremendous phenomena [the folding and subsequent erosion of the Appalachian Mountains] a physical cause which no one now considers necessary." Practical miners and mine managers found such theories difficult to believe; as Lesley observed, "The disputes of the geologists respecting doubtful points, if listened to at all, were regarded as good evidence of the worthlessness of all their theories . . . and the truths in which they agreed seemed . . . like the insanities of exalted imagination or the impious utterances of an irreligious temper."[20] Later experience would show that Rogers's interpretation of the large-scale structure of the Schuylkill coalfield was more nearly correct than that of the local, practical geologists. However, in 1840 a miner found it hard to know which geologist he should listen to for reliable advice.

A third limitation on the utility of the results of the First Survey was that the authors had little to say about the geological structures that were of most direct consequence to the mine operator and the miner, namely, faults and the course of the veins in the vicinity of a mine. Among the 450 pages Rogers wrote on the anthracite districts in his book, he devoted only 3 to faulting, and he included faults in none of his geological sections. Geologists on the First Survey had to guess at the underground course of coal veins; at the time of the Second Survey (Figure 3) they freely admitted that it remained guesswork.[21]

---

19. This evaluation was made by J. P. Lesley in his *Historical Sketch of Geological Explorations in Pennsylvania and Other States,* Second Geological Survey of Pennsylvania, 1874–1876 (Harrisburg, Pa., 1876). The final report was not published until 1858, at which time it was substantially revised and updated. It appeared as Henry Darwin Rogers, *The Geology of Pennsylvania . . . ,* 2 vols. (Philadelphia, 1858).

20. Lesley, *Historical Sketch,* 86, 111.

21. Chance, *Mining Methods,* 50.

FIGURE 4. Job 111, Bethlehem Mines Corporation

A. Eastern Face, August 1988. *Photograph by author.* This open pit mine is located in the Panther Valley, two miles east of Tamaqua. The horizontal lines (forty feet apart) across the face are benches used by vehicles in the excavation. Coal was mined from the veins (dark here, obscured in places by scree) in the nineteenth century by means of breasts driven up from gangways (not visible here).

B. Coal Veins. Sketch of veins in *A*, with apparent displacements caused by the benches removed. Faults have offset the two clusters of folds whose crests are visible.

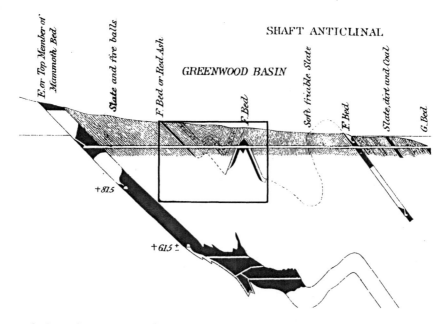

C. Cross Section. *From the Second Geological Survey of Pennsylvania.* Note that the section does not show the faults or the full complexity of the strata that have now been revealed by the open pit excavation.

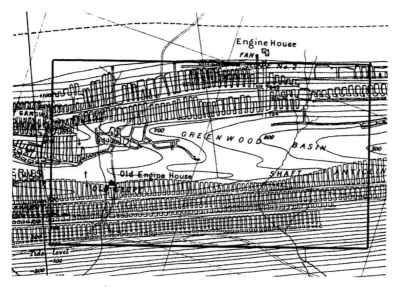

D. The Underground Workings Used in Mining. This map corresponds to C. Contour lines show the elevation of the Buck Mountain coal vein. Where the lines are closely spaced, the inclination of the vein is great (the contour interval is fifty feet). The approximate location of the open pit mine (outlined here) shows how the pit penetrated the old mine workings.

The lack of published results, the difficulty of knowing which geologists to believe, failure to produce the detailed information needed for mine management, and the populist distrust of science then prevalent made it easy for legislators to terminate fieldwork on the first survey in 1841. Some propertyowners' concern over possible diminution in the value of their holdings from unfavorable geological reports contributed to the survey's demise.[22] Even if the survey had continued, miners would still have had to rely primarily on their own skills to carry out successful extraction of coal for many years to come.

THE MINING SYSTEM

The first adventurers on the coal lands took anthracite from small quarries they could work without specialized mining skills. Thomas Bedwell, who had learned about mining by reading an encyclopedia article, managed the opening of the Lehigh coal mine in 1792. Here a thick, nearly horizontal coal vein reached the surface (see Figure 5). Successive operators developed Bedwell's quarry into the Summit Hill mines of the Lehigh Coal and Navigation Company.[23] According to a description in 1833:

> The coal appears to be on average, sixty feet thick, and to follow very nearly the surface of the ground where found. It is overlaid in some parts with stone, in others with decomposed coal and a stratum of yellow soil. The covering where it has been worked varies from ten to twenty feet. . . . This mine is worked by uncovering and quarrying. The excavation now amounts to ten acres.[24]

Abijah Smith introduced an additional component of quarrying technique in 1818, when he recruited John Flannigan, a quarryman from Milford, Connecticut, to experiment with loosening the coal with black powder.[25] However, Smith and others found quarrying methods alone inadequate as soon as they recognized that the Summit Hill vein was unique in the anthracite fields; most of the coal veins cropped out for only a small distance at the surface. They could see steeply inclined veins on the sides of

---

22. Wallace, *St. Clair,* chap. 4.

23. Powell, *Fuel Crisis,* 9–10; Bowen, ed., *Coal Regions,* 23.

24. Pennsylvania, General Assembly, Senate, Committee on Coal Trade, *Report of the Committee of the Senate of Pennsylvania upon the Subject of the Coal Trade* (Harrisburg, Pa., 1834), 53.

25. Hudson Coal Co., *The Story of Anthracite* (New York, 1932), 113.

FIGURE 5. East Workings of the Summit Mine of the Lehigh Coal and Navigation Company. *Drawing from Henry Darwin Rogers,* The Geology of Pennsylvania . . . , *2 vols. (Philadelphia, 1858), II facing 66.* This was the first mine worked by the company, and the coal was shipped to market by way of the Lehigh River. A fold brings the Mammoth coal vein close to the surface here, and the vein plunges downward at the edge of the quarry.

ravines, and, by 1814, miners were working these by driving horizontal tunnels ("drifts") and hauling the coal out in wheelbarrows. Where veins reached the surface,

> shafts were sunk but a few feet, in the *crop* of the vein; and the coal, raised by means of the common windlass and buckets, and so soon as they attained a depth where the water became troublesome, (which seldom exceeded thirty feet,) the shaft was abandoned, and another sunk; and the same process undergone.[26]

In 1825, the demand for coal was growing rapidly. Miners shipped 6,500 tons from Schuylkill County that year; just four years later they had increased production more than tenfold, to 79,973 tons. They had started to substitute horses for men at the windlasses drawing coal from the mines

26. Pa., Gen. Ass., Sen., Com. on Coal Trade, *Report*, 97–98.

FIGURE 6. Coal Vein ("The Fourteen-Foot") at the Nesquehoning Mines. *Engraving from Henry Darwin Rogers,* The Geology of Pennsylvania . . . , *2 vols. (Philadelphia, 1858), II, facing 1011.* Exposed on a hillside and worked by means of a drift (a horizontal tunnel driven into the coal), the coal was brought out in mine cars running on rails and probably pulled by mules. Chutes and screens shown here prepared the coal for market.

by 1823; with this additional power, they were also able to bail water out of the shafts and penetrate the veins to a somewhat greater depth. In 1827, some operators placed rails in their drifts and used wagons drawn by mules to ease the labor of moving the coal to daylight (see Figure 6). By 1833 a few operators had adopted the system of mining based on the slope, gangway, and breast.[27] To develop a mine, they drove a slope (an inclined shaft) to provide access to the coal veins and serve as the main route for hauling the coal to the surface. Next, miners cut gangways into the coal vein and opened chambers called "breasts" out from the gangways. The miners kept the breast partially filled with loose coal so that they could stand within reach of the unmined coal above (see Figure 7). Miners reached their workplaces by climbing up the manways adjacent

27. Hudson, *Story,* 51; Pa., Gen. Ass., Sen., Com. on Coal Trade, *Report,* 7, 84, 98; Bowen, ed., *Coal Regions,* 36–37.

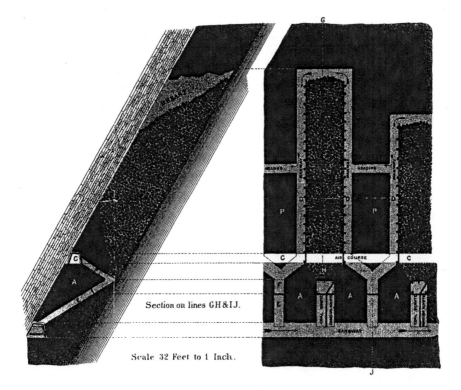

Section on lines GH&IJ.

Scale 32 Feet to 1 Inch.

FIGURE 7. Mining Coal from a Steeply Dipping Vein by Means of a Gangway and Breasts. *From H. M. Chance,* Mining Methods and Appliances Used in the Anthracite Coal Fields, *Second Geological Survey of Pennsylvania, 1883 (Harrisburg, Pa., 1883), unbound sheet.* Miners reached the breast by walking along the gangway and climbing the manway to the top of the broken coal, which provided a footing. Timbers and tools were carried by them up the manways at the start of the working day. The timbering supported the roof once the coal was removed from it. Airflow was directed (arrows) by the mine fans.

to the loose coal. The manways also provided paths for the circulation of air. Miners drilled holes in the coal, packed in black powder, attached and lit fuses, and promptly left the area. Later, they drew enough broken coal from the breast to keep the footing at a convenient height. Mules hauled the coal along the gangways to the shaft, slope, or tunnel that served the mine. As the miners removed coal from the breast, they placed timber props to support the roof of their workplace. Anthracite miners used this system throughout the life of their industry.

After they dug out the coal near the surface, the miners worked deeper. By 1834, some operators were experimenting with shafts extending below water level; by 1839 ten collieries had mines that penetrated the water table. There were seventeen by 1842; and, in 1844, twenty-two collieries had twenty-eight steam pumping and hoisting engines at work. The capital they had to invest in pumping and ventilating systems to work below water level greatly increased mine operators' costs. However, as they increased the complexity of their mines and aboveground plants, they retained the established system of working the coal faces. A miner of the 1830s would have found little different underground in the 1870s; an English miner from 1720 would have been in familiar surroundings.[28]

The system of mining Pennsylvanians had in place by 1834 incorporated familiar elements: the gangway was similar to the drift (shown in Figure 6), and the breast, to the rooms opened out from the coal quarry (shown in Figure 8). It was not a system developed by professional miners. Mine agent William Milnes asserted in 1834 that "the knowledge of mining in the community is very limited." When the miners at Summit Hill reached the bottom of the nearly horizontal coal vein in the 1820s, the proprietors panicked, because none of them realized that the vein plunged below ground. Everyone believed that anthracite existed only above water level until, following a suggestion made by Benjamin Silliman, someone dug to greater depths in 1833. In locating their mines, the adventurers put down shallow test pits that usually failed to reveal the thickness and dip of the vein. For want of rudimentary knowledge of mining, they misplaced their shafts and wasted more money and labor than they usefully employed. Early observers in the anthracite districts, such as Ele Bowen, recognized these deficiencies. Bowen observed that the mining methods adopted in the early stages of the trade were "extremely defective and expensive" and that the quickest mode of mining was adopted by the early, inexperienced miners; the possibility of deep basins of coal "did not occur to the inexperienced adventurers in the Trade, for some time after its commencement in 1826." Bowen also describes how miners dug a tunnel 200 feet below the ridge of the Summit Mine to reach the coal vein being worked from above; they located their tunnel on the assumption that the vein dipped to the south, and had driven it 790 feet before they found the vein did not go that way.[29]

28. C. K. Yearley, Jr., *Enterprise and Anthracite: Economics and Democracy in Schuylkill County, 1820–1875,* Johns Hopkins Studies in Historical and Political Science, LXXIX, no. 1 (Baltimore, 1961), 111, 120.

29. Pa., Gen. Ass., Sen., Com. on Coal Trade, *Report,* 71; Yearley, *Enterprise,* 100–

FIGURE 8. Baltimore Company Mine, Wilkes-Barre, Pennsylvania. *Watercolor from Henry Darwin Rogers,* The Geology of Pennsylvania . . . , *2 vols. (Philadelphia, 1858), II, frontispiece.* Rooms opened out from a quarry to permit extraction of underground coal. Pillars of coal have been left to support the overburden. Breasts were rooms similar to these but opened upwards from a tunnel (called a gangway) driven into the coal.

Inexperienced adventurers aided by British miners established the basic elements of the underground mining system subsequently used in the anthracite districts between 1827 and 1834. The British miners encountered unfamiliar geological conditions when they came to the Pennsylvania anthracite districts. In most of their home coalfields, the coal lay in nearly horizontal seams that rarely outcropped. Britons had anthracite only in a small part of the South Wales coalfield, where they had mined it since the seventeenth century. By 1827, British mine proprietors had replaced the medieval bord-and-pillar method of mining with the longwall system in their mines outside the northern counties. (In the bord-and-pillar system, miners dug coal in rooms called bords separated by pillars of coal left to support the roof. Miners worked alone or with a helper in each

108; Ele Bowen, "A Glance at the Coal Trade," in pamphlet, *The McGinnes Theory of the Schuylkill Coal Formation . . .* (Pottsville, Pa., 1852), 36, 39; Bowen, ed., *Coal Regions,* 24.

bord, and the workplaces were assigned by lot. In the longwall system, miners worked in groups in a common area.) The early Pennsylvania adventurers, in their desire to obtain coal quickly with the smallest investment in mine development, followed the bord-and-pillar system, which was already obsolete in Britain. They cared nothing for the coal they made inaccessible by bad mining technique, which could amount to three to four tons for every ton sent to market. Their methods were not informed by contemporary British practice, even though most of the miners were of British origin.[30]

The abundant coal outcrops made it easy for individuals or partnerships to win coal with a small investment in the early days of the trade. However, two factors helped a few capitalists to gain control of most of the anthracite coal resources quickly. First, the water gaps that pierced the surrounding ridges provided the most economical way of getting coal out of the mining districts. Second, the anthracite fields of eastern Pennsylvania, not being attractive agricultural lands, had been largely wilderness when the exploitation of coal began; individuals who had acquired capital through commercial ventures could easily acquire large tracts. Philadelphians began investing when the Lehigh Coal Mine Company acquired 8,665 acres of "unlocated" (wilderness) land in the 1790s. Subsequently, the Carey and Wetherill families bought large tracts, and Stephen Girard purchased 30,000 acres of the Mahanoy coal region at auction from the First Bank of the United States in 1830.[31] According to Bowen, companies bought up most of the coal land of the Schuylkill district during the "coal rush" of 1829 to get monopoly control of coal production. Thus, while adventurers established a technique of mining in the early years of the coal trade, nonresidents whose principal interests were in the commerce of Philadelphia gained control of the coal lands. They leased their coal properties through agents, often for short terms, to anyone who antici-

30. A. A. Archer, *Geology of the South Wales Coal Field: . . . Gwendraeth Valley and Adjoining Areas* (London, 1968), 166 (also compare his plate 4 with Figure 4, above); Allan R. Griffin, *Coalmining* (London, 1971), 10, 47; Yearley, *Enterprise*, 123. Anthracite mining was still directed by mining agents rather than mining engineers in 1858, according to Volney L. Maxwell, *Mineral Coal* (Wilkes-Barre, Pa., 1858).

31. Bowen, ed., *Coal Regions*, 19, 26; Powell, *Fuel Crisis*, 9; Wallace, *St. Clair*, 54–63; Nicholas B. Wainwright, "The Age of Nicholas Biddle, 1825–1841," in Russell F. Weigley et al., eds., *Philadelphia: A Three-Hundred-Year History* (New York, 1982), 273; more examples of purchase at auction are quoted by Powell, *Fuel Crisis*, 16.

pated a profit from mining. The social usages that became entrenched in the anthracite districts at the same time that the Pennsylvania system of mining was established (1825–1834) grew out of the system of absentee ownership and mine operation by agents under short-term leases.[32]

THE MINER'S WORK

Because we have few records or descriptions of underground mine work in the early years of the anthracite fields, we have to rely on a reconstruction based on the physical requirements of mining and the accounts of mine organization and operation prepared after the system was in place. The underlying principle of the Pennsylvania system of anthracite mining in place after about 1834 was that miners won coal in breasts that opened off one or more gangways in the mine. A miner, paid on piece rate with one or two helpers whom the miner directed and paid, worked each breast. Since an overseer could visit each breast only occasionally, he could not directly supervise any of the miner's work. In the larger mines, a "fire boss" inspected for explosive concentrations of firedamp (methane) at the start of the work day, and an "inside boss," who might visit each breast once in the course of a day, had overall supervision of the underground operations. Individual miners made their own judgments about the working of the breast, the tools to be used, and safety within it.[33]

Once in a breast, the miner and helper had no direct communication with other miners; they worked as an independent team. Although the miner exercised control over the work in his breast and could fully apply his skills in winning coal there, he was dependent on the mine organization for the essential services of hauling out the coal, ventilation, and pumping water. He could be endangered by bad judgment on the part of the mine operator, by the aboveground staff handling the pumps, hoists, and ventilating equipment, or by almost anyone working underground in the mine.

---

32. Bowen, ed., *Coal Regions*, 29–30. The owners of the coal lands in Schuylkill County are described in Yearley, *Enterprise*. The concentration of ownership continued as long as underground mining lasted. By 1901, railroads controlled more than 96% of the anthracite coal lands and 50% of the price of coal at tidewater went to the transportation companies; see Peter Roberts, *The Anthracite Coal Industry: A Study of Economic Conditions and Relations . . .* (New York, 1901), 66. I. Lowthian Bell, *Principles of the Manufacture of Iron and Steel . . .* (London, 1884), 555, describes the efforts of the railroads to control the coal trades and thereby drive up prices in the 1880s.

33. Chance, *Mining Methods*, 169, 401.

Other miners might, for example, block the ventilation system, cause an explosion by using candles when safety lamps were called for, or cause a crush by robbing too much coal from the pillars. Each miner relied on the managers and owners of the mine for adequate ventilation and drainage systems and for the overall plan of mine development. He could be endangered by the failure of past mine operators to keep adequate records of their work, thereby creating the possibility of inadvertently breaking into old, unmapped workings filled with gas or water, or by the operators of adjacent mines who, through greed or incompetent surveying, extended their underground workings beyond their property lines.[34]

The organization of work that emerged in the initial years of underground anthracite mining in Pennsylvania differed markedly from that in other nineteenth-century American industries. Workers had nearly total control with minimal supervision within the breast. Here the miner decided how the work was carried out on the basis of his skill and experience. However, the miner and his helper depended on others for the essential services of pumping, ventilating, hauling, and hoisting. Three factors set a coal mine apart from other workplaces. The miner and helper worked in an isolated room out of view of supervisors and out of communication with other miners. The mine itself, unlike a textile mill or ironworks, was invisible to people on the surface who might have taken an interest in the underground proceedings had they been able to see them. The mine was at best a dangerous place where any person within could endanger all others. Others, such as sailors, worked in inherently hostile environments. However, on a sailing vessel the sailors worked outside in full view of one another, and everyone accepted the need for a chain of command. The management structure of a coal mine did not provide for either a clear chain of command or the fixing of responsibility above the level of the inside boss. Other American industries encountered difficulties in developing responsible and effective supervision at this time.[35] However, the consequences were particularly severe in coal mining.

THE MINER'S SKILLS

The geological complexity of the anthracite coal deposits was much greater than that of the bituminous coals in Britain and North America.

---

34. *Ibid.*, 293.

35. Thomas C. Cochran, *Frontiers of Change: Early Industrialism in America* (New York, 1981), 124.

In the years before 1850, geological science provided little guidance to the solution of the practical problems of mine operation. Both miners and mine operators contended with a variable and uncertain environment that offered few clues that they could use for guidance in the development and operation of a mine. Additionally, those who established the mining practices and associated social usages in the anthracite regions had worked in an environment where ignorance and greed were more common than mining knowledge or a desire to develop the coal resources systematically. All of these conditions left individual miners dependent on their own skills in the winning of coal.

An important component of skill is an individual's capacity to carry on work with only incomplete information available to guide decisions. The many geological complexities that give few clues, the need to reach decisions without the presence of supervisors, and the likelihood of serious accident due to misjudgment demanded a high level of skill from the miner. Additionally, work in the anthracite mines was arduous.[36] The miner and his helper climbed to the working face of their breast through narrow manways carrying their tools and the heavy timbers needed to support the roof. They drilled holes for the powder charges, placed the roof props, and moved the broken coal entirely by hand.

Miners learned to interpret the geological structures near their work areas and, drawing on their general knowledge, adapt their mining methods to them. Both the mine operator and the individual miners depended on their capacity to adapt to the conditions they encountered in the property they worked. Since the visible clues about the geological structure of a mine were sparse, miners and mine operators needed a prolonged learning period in which to discover how to deal with the geological characteristics of their mine. Much of this knowledge was of value only at that locality. Henry Darwin Rogers emphasized this in his report on the First Geological Survey of Pennsylvania:

> It is of the greatest importance in mining that the collier should make himself familiar with the derangements of the strata in his particular neighborhood, so as to determine for himself, if possible, the prevailing

---

36. J. R. Harris shows that coal mining in Britain was highly skilled work but was not so recognized outside the trade, because the skills had developed gradually and little was written about them, in "Skills, Coal, and British Industry in the Eighteenth Century," *History*, LXI (1976), 167–182. On the arduousness, see the description of the difficult physical tasks in Chance, *Mining Methods*, 425–426.

character and direction of the displacements. An intimate acquaintance with the underground workings of the adjoining mines, will very frequently show him that these follow a certain rule or law, and it will teach him to infer, from the presence of peculiar signs of irregularity, not only the nature of any fault or derangement which he may be approaching, but the readiest means for either avoiding or passing through it.[37]

Because they needed local knowledge, and could gain it only by experience, the value of the skills of the mine operator and the individual miners increased with the duration of employment at a given mine. Their specialized knowledge had less value at nearby mines, and even less in other mining districts. While working in a mine, the miner and the operator made an investment in knowledge they could not use elsewhere. In contrast, nineteenth-century mechanics in metal- and woodworking learned skills that they could apply to many different products made by numerous factories. They frequently moved between different plants, often making distinctly different products. As they moved about and exchanged information with their peers, they contributed to discovery of the best way to carry out any given task, to the benefit of both the entrepreneurs and the artisans. Miners had fewer opportunities for this sort of incremental improvement.

The skills of the coal mine operator and of the individual miner contributed most to the national economy when they were fully developed at a particular mine over the learning time needed to uncover all its particular characteristics. However, the practice of moving about among mines was established among miners early in Pennsylvania. John White, a Philadelphia coal merchant, reported how,

> my new partners and myself furnished to a person who went to England for that purpose about one thousand dollars to pay the passage and expenses of miners, who had been brought up to that business. They arrived early in 1827, under an agreement that they would repay the advances made by a weekly per centage from their pay until the amount was refunded. We, however, had no benefit from the importation . . . as some broke their agreement and deserted, and the others were dismissed for disorderly conduct etc.[38]

37. Rogers, *Geology*, 238.
38. Pa., Gen. Ass., Sen., Com. on Coal Trade, *Report*, 83–84.

Human and natural resources would have been best used in anthracite mining by keeping operators and miners at work in one mine until they had dug out all the coal they could get from it. Instead, agents representing absentee owners arranged short-term leases. Working on these terms, operators could not develop a mine systematically to ensure long-term productivity. The tenure of an operator was short compared to the learning time he needed to make the most of a mine. Neither the operator nor the miners had enough time to learn the peculiarities of the mine they worked and so could not gain the full reward of the exercise of their skills.[39]

This system of working the mines wasted both the natural and the human resources of the coalfields. It also left both the mine operators and the individual miner vulnerable to exploitation. In learning how best to work a mine, both acquired intellectual property. However, they had neither ownership of the physical property nor a market that could value their intellectual property. Consequently, the miners and the operator had little incentive to investment in either fully learning the geology of a mine or in the timbering and ventilation system needed for its safe operation.

Environmental and Social Consequences

As early as 1834, knowledgeable Pennsylvanians saw that the mining adventurers in the coalfields lacked the knowledge they needed to make the best use of the state's anthracite resources.[40] By midcentury, authorities such as Samuel Daddow and Ele Bowen were emphatically condemning the poor mining practices used by miners and mine operators in the anthracite districts. They saw capital wasted in ill-planned ventures and the coal forever lost by poor mining practice. The adventurers' inadequate mining techniques led to environmental damage and a high accident rate. These, in turn, contributed to social distress in the mining districts. While many Pennsylvanians noted the environmental damage, they were not much concerned about it at that time. They routinely ascribed accidents to the carelessness of the miners.

Many of the costs of industrial wealth are not direct costs charged to those who receive goods and services. Instead, they are externalities, costs

39. See, for example, Roberts, *Coal Industry*, 17.
40. Pa., Gen. Ass., Sen., Com. on Coal Trade, *Report*, 71.

(or benefits) distributed without the award of corresponding benefits (or costs). Coal users paid part of the cost of mining and transporting coal in the purchase price. However, consumers did not pay the social and environmental costs—death and injury from mine accidents, exploitation of workers and mine operators, social strife, and environmental degradation—of mining. Residents of the mining regions bore these burdens; distant consumers knew little about them. Consumers gathered the benefits of using anthracite as canals and railways carried coal away from the mines for the production of iron and the manufactured goods. Such separation of costs from benefits was new in North America.

ENVIRONMENTAL COSTS

Pennsylvanians could easily see some of the environmental effects of mining in the nineteenth century. Ele Bowen wrote of the mining regions in 1848:

> There never was a more grand, picturesque region—beautiful in all its seasons—grand in all eyes,—precious to the man of science, the capitalist, and to the whole world of business. But if it be wild and beautiful now, when jealous art has despoiled it, somewhat, of its wild aspect— stripped the mountains of their gaudy foliage, and levelled the venerable and sturdy forest trees to the earth—with here and there one remaining, stripped of bark and branches—as if intended for monuments of their perished fellows;—what must it not have been when the howls of the wild beasts went forth in the solitary depths of the woods,—in the deep ravines and mountain passes until then unexplored by man?[41]

Through most of the nineteenth century, Americans observed and commented upon environmental degradation. They accepted it as an unavoidable component of creating wealth, much as we accept highway accidents today as an unavoidable component of the motor vehicle transportation system. The one environmental consequence of anthracite mining that did worry Pennsylvanians by midcentury was the possibility of running out of coal prematurely through the large waste inherent in their mining system. Of the coal in a vein, miners left, on average, 45 percent in pillars supporting the roof and broke up 8 percent to a size too fine to be used; additionally, 7 percent adhered to shale to be lost when the coal was cleaned. Hence, miners wasted 60 percent of the coal in a vein before

---

41. Bowen, ed., *Coal Regions*, 26.

they got any to the surface. Then the breakermen lost another 16 percent in crushing, picking, and cleaning the coal. When miners finished with a vein, they had sent only 34 percent of its coal to market.[42] Successive generations of owners, operators, and miners made few improvements in this system of underground mining.

Much mine waste accumulated in the anthracite districts because the breakermen crushed much of the coal smaller than three-eighths of an inch, the finest anyone could burn successfully. Mine operators found no uses for this small coal and dumped it together with waste rock in piles next to their breakers. The piles of "culm," fine coal mixed with rock and other debris, remain marking the landscape near the sites of most underground anthracite mines today. After miners left, water flowed through underground workings, encountering the remaining coal. Anthracite usually contains about .6 percent sulfur. Reacting with water, it forms dilute sulfuric acid that can kill plants and animals when it reaches rivers and lakes. Additionally, rain washes culm into streams and, leaching through culm piles, adds to the acidic water.[43] Today, water continues to flow through many old workings and waste piles.

As miners pumped water from mines to keep them dry, they lowered the level of surrounding ground water, often causing long-established wells to dry up. If someone built a house near a mine when the water table was depressed by pumping and mine operation subsequently stopped, rising groundwater often flooded the house. To prevent such damage, pumping would have to be continued at a worked-out mine for the indefinite future.

Miners left coal pillars to support the mine roof. However, they rarely knew just how big the pillars had to be. Operators began to rob the pillars after as much coal as could be safely taken from the breasts had been extracted. Miners commenced robbing at the distant parts of the mine

42. Eckley B. Coxe *et al.*, *Report of Commission Appointed to Investigate the Waste of Coal Mining* (Philadelphia, 1893); Chance, *Mining Methods*, 475.

43. Franklin Platt, *Causes, Kinds, and Amount of Waste in Mining Anthracite* (Harrisburg, Pa., 1881), 23; Chance, *Mining Methods*, 475; Y. K. Rao, *Stoichiometry and Thermodynamics of Metallurgical Processes* (Cambridge, 1985), 97. Possible uses for culm were explored by the commission on waste. Only in the last two decades has the demand for fine coal become great enough to justify reworking old culm banks. The commission on waste in mining was concerned also about the loss of coal in streams but noted that, since most of it filled in ponds along the stream routes, it could be recovered by dredging (Coxe *et al.*, *Waste*, 46).

and worked toward the entrance, allowing the roof to collapse as they retreated. Subsidence at the surface resulted.

Nineteenth-century holders of mining estates often sold building lots to increase the return on their investment. They retained the mineral rights and wrote deeds containing provisions releasing themselves from responsibility for damage that might result from extraction of any coal that happened to be under the property. Subsidence might occur while mining was in progress or many years after the miners had left, when residents might have forgotten the restrictive clauses in their deeds. In cities such as Pittston, where abandoned mines extend under much of the urban area, subsidence damaged structures and roads and occasionally caused loss of life, long after mining ceased.[44]

Anthracite does not ignite by spontaneous combustion. However, boiler tenders at mines sometimes inadvertently ignited culm banks when they dumped hot ashes on them. Underground fires could be a more serious problem. Abandoned anthracite mines contain residual coal exposed to air in any workings above the water table. It is usually impossible to seal all of the openings of older mines. Often, there are no records of these openings, particularly where miners dug bootleg holes to take coal without paying royalties. Consequently, extinguishing a mine fire by cutting off its air supply may be impossible. Fires may be started long after mining operations have ceased. In 1962, the volunteer fire department of the town of Centralia, Pennsylvania, inadvertently ignited coal as they burned trash near old mine workings. Gradually, the fire spread under the town, forcing its evacuation. Years later, the state had to buy all the private property in the village and raze the town.[45]

One estimate of the cost of cleaning up the environmental damage done by anthracite mining is fifteen billion dollars.[46] A surcharge of less than two cents per ton added to the price of the total of nine billion tons of

44. Roberts, *Coal Industry,* 10.

45. This fire will probably continue to burn for many years, because it has proved to be impossible to cut off its air supply. Because carbon monoxide gas from the fire entered homes in Centralia, most residents of Centralia have been relocated through a federal project that has cost $42,000,000. See David DeKok, *Unseen Danger: A Tragedy of People, Government, and the Centralia Mine Fire* (Philadelphia, 1986); Renée Jacobs, *Slow Burn: A Photodocument of Centralia, Pennsylvania* (Philadelphia, 1986); *Reading Times,* Oct. 24, 1988.

46. DeKok, *Unseen Danger,* 33.

coal mined, and put in a trust fund would have generated the fifteen billion. Charging the users of coal the cost of environmental damage would probably not have been a significant drag on nineteenth-century economic growth.

SOCIAL COSTS—THE LIMITS OF SKILLS

As soon as coal miners began working underground instead of in quarry pits, they faced increased dangers. Blasting coal out of a vein with explosives in a confined, underground space was inherently dangerous work. Today we use many technologies with a high potential for danger, such as flight in aircraft, with little risk because a combination of market forces and regulation makes operators devote substantial resources to the necessary safety precautions. In anthracite coal mining, neither markets nor regulations had much influence on the safety of mine work, and the accident rate increased as the industry grew. The system of mining in breasts, in addition to wasting much coal, had a great potential for accidents. For example, mine operators had little technical knowledge or experience to guide them in deciding how thick the pillars supporting the mine roof had to be; and, even when they left adequate pillars, exposure of the coal to air reduced its strength, with roof falls resulting. In spite of the waste and frequent accidents, mine operators continued with this system of mining through custom, ignorance, and royalties paid on the basis of coal removed regardless of the amount wasted within the mine.[47]

No one kept systematic data on mine accidents in Pennsylvania until 1870. Nevertheless, most nineteenth-century observers reported that accidents were common and became more common as the miners went deeper.[48] Since mine operators made few changes in their basic system of operating the underground portions of their mines after 1833, the detailed records of accidents kept after 1870 are probably a good guide to the types of accidents that occurred in the earlier years of mining. Accounts of mine explosions dominate the descriptive and popular literature on coal mine accidents. However, the records show that roof falls close to the working

47. Yearley, *Enterprise,* 123.

48. *Ibid.,* 167, 172. Systematic collection of data on mine accidents began in response to the requirements of the act of Apr. 5, 1870, passed by the Pennsylvania legislature. The first data appear in Pennsylvania, Inspector of Mines, *Reports of the Inspectors of Coal Mines of the Anthracite Coal Regions of Pennsylvania for the Year 1870* (Harrisburg, Pa., 1871). Reports were issued annually thereafter.

faces in the breasts, the areas of the mine under the immediate control of the individual miners, caused most of the fatal accidents in anthracite mines.[49] Since they killed or injured only a few men at a time (albeit day in and day out), they did not attract the attention accorded the drama of the less frequent gas explosions. The remedy for roof falls was generous use of props to support the roof. Each miner decided how much effort and wood to put into timbering his breast on the basis of his experience and local knowledge.

To find why accidents from roof falls were so frequent, we need to decide whether they arose because mining skills were ineffective, because miners were induced not to use their skills to the full, or because miners and owners were careless. Professional writers about mining at least as far back as the time of Georgius Agricola (Georg Bauer) in the sixteenth century routinely ascribed mine accidents to carelessness rather than the inadequacies of miners' skills. Anthony F. C. Wallace has emphasized how mine inspectors and owners avoided confronting other causes of accidents by blaming accidents on the carelessness of miners. H. M. Chance, in his report on mining methods for the Pennsylvania Geological Survey, attributes the frequent roof falls to a combination of carelessness, poor judgment, and reluctance to take precautions because timbering was "dead time," work for which there was no direct payment to the miner or the mine lease–holder.[50] In fact, the system of mine organization and inspection failed to provide either effective enforcement or inducements to miners to put up adequate roof support. The situation seems similar to that in the trucking industry today, where individual drivers make their own decisions about safe operation without supervision in an environment where enforcement of regulations is lax and there are no economic rewards for following safe procedures. Anyone can observe the range of responses among truck drivers on the highways. Miners probably behaved in much the same way.

Neither mineowners, mine operators, nor state authorities overcame the social and technological barriers to safe mining in the nineteenth century. In 1899, the annual death rate was 1 in 305, and the injury rate 1

---

49. Chance, *Mining Methods*, 409.

50. Georgius Agricola, *De Re Metallica* (1551), trans. Herbert Clark Hoover and Lou Henry Hoover (1912; New York, 1950), 6; Wallace, *St. Clair*, 265; Chance, *Mining Methods*, 400. Roberts and other writers in the late 19th and early 20th centuries rather glibly blame nearly all mine accidents on the careless miner.

in 136. (For comparison, the death rate among workers on railroads in Pennsylvania was 1 in 434, although the injury rate was higher than in the anthracite mines.) Between 1870 and 1899 the death rate paralleled the production rate, showing that no improvements in mine safety had been made.[51] Since the 1870 mining system was the one established decades earlier, the late-nineteenth-century accident data represent a characteristic of the early mining system. Even these data do not fully measure the actual risk faced by the underground miners. First, only about half of the mine labor force worked underground, where nearly 90 percent of the deaths occurred. Second, deaths resulting from injuries were not counted in the official reports; and, third, there was some undercounting of accidents. Wallace suggests correcting by a factor of 1.3 for the second and third errors. All of these factors together suggest a death rate at midcentury for underground work of about 1 in 130 per year. At some mines the rate was significantly higher, approaching, as described by Wallace, that of military losses in combat.[52]

Safe operation of an anthracite mine depended on the skills of individual miners. However, the established organization of mine operation discouraged miners' exercise of prudence. As the pace of American industrialization quickened in the nineteenth century, artisans found their craft skills increasingly taxed in tasks such as puddling iron, casting brass, or navigating steamships. Artisan skills alone could not provide the degree of reliability the public expected. In an anthracite mine the geological uncertainties were so great and the clues about them so limited that it is unlikely that exercise of craft skills alone, even if given free play, could have made anthracite mining safe.

One part of the social context in which a technology is practiced is the way that the limitations of craft skills are dealt with. Industrial societies impose regulations and discipline and provide support to the victims of accidents. The attitudes of miners, operators, and owners in the anthracite districts discouraged those who would emplace these social structures. The result was a progressive deterioration of the quality of life in the mining communities that bred violence, intransigence, and despair. By the dawn of the twentieth century, these could not be put right.

We do not have a calculation of the cost of reducing the frequency of accidents in underground mining of anthracite to the lowest level that

51. Roberts, *Coal Industry*, 168, graph facing 159.
52. Wallace, *St. Clair*, 249–253.

could have been attained with existing technology.[53] It seems likely, however, that in the nineteenth century, when there was little competition from other fuels, these costs could have been passed on to the users of coal with little harm to the national economy. However, they could not be charged to coal users unless safe mining practices were required by enforceable regulations imposed by outside authority. Other nineteenth-century industries besides coal mining created social costs because of inadequate technology or regulation. However, anthracite mining was nearly unique in the complexity of its social structure and the high degree to which its human and environmental costs were confined to the mining communities. The geographical isolation of the anthracite basins kept the costs of mining out of the direct purview of both the owners of the coal lands, most of whom were located in Philadelphia, and the users of anthracite, who were found wherever railways and waterways were capable of delivering coal.

AFTER 1825, Americans rapidly overcame the technical difficulties that had inhibited their use of anthracite. Factory proprietors and ironmasters who substituted energy from this coal for energy from wood and the fall of water reduced their energy costs and avoided the constraints of location inherent in the other energy sources. A technological unbalance accompanied the rapidly growing demand for the new fuel because no one made corresponding advances in the technology for the safe and environmentally benign mining of this coal. The mining methods and the social uses among miners, mine operators, and mineowners put in place by inexperienced adventurers and a few English miners between 1827 and 1834 soon proved to have unfortunate social and environmental consequences. These were exacerbated by the peculiar geology of the anthracite deposits, the

53. In 1901 Roberts calculated the economic cost of mine accidents for the decade 1890–1899 to be $60,000,000. The average annual production was about 50,000,000 tons per year in this decade. The economic cost of accidents would then have increased the cost of producing coal by $.12, or about 10% of the direct cost of mining (about $1.25 per ton) (Roberts, *Coal Industry*, 55, 170, graph facing 159). If allowance were made for the undercounting documented by Wallace, the cost would be nearer 23%. By today's standards, Roberts's estimate of the cost of accidents is incomplete because it makes little provision for medical attention or the support of miners' dependents. Nor does it count the costs of debilitating diseases, such as black lung. While the direct economic cost of mine accidents may then have been more than 25% of the cost of mining coal, it seems likely that it was not high enough to retard significantly the national or state economy. This is one reason that little was done to reduce the accident rate.

system of mine ownership and control, and new technology in transportation of coal that made it possible to isolate most of the environmental and social costs of producing anthracite in the mining districts while the fuel was carried elsewhere to use in producing wealth.

To explain the ill distribution of the costs of anthracite wealth, we have to consider both the geological and technical complexities of mining the coal and the social uses and practices established in the early years of the industry. These practices included partial worker control, managers who were not owners, and owners who did not manage. The complex relations between these groups are not easily described by the existing terms of discourse in business and labor history, and their analysis remains a challenge. The subject is worthy of study because experience has shown that, as new energy technologies involving potential environmental and social costs replace the energy system based on coal, many aspects of the anthracite story are being repeated.

CAROLYN C. COOPER

A Patent Transformation

Woodworking Mechanization in Philadelphia,

1830–1856

One of the earliest nationwide institutions of the new American republic was the patent system, established by the Constitution in 1790 and revised substantially in 1793 and 1836, intended to promote "the Progress of Science and useful Arts, by securing for limited Times to Authors and Inventors the exclusive Right to their respective Writings and Discoveries." Even as the Patent Office building loomed large on the landscape in Washington at the time of its construction in the 1830s and 1840s, so did the burgeoning system of interaction by inventors, businessmen, patent examiners, lawyers, users of the inventions, and, ultimately, the public loom large in the national consciousness of Americans as doers. An English parliamentary commissioner sent to the New York Exhibition of 1853 to study the American system of manufactures observed that,

My thanks go to Robert Gordon for his moral support as well as drawings, to Judith McGaw for her entrepreneurship, to the National Science Foundation and the National Endowment for the Humanities for funding, to Diana Long and Tom Patton for providing a home in Philadelphia, and to the members of the Philadelphia Transformations Project, in particular Donna Rilling, for helpful discussion and information about carpenters in 19th-century Philadelphia.

for "the American working boy," the examples "of ingenious men who have solved economic and mechanical problems to their own profit and elevation, are all stimulative and encouraging."[1]

Invention was one thing; its actual adoption and use, necessary for the "profit and elevation" of the inventor, was another. Reward did not come automatically, but only through skillful management of the patent, which involved an inventor in the above-mentioned interaction with businessmen, lawyers, patent officials, and so on. The results of such management interactions in specific localities, as well as on a national level, not only determined how quickly or widely an invention was adopted but also sometimes had the surprising effect of reshaping the invention, as perceived and further acted upon by the persons who were interacting. Patent management, then, was one process by which technology was socially constructed.[2] This essay attempts to show how the patent management interactions that took place during mechanization of woodworking in a specific locality—Philadelphia—resulted in the social reconstruction of certain woodworking machines as well as the restructuring of the existing arrangements for woodworking. It first sets the stage for these interactions by discussing the structure of the woodworking industry in early nineteenth-century Philadelphia before these machines appeared on the scene.

American Wood and Its Workers

As emphasized in Brooke Hindle's book *America's Wooden Age,* Americans have enjoyed an abundance of wood. Through the first half of the nineteenth century Americans were not only exporting lumber to England and elsewhere overseas but were themselves consuming from three to six times more of it per person (even aside from their use of wood for fuel) than were their English cousins. They also became prolific inventors of

---

1. United States Constitution, Article 1, Section 7; George Wallis, *New York Industrial Exhibition: Special Report of Mr. George Wallis* . . . (London, 1854), reprinted in Nathan Rosenberg, ed., *The American System of Manufactures* (Edinburgh, 1969), 204.

2. Carolyn C. Cooper, "Social Construction of Invention through Patent Management: Thomas Blanchard's Woodworking Machinery," *Technology and Culture,* XXXII (1991), 960–998.

machinery for working wood into useful form. Knowledgeable visitors from England commented in midcentury on "the wonderful energy that characterizes the wood manufacture of the United States."[3]

In Philadelphia, two important woodworking processes—irregular shaping and planing—underwent mechanization by the middle of the nineteenth century. In twenty years, two patented machines not only replaced hand tool methods of shaping and planing wood for certain purposes but also prevailed over alternative machines for accomplishing those two tasks. The Blanchard (or irregular turning) lathe, patented in 1820, shaped wood into tool handles, shoe lasts and boot trees, hat blocks, wheel spokes, and gun stocks. In Philadelphia, as elsewhere, these irregularly shaped wooden objects in turn would feed into the manufacture of hand tools, hats, shoes, carriages and wagons, and firearms. The Woodworth (or cylinder) planing machine, patented in 1828, made wooden boards and planks flat and smooth and of uniform thickness. It also cut tongues and grooves on the edges of boards and could cut moldings useful in making doors and window sashes. These wooden building components in turn would feed into Philadelphia's construction industry. Thus, wearing shoes made on Blanchard-shaped lasts and treading on Woodworth-planed floorboards, Philadelphians continued the nineteenth century on a new footing. The advent of the two machines had also placed Philadelphia's woodworking on a new footing.

Both the Woodworth planer and the Blanchard lathe originated elsewhere. William Woodworth lived in Hudson, New York, at the time of his invention, Thomas Blanchard lived in Millbury, Massachusetts, and neither man migrated to Philadelphia himself. However, an inventor did not need to use his own patented invention, but could assign (give or sell) his patent rights for any specific geographic territory to a local innovator who would then use it.[4] Through the patent system local innovators learned of the Blanchard and Woodworth machines and introduced them

3. Nathan Rosenberg, "America's Rise to Woodworking Leadership," in Brooke Hindle, ed., *America's Wooden Age: Aspects of Its Early Technology* (Tarrytown, N.Y., 1975), 40, 56; John Anderson et al., *Report of the Committee on the Machinery of the United States of America* (London, 1855), reprinted in Rosenberg, ed., *American System*, 171.

4. For the distinction between innovation and invention, see F. M. Scherer, "Invention and Innovation in the Watt-Boulton Steam-Engine Venture," *Technology and Culture*, VI (1965), 165–187.

into Philadelphia. But the innovators of these machines in Philadelphia met opposition that sparked contention in legal and political arenas.

What kind of persons became the innovators, and what kind of persons became their opponents? Were the innovators already experienced woodworkers who wanted to improve their productivity with the new machines? Were they machinists who wanted to make and sell the machines? Or were they persons with extra money, interested only in making a return on their investment? Were their opponents traditional hand tool–wielding woodworkers who felt their jobs were threatened by the new technology? To answer such questions, one should first understand the relationships of persons who dealt with wood for Philadelphia. In the order of transactions from tree to consumer, these were, by and large, (1) the lumbermen who cut down trees and transported lumber, (2) sawmillers or sawyers who cut the lumber into convenient dimensions, (3) lumber merchants in the city who bought, stored, and sold the dimensioned lumber, (4) carpenters/joiners, wheelwrights, lastmakers, and other producers of wooden objects, and (5) householders and other consumers of wooden objects. Persons in any of these categories might have their own reasons to welcome or resist the introduction of a new woodworking technology, especially one that was patented. Who welcomed or resisted depended in part on who else was doing the introducing.

Lumber Supply, Sawmills, and Lumber Merchants

By 1860, lumber ranked second among United States manufactures, after cotton goods, in value added.[5] As a large port city, Philadelphia was an important user as well as shipper of lumber. Situated on a peninsula between the Delaware and Schuylkill rivers, Philadelphia was readily supplied by water with lumber from the interior of New York and Pennsylvania (see Figure 1). Both because of lumber's arrival by water and its use in building wooden ships, at the beginning of the nineteenth century "the bulk of the lumber business was conducted along the Delaware River front. . . . Nearly, if not all, of the lumber or timber reached the city in rafts [that] came down the Delaware from the Pocono Mountains . . . or down

5. Rosenberg, ed., *American System*, 26. Value added equals value of product minus cost of materials from which it is made.

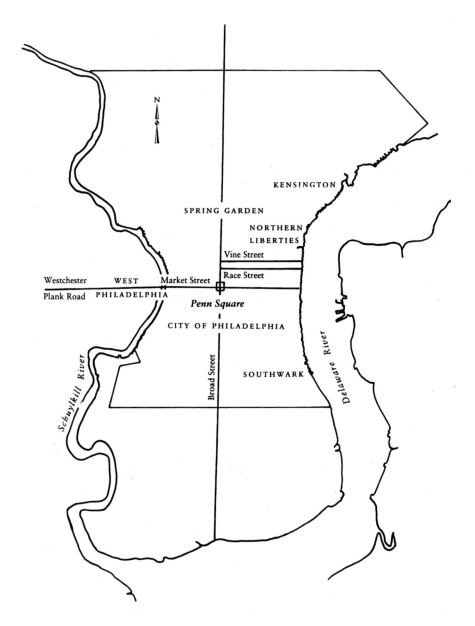

FIGURE I. Philadelphia in 1852. *Adapted from "Plan of the City of Philadelphia and Environs," in* Philadelphia as It Is in 1852 *(Philadelphia, 1852). Drawn by Robert B. Gordon*

the Susquehanna and into the Delaware by way of the old canal, from the Moosic Mountain region" just northeast of Scranton. "Board yards" and wharves for fuel wood were interspersed with other businesses all along the Delaware riverfront (see Figure 2). Once the Schuylkill Navigation Company began operation in the 1820s, more fuel wood and lumber also arrived on the less heavily populated but growing western side of Philadelphia. West Philadelphia on the other side of the Schuylkill acquired a lumberyard at present-day Thirty-fourth and Market streets, run by a long-established family of lumber merchants.[6]

Port and canal company records give some idea of this waterborne traffic in lumber. In the years 1832–1850 between 8,400 and 14,800 tons of lumber annually descended toward Philadelphia on the Schuylkill Canal. In 1851, lumber ranked fifth in the number of cargoes entering Philadelphia in the coastal trade from the four nearest states and third among the more remote coastal cargoes. It is difficult, however, to tell how much lumber from Philadelphia's own more immediate hinterland was cut and sawed to dimensions in rural sawmills and entered the city by wagons, a much less economical method of transport. (In 1830 lumber could travel to Philadelphia 280 miles by canal as cheaply as 39 miles by road.) At least one waterpowered sawmill in Haverford Township whose ledger has survived seems in the period 1837–1860 to have supplied both a nearby rural clientele and a number of customers in Philadelphia. To Penn Square in Philadelphia from Samuel Leedom's grist- and sawmill on Darby Creek was only about eight miles by way of the West Chester Plank Road and the bridge across the Schuylkill. Such a sawmill could produce perhaps two to five thousand feet of boards in a day.[7]

6. James Elliott Defebaugh, *History of the Lumber Industry of America*, II (Chicago, 1907), 577, 584. See also Ross McGuire and Nancy Grey Osterud, *Working Lives: Broome County, New York, 1800–1930* (Binghamton, N.Y., 1980), 5–13, for lumbering in New York; Diane Lindstrom, *Economic Development in the Philadelphia Region, 1810–1850* (New York, 1978), 25–26, 101; John F. Watson, *Annals of Philadelphia . . .* (Philadelphia, 1830), 208; William Barton Marsh, *Philadelphia Hardwood, 1798–1948: The Story of the McIlvains of Philadelphia and the Business They Founded* (New York, 1948).

7. Spiro G. Patton, "Transportation Innovation and Market Expansion for an Industrial City: Reading in the Nineteenth Century," *Canal History and Technology Proceedings*, VII (1989), 141–160, esp. tables 1 and 2; Lindstrom, *Economic Development*, 98–99, 107, 118; Samuel Leedom's account book, Archives, Historical Society of Pennsylvania, Philadelphia (hereafter HSP); Defebaugh, *Lumber Industry*, II, 582.

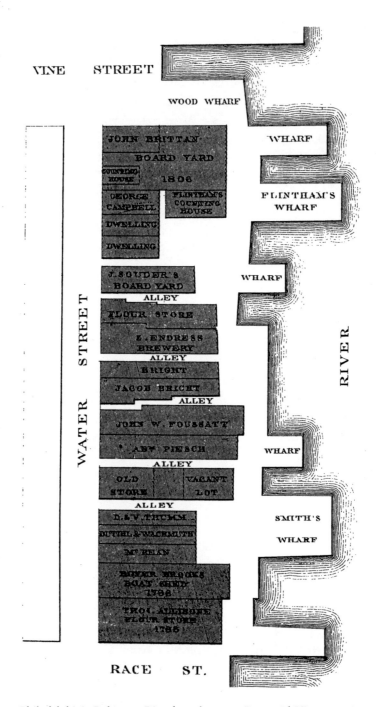

FIGURE 2. Philadelphia's Delaware Riverfront between Race and Vine, ca. 1800. *From Abraham Ritter,* Philadelphia and Her Merchants as Constituted Fifty @ Seventy Years Ago . . . *(Philadelphia, 1860)*

Within Philadelphia itself, however, where adequate waterpower was not available, the establishment of sawmills had to await the application of steam power. Oliver Evans, Philadelphia's most famous inventor, built high-pressure steam engines that found use by 1812 in a sawmill on the lower Mississippi, but Philadelphia's first steam-powered sawmills got under way in the 1820s, in part to cut veneers for cabinetry. Three sawmills in nearby Kensington, of which two were explicitly "steam sawmills," appeared on the Philadelphia County tax rolls in 1826, owned by James Gibson, John Naglee, and the Lehigh Coal and Navigation Company. That same year, mahogany sawyer William White offered his "circular saw mill and fixtures" for sale, and the works of well-to-do cabinetmaker Michael Bouvier's City Saw Mill were "carried on by steam" in 1833. According to an old-timer's memory in 1897, the sawmills were few but large.[8] The paucity of listings in Philadelphia city directories of the 1830s to 1850s for sawmills and sawyers suggests there were indeed few of them in the city proper (see Table 1). Ten steam engines were reported in use at Philadelphia sawmills in 1838. Block- and pump-maker Jonathan Wainwright established one near the waterfront in Kensington in partnership with a Gillingham who was presumably related to Samuel H. Gillingham, a nearby lumber merchant. In 1860 there were only seven sawmills in Philadelphia, according to the census conducted by the Philadelphia Board of Trade.[9]

Philadelphia lumber merchants were much more numerous than sawmills or sawyers (Table 1). One study counted more than 250 lumber

8. Dorothy Bathe and Greville Bathe, *Oliver Evans: A Chronicle of Early American Engineering* (Philadelphia, 1935), 78, 135, 184; Charles E. Peterson, "Sawdust Trail, Annals of Sawmilling, and the Lumber Trade . . . to the year 1860," Association for Preservation Technology, *Bulletin,* V (1973), 87–88; Kathleen Matilda Catalano, "Cabinetmaking in Philadelphia, 1820–1840" (master's thesis, University of Delaware, 1972), 26, 39; Thomas Wilson, ed., *Philadelphia Directory and Stranger's Guide for 1825,* 151; Member of the Philadelphia Bar, *Wealth and Biography of the Wealthy Citizens of Philadelphia . . .* (Philadelphia, 1845), 6; Defebaugh, *Lumber Industry,* II, 582, 583. On Kensington sawmills, I am indebted to Donna Rilling, personal communication.

9. Carroll W. Pursell, Jr., *Early Stationary Steam Engines in America: A Study in the Migration of a Technology* (Washington, D.C., 1969), 87; pattern of entries for Jonathan Wainwright, Wainwright and Gillingham, and Samuel H. Gillingham in Philadelphia city directories for 1831, 1837, 1842, and 1847 (for Wainwrights, see Abraham Ritter, *Philadelphia and Her Merchants as Constituted Fifty @ Seventy Years Ago* [Philadelphia, 1860], 82); Philadelphia Board of Trade, *Manufactures of Philadelphia* (Philadelphia, 1861), 19.

TABLE 1. *Philadelphia Occupations, 1831–1853*

	Percentage of Persons Listed in Philadelphia City Directories				
Occupational Groups	1831	1837	1842	1847	1853
Primary processors of lumber					
Sawmills	0	0	0	.09	0
Sawyers (includes mahogany sawyers)	.11	0	.13	.05	.11
Storers of lumber					
Lumber merchants	.32	.16	.31	.50	.34
Constructors of large wooden structures					
Carpenters (includes joiners and measurers)	4.14	4.13	3.32	5.48	3.80
Ship carpenters (includes ship joiners, shipwrights, and boat builders)	.83	.88	.61	.73	.75
Suppliers of tools and components for construction					
Sawmakers	.04	0	0	.09	.11
Plane makers	.18	.06	.04	0	0
Sash and blind makers	0	.20	.04	.05	.11
Makers of smaller wooden objects					
Lastmakers	0	.03	.04	0	.08
Wheelwrights	.29	.20	.18	.32	.23
Blockmakers	.07	.20	.04	.14	.04
Machine makers					
Machinists (includes mechanics, engineers, machine makers, engine manufacturers)	.18	.26	.96	1.00	1.84
Other occupations					
Cordwainers (includes boot- and shoemakers)	4.29	4.75	4.55	6.03	4.92

TABLE 1. Continued

Occupational Groups	Percentage of Persons Listed in Philadelphia City Directories				
	1831	1837	1842	1847	1853
Weavers (includes carpet weavers)	2.41	2.65	1.79	4.11	1.96
Laborers	1.04	4.95	4.90	6.21	5.64

Sources: Samples from Philadelphia city directories (size of sample and basis in parentheses): 1831: *Desilver's Philadelphia Directory and Stranger's Guide* (2,226, every 10th page); 1837: *Desilver's Philadelphia Directory and Stranger's Guide* (3,052, every 10th); 1842: *McElroy's Philadelphia City Directory* (2,285, every 15th); 1847: *McElroy's Philadelphia City Directory* (2,189, every 20th); 1853: *McElroy's Philadelphia City Directory* (2,660, every 20th).

Note: Although the 1850 U.S. census identified respondents' occupations, censuses of 1830 and 1840 did not. To obtain a rough idea of trends in certain wood-related occupations, I used Philadelphia city directories. These directories in themselves are by no means comprehensive for the actual population of Philadelphia, or even for heads of households, and are probably biased toward including more visible and stable residents. Thus these data should not be taken as valid for the actual work force of Philadelphia, but as merely *suggestive* for the purpose of comparison. (I am assuming that the bias of the directories did not change radically from one edition to the next, so that the relative proportions of the different occupations and their general direction of growth or decline are not misleading.)

dealers operating in Philadelphia at some time during the 1820s and 1830s. So we can infer that, unlike Gillingham, few lumber merchants had their own sawmills, and most bought sawn lumber from the sawmillers. Lumber merchants then stored the drying lumber in scattered yards throughout the city, where their customers, such as carpenters or cabinetmakers, paid for seasoned lumber at convenient locations. Samuel Leedom's rural sawmill sold a wide variety and large quantity of lumber over the many years from 1844 to 1861 to customer Samuel Davis, who was probably Samuel H. Davis, a lumber merchant at Fourth and George streets in Philadelphia. To judge by Davis's account with Leedom's sawmill, a lumber merchant bought timber for house construction already cut to proper dimensions for girders, joists, lintels, floorboards, rafters, and so forth. He also bought scantlings, or pieces of wood that would be suitable for working into such items as parts of bedsteads or axle trees for wagons. He would then presumably sell them to carpenters, wheelwrights, furniture makers, and other customers. For instance, cabinetmaker George Ritter bought lumber in log or plank in small or large lots from various lumber dealers in the 1820s and 1830s. And in one nine-month period lumber merchant Davis bought from sawmiller Leedom more than six thousand board feet of poplar scantling intended for Biddle Reeves, a Philadelphia bedstead maker, of whom we learn more below (see Figure 3).[10]

Presumably the steam-powered city sawmills also cut lumber to different dimensions for the carpenters and other workers of wood. At least one sawmill also doubled as a door, sash, and blind mill: when John Naglee, one of the few combined lumber merchant-sawmillers in Philadelphia, died in 1852, the inventory of his Union Mill on Front Street above Maiden showed the stock on hand consisted of not only some twenty-six thousand feet of boards but also a wide variety of different sizes of shutters, doors, blinds, and window sash with and without glass.[11]

Lumber merchants varied in size of operation, but John Naglee was worth $100,000 in 1845 and was also one of two lumber merchants among directors of the Bank of the Northern Liberties. Lumber merchants located in Kensington, likely perhaps to be large-scale operators in the transshipment of lumber, apparently also had sufficient wealth and status to be heavily represented on the Kensington Bank's board of directors, whose president for many years was blockmaker and sawmiller Jonathan Wainwright. Only one of the fourteen directors of the Philadelphia Bank was a lumber merchant in 1837 and none in 1847, but the Kensington Bank had three lumber merchants among its fourteen directors in each of those years, plus two in each year who were in related occupations: sparmaker, ship joiner, shipbuilder, pump maker. If Philadelphia banks operated as banks did in New England at this time, directors could expect easy access to business loans from their own bank.[12] We can suppose capital was

10. Catalano, "Cabinetmaking," 14; Samuel Leedom's sawmill ledger, entries August 1857 and April 1858 in accounts of Samuel Davis, HSP.

11. *McElroy's Philadelphia City Directory* for 1851; inventory of Union Mill, Feb. 3, 1852, in account book for estate of John Naglee, Archives, HSP.

12. *McElroy's Philadelphia City Directory* for 1847; Member of Bar, *Wealthy Citizens of Philadelphia*, 17; Philadelphia city directories, 1837 and 1847. For remarks upon the wealth of Jonathan Wainwright and other Kensington lumber merchants, see David Montgomery,

BIDDLE REEVES & SON,

BEDSTEAD MANUFACTORY,

Nos. 89 & 91 ST. JOHN STREET,

ABOVE WILLOW STREET.

Cabinet-Makers, Upholsterers, Hotels, Private Families, and Boarding-Houses can be supplied at this establishment with

PATENT SCREW, AND OTHER BEDSTEADS,

ALSO,

BUREAUS AND WASH-STANDS,

OF EVERY KIND AND QUALITY,

OF THE MOST FASHIONABLE AND SUBSTANTIAL MANUFACTURE.

BIDDLE REEVES & SON,

FURNISH ALSO FROM THEIR STORE,

No. 216 Second Street, above Vine Street,

EVERY DESCRIPTION OF

FURNITURE,

And offer to Wholesale Purchasers a LARGE VARIETY OF GOODS, suitable for City and Country Trade, which they will dispose of

ON REASONABLE TERMS.

☞ Turning in Wood done at the Shortest Notice. ☜

FIGURE 3. Bedstead Manufactory Advertisement, 1852. *From* Philadelphia as It Is in 1852 *(Philadelphia, 1852), 386.* Biddle Reeves and Son used steam-powered machinery to manufacture furniture.

available to John Naglee and other well-to-do lumber merchants either
for expansion of business as usual or for investment in new technology
like the Woodworth planing machine. Which course did they choose?

Mechanics and Carpenters

Philadelphia was shifting its economic emphasis from commerce to
manufacturing in this period. The Franklin Institute, founded in 1824 to
foster study and dissemination of new technology, was both a manifestation and promoter of this change. The number of machinists, engine
makers, and engineers rose tenfold between 1831 and 1853, from fewer
than two to nearly twenty per thousand directory names—that is, to 2 percent (Table 1).[13] They developed their mechanical skills in the industrial
establishments of various sizes that Philadelphia was acquiring—dye and
chemical works, textile mills, papermakers, steam engine builders, locomotive and car works, machine shops, and all the smaller industries to
supply their needs as well as a multiplicity of consumer goods. By 1852
Philadelphia booster Job R. Tyson boasted to the British consul William
Peter, "So much are manufactories the order of the day, that it is said in no
community of the same size on the globe, are there so many steam-engines
at work, nor any in which they are applied to so great a variety of purposes
as in Philadelphia." Late in the 1850s Edwin T. Freedley listed more than
seven hundred separate items, from accordions to zinc paints, "now made
in Philadelphia, with the address of one or more manufacturers of each."[14]

"The Shuttle and the Cross: Weavers and Artisans in the Kensington Riots of 1844," *Journal
of Social History,* V (1971–1972), 412.

On banks, see Naomi Lamoreaux, "Banks and Insider Lending in Jacksonian New
England: A Window on Social Structure and Values," Colloquium on Early New England
Society and Culture, Old Sturbridge Village, March 1989. See also Lamoreaux, "Banks, Kin-
ship, and Economic Development: The New England Case," *Journal of Economic History,*
XLVI (1986), 647–667.

13. For descriptions of this growing community of Philadelphia mechanicians, see
Eugene S. Ferguson, ed., *The Early Engineering Reminiscences, 1815–40, of George Escol
Sellers,* Smithsonian Institution, United States National Museum Bulletin no. 238 (Wash-
ington, D.C., 1965); Bruce Sinclair, *Philadelphia's Philosopher Mechanics: A History of
the Franklin Institute, 1824–1865* (Baltimore, 1974); Anthony F. C. Wallace, *Rockdale . . .*
(New York, 1978), 214–239.

14. Philadelphia Board of Trade, *Letters on the Resources and Commerce of Phila-*

Already in the 1830s Rush and Muhlenberg, successors to Oliver Evans at the Mars Works, advertised, "Machinery of every description made and repaired," and locomotive builder Matthias Baldwin was making turning lathes and calico-printing machines.[15] Did these or other Philadelphia machinists branch out into production of Blanchard lathes or Woodworth planers to become woodworking innovators?

Machinists and engineers were on the rise in industrializing Philadelphia, but carpenters were much more numerous than those mechanicians, comprising 3–5 percent of the directory listings (Table 1). More carpenters were listed than weavers, and about the same number as cordwainers, although textiles and shoes were two important manufactures in Philadelphia at the time. The high proportion of carpenters in the directories undoubtedly reflects the rapid population expansion for the city and county of Philadelphia—nearly tripling from 137,097 in 1820 to 408,762 in 1850—and the need of these new residents for houses, the number of which rose from 53,078 in 1840 to 61,278 in 1850 and to 89,979 in 1860. Even buildings of brick and stone required large amounts of carpentry. For instance, for his new twenty-seven-foot-wide brick house on Fourth Street, banker Joseph Trotter in 1829 paid somewhat more for carpenters' work and lumber than for bricks, stone, and bricklayers' work. Master carpenters like Moses Lancaster acted as builders, or general contractors, undertaking to provide other types of supplies and labor on a house besides carpentry.[16] Of the 255 names listed under "Carpenters" in *Boyd's Philadelphia City Business Directory* for 1859–1860, 159 were "also Builders."

Carpenters were highly organized in Philadelphia. The well-heeled, prestigious, and cohesive Carpenters' Company of the City and County of Philadelphia, to which master carpenters like Moses Lancaster might

———

delphia: From Job R. Tyson, LL.D., to William Peter, Her Britannic Majesty's Consul for Pennsylvania, with Mr. Peter's Answer Prefixed (Philadelphia, 1852), 38, 39; Edwin T. Freedley, *Philadelphia and Its Manufactures . . .* (Philadelphia, 1859).

15. Sinclair, *Philosopher Mechanics*, 77, 82.

16. Lindstrom, *Economic Development*, 25; Elizabeth M. Geffen, "Industrial Development and Social Crisis, 1841–1854," in Russell F. Weigley *et al.*, eds., *Philadelphia: A Three-Hundred-Year History* (New York, 1982), 309; Joseph Trotter, account book, Archives, HSP; Philadelphia directories, 1830s–1850s. Lancaster's receipt book for 1812–1822 shows payments for deliveries not only of nails but also of such items as stone, bricks, and plaster and for carpenters', paperers', turners', and tinworkers' labor (Archives, HSP).

aspire to be elected, owned and rented out property in its own right and provided welfare to widows and orphans of its members. It also had a special relationship to the city government, in that official "measurers," who determined the costs of given construction jobs and settled any disputes between carpenters and their customers concerning those costs, were chosen only from Carpenters' Company members. The Carpenters' Company had a standing committee whose task it was to review and establish the prices for various carpenters' operations, including new ones arising from changes in architectural fashion. These prices were listed in closely held price books, issued in 1786, 1805, 1831, and 1852. As the number of carpenters in Philadelphia grew, the membership of the Carpenters' Company became increasingly exclusive, dropping from one-fourth of the master carpenters in pre-Revolutionary times to one-sixth after the Revolution. Despite the tripling of Philadelphia's population by the mid-nineteenth century, the Carpenters' Company membership was held to about one hundred between 1813 and 1853.[17]

In protest of the secrecy and exclusivity of the Carpenters' Company, carpenters also formed other societies in Philadelphia, such as the Practical House Carpenters' Society, founded in 1811, which published its own constitution, rules, and regulations for carpentry work in 1812. On the whole, carpenters, many of whom were members of no such society, were undergoing strains in their organization as a craft. Journeymen carpenters went on strike unsuccessfully in 1827 and joined the general strike for a ten-hour day in 1835, which was at least temporarily successful.[18]

17. Minutes of the Managing Committee, Carpenters' Company of the City and County of Philadelphia, 1838–1851; minutes of the Price Book Committee of the Carpenters' Company of the City and County of Philadelphia, Apr. 27, 1827, which mention that would-be measurers were to apply to the Managing Committee for a certificate; Carpenters' Company bylaws adopted Mar. 12, 1827, Price Book Committee minute book, April 1827–May 1829, Carpenters' Company Collection, Archives, American Philosophical Society, Philadelphia; Charles E. Peterson, ed., *The Rules of Work of the Carpenters' Company of the City and County of Philadelphia, 1786* (New York, 1971), xvii–xviii. On exclusivity of membership, see Roger William Moss, Jr., "Master Builders: A History of the Colonial Philadelphia Building Trades" (Ph.D. diss., University of Delaware, 1972); also, lists of members in *An Act to Incorporate the Carpenters' Company of the City and County of Philadelphia; By-Laws, Rules, and Regulations; Together with Reminiscences of the Hall, Extracts from the Ancient Minutes, and Catalogue of Books in the Library* (Philadelphia, 1866).

18. Peterson, ed., *Carpenters' Company,* xvii; Nicholas B. Wainwright, "The Age of Nicholas Biddle, 1825–1841," in Weigley *et al.,* eds., *Philadelphia,* 279–280.

Under such circumstances we might expect master carpenters to turn to mechanization as a technological fix to a labor problem and start using the Woodworth planer. Did they?

Carpentry remains today a relatively unmechanized technology; we still see carpenters at building sites sawing and hammering boards to fit. But before the early nineteenth century, they had to prepare the boards also, by hand planing them. In particular they had to plane floorboards to a uniform thickness if they were to lie evenly on the joists of a house. To make a strong and tight fit they also made tongues along one edge of the floor boards and grooves along the other. They did so with jack and trying planes for thinning and smoothing the boards and matching planes for tonguing and grooving (see Figure 4). For this job the master carpenter had the assistance of his journeymen and apprentices. With hand planes, a really energetic journeyman carpenter working flat out was estimated in 1833 to be able to plane, tongue, and groove twenty-five boards a day, each, say, 12½ feet long and 9 inches wide. At that rate, to prepare the floorboards for a small, one-and-a-half-story house, only one room deep, would take him seven days at ten hours a day constant and vigorous labor.[19] To save precious good-weather summer time for on-site construction, a carpenter able to spare the storage space and investment might attempt to stock up on planed, tongued, and grooved floorboards in advance, by preparing them during the winter, the slack season for construction.

The Woodworth Planer

In the 1830s in Philadelphia, two men running a Woodworth planing machine would be able to do that same job in 1¾ hours. House carpenters no longer needed to prepare their own floorboards long in advance for a job, but could take them from a lumberyard to a planing mill and get them back to the house site the same day.[20] Put another way, a planer-matcher did in fifteen minutes what a man with hand planes could do in a day. Furthermore, incremental improvement in the construction of Woodworth planers soon more than doubled their output from an adver-

19. Gregory Clancey, "The Cylinder Planning Machine and the Mechanization of Carpentry in New England, 1828–1856" (master's thesis, Boston University, 1987), 209.

20. Planing machine output time for 1830s adapted from Clancey, who calculates it for a later, faster machine in "Cylinder Planing Machine," 211.

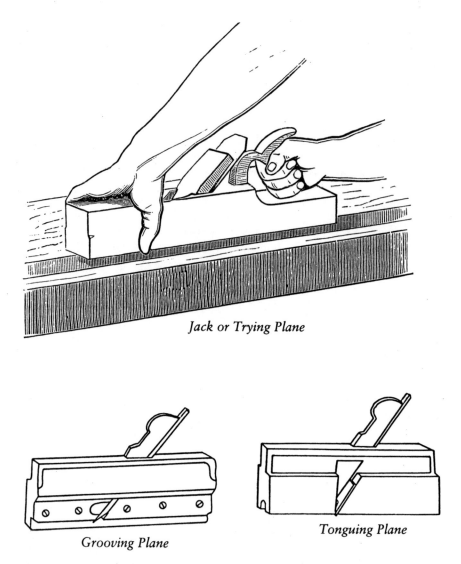

Jack or Trying Plane

Grooving Plane

Tonguing Plane

FIGURE 4. Planing, Tonguing, and Grooving by Hand Tools. *Adapted from* Wood-working Tools; How to Use Them: A Manual *(Boston, 1882), 31. Drawn by Robert B. Gordon*

tised 625–1,250 board feet per hour in 1839 to 3,000 feet per hour by 1849 (see Figure 5). Such dramatic technological change in the ease and speed of constructing wooden floors was sure to have far-reaching effects not only on carpenters but on nearly everyone, for nearly everyone wanted a house with wooden floors. After twenty years' experience, Woodworth's planer was hailed in Congress in 1850 as "next to Whitney's cotton

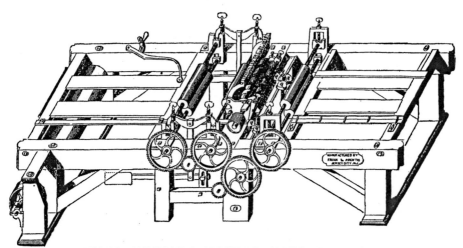

WOODWORTH'S PATENT PLANEING MACHINE.

F I G U R E 5. Woodworth Planing Machine, ca. 1847. *Advertisement of manufacturers Frink and Prentis, Jersey City, N.J.,* Scientific American, *II (1847), 407.* Cylindrically-rotating plane irons planed a board (not shown) that was fed to it between the rollers.

gin . . . the greatest labor-saving invention which has been produced in this country."[21]

William Woodworth's machine, for which he received a patent December 27, 1828, did not appear from a technological vacuum. Machines for planing had been attempted both in Europe and America for several decades. In England, for instance, the planers of Joseph Bramah and Malcolm Muir received patents in 1802 and 1827, respectively, and in the United States fourteen planing machine patents were issued—four of them to Pennsylvanians—between 1805 and 1828.[22] They were of three usual types: having stationary planing cutters, or having planing cutters

21. Advertisements by Samuel B. Schenck and John Gibson in Warshaw Collection, National Museum of American History, Washington, D.C.; Committee on Patents and the Patent Office, 31st Cong., 1st sess., H.R. Report 150, on H.R. 168 to extend the Woodworth patent for 14 years, Mar. 12, 1850, Legislative Records, National Archives (hereafter NA), Washington, D.C.

22. Bennet Woodcroft, *Alphabetical Index of Patentees of Inventions . . . ,* ed. A. M. Kelley (London, 1969 [orig. publ. London, 1854]), 63, 391; I. L. Skinner, ed., *The American Journal of Improvements in the Useful Arts and Mirror of the Patent Office in the United States* (Washington, D.C., 1828), 161, 224, 238, 252–255, 261, 536; M. D. Leggett, *Subject-Matter Index of Patents for Inventions . . . 1790–1873 . . .* (Washington, D.C., 1874), 1057, 1058, 1059.

on horizontally or on vertically rotating disks (Figure 6). In Bramah's machine, cutters rotated horizontally in a disk-shaped path (see Figure 7). But Woodworth's cylindrical planing machine succeeded better and eventually became the standard in Europe as well as America.

William Woodworth was born in 1780, probably in Massachusetts, but had moved to Hudson, New York, by the time of his invention.[23] Later identified variously as a carriage maker or a carpenter, he apparently had experience working with wood. Woodworth's machine design mounted knives around a rotating hollow cylinder and used pressure rollers to hold a vertical board firmly as it passed the planing cylinder, to prevent its being drawn back into the knives and chewed up. Small rotary "duck-billed" cutters on the sides simultaneously cut tongues and grooves into the edges (see Figure 8).

After an invention was patented, the task of bringing about innovation, or adoption in localities, did not fall solely on the inventor-patentee, but could be divided up and parceled out to many small investors. This act of patent management, called assignment, could occur even in anticipation of a patent, by advance sale of future patent rights. In the 1820s Woodworth was superintendent at a pulley-block mill on the Livingston family estate downriver from the town of Hudson but lacked sufficient funds of his own to get a working machine built and tested and patented. So he assigned half of his patent rights to his local congressman, James Strong, who put up the fifteen hundred dollars needed for "building a model and first machine and for the taking out of Patent."[24] (Of this, only thirty dollars was needed for the patent fee.)

Research and development expenses were not trivial. After Hudson machinist David Dunbar built the first machine for them, Woodworth and Strong took it to New York City and ran it at the Dry Dock, on the East River. There it underwent some significant revision: the board was

23. Loren Johnson, affidavit, *Artemas Brooks et al.* v. *John Fiske and Nicholas G. Norcross,* Transcripts of Records, Supreme Court of the U.S., 1853, II, Judicial Records, NA, Washington, D.C.; 1820 population census manuscript returns for Hudson, Columbia County, microfilm roll 70 for New York, NA; death notice, *New York Post,* Feb. 9, 1839.

24. William W. Woodworth, papers headed "Receipts" and "Account of expenditures and losses referred to in the annexed affidavit," in case file for *William W. Woodworth et al.* v. *Hiram Casley et al.,* U.S. Circuit Court, Massachusetts District, October term 1849, N.A., Boston Branch, Waltham, Mass.; assignment of William Woodworth to James Strong, Dec. 4, 1828, *Digests of Patent Assignments,* W-1, 4, Civil Records, NA, Washington, D.C.

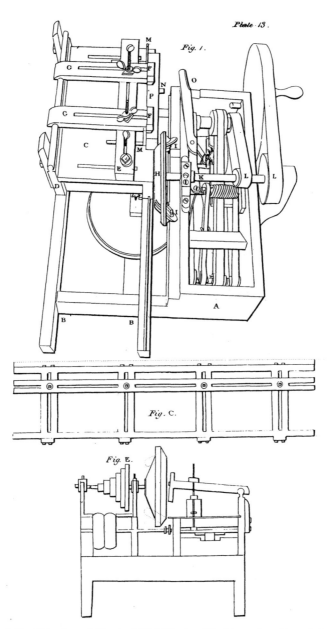

FIGURE 6. Pre-Woodworth Vertical-Disk Planing Machine, 1827. *From I. L. Skinner,* ed., The American Journal of Improvements in the Useful Arts, and Mirror of the Patent Office, *I (Washington, D.C., 1828), plate 15, 159–161, 224.* Josiah Reihm's machine, patented Nov. 1, 1827, did not tongue and groove, but planed a piece of lumber (P) by sliding it past the rotating two-feet-diameter cast iron disk on which "flat bits, or cutters" were fastened. In operation at the Savage Factory in Anne Arundel County, Md., it could plane three-fourths of an inch in depth in one pass.

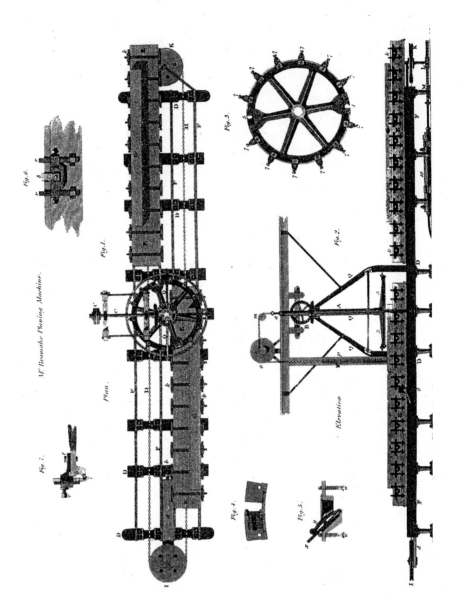

FIGURE 7. Bramah's Horizontal-Disk Planing Machine, ca. 1819. *From Abraham Rees*, The Cyclopaedia . . . (*London, 1819*), XXVII, s.v. "*Planing Machines.*" Patented in England in 1802, Joseph Bramah's planer moved a timber past cutters on a horizontally rotating iron wheel, shown from above at *Fig. 1*, from the side at *Fig. 2*, and in detail at *Fig. 3*. As the timber passes, it is cut first by gouges (7), then by small planes (4 and 5, shown in further detail in *Figs. 4, 5,* and 7).

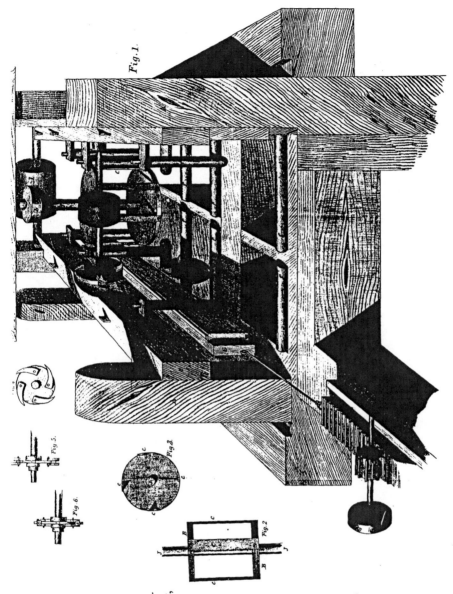

FIGURE 8. Woodworth's First Planing Machine, 1828. *National Archives, Philadelphia.* This drawing appeared in many Woodworth planer lawsuits in the 1840s and early 1850s, including *Jacob P. Wilson v. Daniel Barnum*, U.S. Circuit Court, Eastern District, Pennsylvania, April sess. 1849, Equity Cases, NA, Philadelphia. The rack and pinion at left drew an upright board fastened to the carriage G past the blades C of the rotating planing cylinder J, B (shown also at *Fig. 2*). Simultaneously, the cutters for tonguing (details in *Figs. 4 and 5*) and for grooving (details in *Figs. 3 and 6*) cut the edges of the board.

reoriented from a vertical to a horizontal position, and the cutters re-arranged accordingly (see Figure 9). Their subsequent "expenditures and loss in further experimenting and putting up and running" two machines in New York City cost an additional eight thousand dollars. They lost some thousands more in building four more successful and unsuccessful machines and three models for exhibition. At this time an upstate New Yorker named Uri Emmons claimed he had already invented essentially the same machine, and obtained financial backing for his future patent, which was issued April 25, 1829. Later lawsuit testimony after Woodworth and Emmons were both dead provided strong evidence that Emmons's claim was fraudulent and inspired by his assignees William Tyack, Daniel H. Twogood, and Daniel Halstead, who had considered buying assignments from Woodworth and Strong but apparently decided to muscle in on them instead. At the time, to avoid trouble, Woodworth and Strong agreed with Emmons and his backers to share rights to both patents and divided the country between the two groups. The two patents were then assigned together to purchasers in various territories all over the country.[25]

On his way to Washington to take out a patent in December 1828, Woodworth stopped in Philadelphia. There he promoted his machine by submitting "a working model and specimens of the work executed on a large scale" to the Committee on Inventions of the Franklin Institute, which examined them and reported favorably. The *Journal of the Franklin Institute* accordingly praised the Woodworth planer in early 1829: "The machine for which this patent was obtained, is in actual operation, and executes good work. We have seen a piece of cross grained white pine plank, which had been planed, tongued, and grooved, by it, with a finish which it would have been difficult to have given with the ordinary fore and match planes. It differs essentially from all the planing machines which we have heretofore known." The editor of the *Journal*, Thomas P. Jones, was at this time superintendent of the Patent Office in Washington, so this opinion was presumably quasi-official.[26]

25. William W. Woodworth, "Receipts" and "Account"; *Digests of Patent Assignments*, W-1, 4. For the use of patent assignment records, see Carolyn C. Cooper, "Thomas Blanchard's Woodworking Machines: Tracking Nineteenth-Century Technological Diffusion," *IA: Journal of the Society of Industrial Archeology*, XIII (1987), 41–54.

26. Leonard Chester, deposition in *Jacob P. Wilson* v. *Daniel Barnum*, U.S. Circuit Court, Eastern District, Pennsylvania, April sess. 1849, Equity Cases, NA, Philadelphia; *Journal of the Franklin Institute*, N.S., III (1829), 199; Sinclair, *Philosopher Mechanics*, 71.

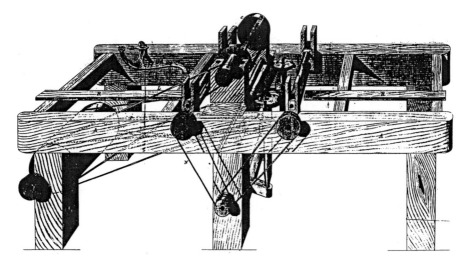

FIGURE 9. Woodworth's Second Planing Machine, 1828. *National Archives, Philadelphia.* This drawing accompanied that in Figure 8 in Woodworth patent lawsuits, including that of Philadelphia assignee Jacob P. Wilson against Daniel Barnum and his rotary planing machine, in 1849. Two sets of horizontal rollers, the lower ones driven by belts around pulleys M, fed the board R past the planing, tonguing, and grooving cutters.

Innovation of the Woodworth planer in Philadelphia—its adoption and application to the needs of woodworkers there—did not occur automatically. As usual, adoption required the "interest," meaning finance and promotion, of innovators, as distinct from the inventor himself. Investors in Philadelphia fairly promptly acquired rights in 1830 to make, use, and sell the Woodworth planer in Philadelphia City and County and sold and resold these rights thereafter. Who were these innovators in Philadelphia of the new planing technology? Were they persons who wished to use the machines themselves to plane wood, or were they capitalists seeking financial return only? One might expect these earliest investors to be nationally famous commercial or financial giants like Philadelphians Stephen Girard or Nicholas Biddle or perhaps the locally well-to-do lumber merchants who could get loans from the Kensington Bank. Or, perhaps, the carpenters of the venerable and affluent Carpenters' Company. Or, some mechanically knowledgeable members of the Franklin Institute might like to build and install Woodworth planers at, say, lumberyards or sawmills.

But the early Woodworth patent assignees in Philadelphia were relatively obscure and unlikely characters—Martha Inslee, Joseph Inslee (who ran a small hotel on Chestnut Street), George McClelland (who became Martha's financial trustee), and a soap- and candlemaker named Tobias Huber. Joseph Inslee's National Hotel at 116 Chestnut, on a corner across from the Bank of the United States, was housing twenty-five persons aged ten to sixty on census day in 1830, and kept a bar, where a pint of Madeira cost two dollars. Martha's trustee George McClellan[d] was possibly a professor of surgery or (more likely) a merchant of bonnets and shoes, later listed as a wealthy citizen worth fifty thousand dollars. Huber and McClelland and the Inslees raised six thousand dollars—two thousand of it by mortgaging some of Martha's property—to pay Woodworth and his partner for the right to make, use, and sell planing machines in the city and county of Philadelphia. They did not join with any lumber merchants or sawmillers. Between February and October 1830 from various persons they leased a steam engine in Kensington, a building, and a wharf lot and contracted for a supply of lumber. Among their first customers was a brewery. With a wharf lot in Kensington, they could probably expect orders from shipbuilders for decking, but in 1831 they moved to another location, at the corner of Arch and Broad streets, with new partners.[27]

The Patent Floor Board Company was joined by three other men, who expanded its operation and eventually bought out first Huber and then the Inslees. John Hemphill, Edgar H. Richards, and Mark Richards were apparently also newcomers to dealing in wood but were partners with one another in various other speculative commercial and industrial ventures. Edgar and Mark Richards were involved in iron mining and smelting in Pennsylvania and adjacent states. In 1831 John Hemphill, Mark Richards, and Joseph Inslee also bought half of the Woodworth-Emmons patent right for the two Floridas, Georgia, and South Carolina, which they sold again a year later. Mark Richards joined Hemphill in 1835 in "the adventure to Batavia, by the ship Walter," involving 1,089 bags of Java coffee. By 1835 Joseph Inslee was out of the hotel business and into real estate brokering, in partnership with his son and later alone, at least through

27. 1830 population census for Pennsylvania, roll 159, NA; *Philadelphia in 1830–1* . . . (Philadelphia, 1830), 226; bill from National Hotel, Archives, HSP; City directories, 1830–1853; Member of Bar, *Wealthy Citizens of Philadelphia*, 16; *Transfers of Patent Rights*, Liber E, 473, 478–480, 521, F, 23, 92, Civil Records, NA, Washington, D.C.; Leonard Chester, deposition in *Wilson v. Barnum*, 1849; James E. Johnston, deposition, Nov. 5, 1840, case file for *E. H. Richards v. Jacob Swimley*, U.S. Circuit Court, Eastern District, Pennsylvania, April sess. 1841, NA, Philadelphia.

1847. All Inslees apparently departed Philadelphia by 1851. By 1840 E. H. Richards opened a lumberyard at Market Street near the Schuylkill River, in which Woodworth machines planed boards, and Mark Richards was the active superintendent of the operation.[28]

Although one might well expect already-established lumber merchants also to take an "interest" in the planing machine, it was not until the mid-1840s, long after it was clearly the dominant technology, that any established Philadelphia lumber merchants—grudgingly—bought rights to use the Woodworth patent. James Patton, lumber merchant, bought a license for three Woodworth machines in Kensington in 1846, but only in the course of a long series of law suits *against* the Woodworth monopoly in Philadelphia. In early 1847 lumber merchants Thomas Manderson and Andrew Manderson, Jr., and merchant James Manderson bought the right to construct and use three Woodworth machines in Kensington until the end of 1849. But carpenters did not buy Woodworth patent rights at all, although they were not generally averse to buying patent rights for wood-working machines. Moses Lancaster, for instance, owned one-tenth of the right to a patented mortising machine.[29] Instead of Woodworth's machine, the lumber merchants and carpenters fostered other planing machines and provoked lawsuits by the Woodworth assignees.

So miscellaneous newcomers to the business of lumber processing were the innovators in Philadelphia of the machine that became not only the worldwide standard for planing machines but also the subject of numerous lawsuits that overloaded the federal judicial system on up to the Supreme Court. Instead of adopting the improved planing technology themselves, the carpenters and lumber merchants who dominated Philadelphia woodworking seemingly resented the upstart newcomers to the working of wood who had introduced the Woodworth machines. But they met their match in the courtroom. Most of the patent lawsuits ensued in the 1840s, after Woodworth himself had died and James G. Wilson, one of

28. Edgar H. Richards, deposition, Sept. 11, 1838, in case files for *Edgar H. Richards v. Thomas S. Richards and Thomas M. Smith,* U.S. Circuit Court, Eastern District, Pennsylvania, 1838, NA, Philadelphia; Philadelphia city directories, 1830–1853; assignments, Oct. 6, 1831, Mar. 15, 1832, *Digests of Patent Assignments,* W-1, 5; papers of *Hemphill v. Cochran,* in Cadwalader Collection, HSP; James E. Johnston, deposition in *Richards v. Swimley,* 1841.

29. *Transfers of Patent Rights,* Liber F1, 108, G1, 351; Philadelphia directories, 1842, 1847; Moses Lancaster's receipt book for 1828–1835, 6, Archives, HSP.

his early New York assignees, gained control of the renewed and reissued patent rights and formed an interregional cartel that fixed the price of planed boards. The frequently "odious" litigation of Woodworth patent managers eventually provoked a protest movement that generated thousands of signatures on impassioned petitions to Congress from all over the country.

In Philadelphia, Wilson supplanted the older generation of assignees in several stages, installing a second-generation triumvirate consisting of his own two sons and his son-in-law, who was the son of assignee and millwright Jonas Sloat. Henry R. Wilson, twenty-four years old, Jacob P. Wilson, twenty-one, and George B. Sloat, twenty-six, were all born in the state of New York and were relative newcomers to Philadelphia when they became Woodworth assignees in 1846 for the term 1849–1856. In 1850 they were residing in the districts of Spring Garden, Southwark, and Kensington, respectively, each within the territory delineated for his patent rights.[30] In taking on the management of the planing mills, they also took on the management of their patent rights in the law courts and continued the fight against the less effective planing machines that were backed by carpenters and lumber merchants.

Social Construction of Invention through Patent Litigation

Even if "odious," the lawsuits offered an unintentional channel for communication of technical information, in addition to the intended channel of the patent specification, which was to become public property after the patent period. Courtrooms and lawyers' chambers seem unlikely places for practical mechanics and engineers to hold seminars in identifying mechanical equivalents in machines, but something like a seminar had to be going on, when such expert witnesses would be asked to read specifications, look at models, and remember and describe machines they had seen elsewhere, in forming their opinions about relevant differences and similarities in the machines. In patent litigation, expert witnesses came from long distances to testify or at least sent written and pictorial depositions that were entered into the court record for response by other

---

30. 1850 population census manuscript returns for Philadelphia, NA; Philadelphia city directories of the 1840s; *Transfers of Patent Rights,* Liber D1, 318, E1, 140, 142, 144, T1, 405, 408, 412, 416; *Digests of Patent Assignments,* W-1, 128, 129, 130, 172.

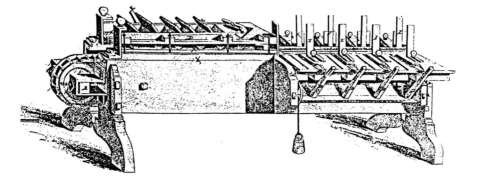

FIGURE 10. Kugler's Planing Machine, 1833. *National Archives, Cartographic Section, Alexandria, Va.* This drawing accompanied the specification of its patent, granted June 3, 1833. A series of successively deeper plane irons planed a board drawn past them.

expert witnesses. Over time, these experiences must have helped shape and spread the criteria for mechanical equivalence that came to be accepted by the technically knowledgeable community of mechanicians in the United States. In the course of these experiences, legal constructions in infringement suits gradually became social constructions of the patented inventions at issue.

The Woodworth planing patent trials held in Philadelphia demonstrate this gradual process of social construction. During the first term of Woodworth's patent, 1829–1842, thirty-two United States patents for planing machines were issued, including two to Uri Emmons in 1829 and one more to Woodworth in 1836. Three of the planing patentees early in this period were Philadelphians: one J. Percival in 1831, carpenter Moses Lancaster in 1832, and lumber merchant Benjamin Kugler in 1833. Kugler's patent was for a planer of the type having stationary rather than rotating plane irons. He spent twenty thousand dollars developing it and made "diligent inquiry" of the Franklin Institute's Thomas P. Jones in Washington (who was no longer head of the Patent Office but still editor of the *Journal*) to avoid the possibility of infringing other patents (see Figure 10).[31] As mentioned above, Jones had described Woodworth's machine as "essentially"

31. Leggett, *Subject-Matter Index*, 1057, 1058, 1059; Benjamin Kugler and James M. Patton, depositions, case files for *Mark Richards v. James M. Patton*, U.S. Circuit Court, Eastern District, Pennsylvania, April sess. 1839, NA, Philadelphia; Sinclair, *Philosopher Mechanics*, 200–201.

different from previously known planing machines, such as those with stationary plane irons.

Benjamin Kugler's 1833 stationary plane machine apparently worked passably well, for he ran one himself for three or four years in Philadelphia on a Delaware River wharf above Noble Street, and another was used by an assignee in Troy, New York. In 1837 Kugler sold all his remaining patent right to James M. Patton, who improved the machine and ran it at the same wharf in Philadelphia. Woodworth planer assignee Mark Richards noticed this competition and proposed, according to Patton, that they "unite their interests . . . to control the market." When Patton declined to do so, Richards brought a patent infringement suit against him. He didn't, however, claim infringement of Woodworth's patent, apparently deeming it too different from Kugler's. Instead, he went to Erie County in New York and paid one Caleb Taylor sixty dollars for his otherwise unused 1829 patent for a stationary-iron planing machine, even though the *Journal of the Franklin Institute* had found Taylor's machine to have "numerous and insurmountable" difficulties and expressed doubt that it would work. Returning to Philadelphia as Taylor's assignee, Richards sued Patton for infringement of Taylor's patent. Patton remarked that the low price of Taylor's patent right showed it was not a good machine and said Richards had brought the lawsuit "solely and entirely in hope and desire to vex, harass, embarrass and disturb" him and Kugler. Whether or not he was vexed, Kugler went out of the lumber business for good. He became a prosperous and civic-minded medical doctor instead. In 1850 he was active on a committee to set up a free reading room in Spring Garden for working young men and apprentices.[32]

James M. Patton remained in the planing business, however, using various kinds of machines at shifting locations. He continued to be sued in the 1840s and early 1850s for patent infringement by the Philadelphia proprietors of the Woodworth patent during its two extensions.[33] In these later

32. *Journal of the Franklin Institute*, N.S., III (1829), 336; James Patton, deposition in *Richards* v. *Patton*, 1839; Philadelphia directories, 1837, 1847–1853; 1850 census manuscript returns, microfilm roll 818, p. 481; *Pennsylvania Inquirer*, Apr. 6, 1850.

33. See Patton's various addresses in the Philadelphia directories for 1837, 1842, 1847. On lawsuits, see records of U.S. Circuit Court, Eastern District, Pennsylvania, NA, Philadelphia: *William W. Woodworth* v. *James Patton*, April sess. 1844; *Barzillai C. Smith* v. *James M. Patton*, October sess. 1846; *George B. Sloat* v. *James M. Patton*, April sess. 1847; *George B. Sloat* v. *James M. Patton*, April sess. 1851.

suits, the planing cutters compared with one another were, not stationary like Kugler's, but rotary cutters of various types. Regardless of how
Patton changed his machines, all of them raised accusations by Woodworth assignees that he had infringed the Woodworth patent. In 1846,
for instance, Patton agreed to pay Woodworth assignee George B. Sloat
a seventy-five-cent royalty per thousand board feet, but then instead took
the cutter out of his excessively Woodworth-type machine, substituted a
disk-type cutter "like that of Bramah" (that is, horizontal), and resumed
operation without paying royalties. Justices Robert C. Grier and John K.
Kane enjoined Patton in April 1847 to refrain forever from construction,
use, or sale of that particular machine and of any Woodworth machine,
"under penalty of $10,000." But by 1851 he was in court again for using a
machine that had a disk-type "face plate" (that is, vertical) cutter, whose
use was enjoined in December 1852. A complaint in January 1853 accused
him of using it again.[34] These court decisions constituted social constructions of the law and were cumulatively defining the different cutters in
Patton's machines to be "the same."

In these and the numerous other Woodworth patent cases in Philadelphia, witnesses were called upon repeatedly to decide whether specific
features of any two planing, tonguing, and grooving machines at issue
were "the same" or "different." For instance, the planing machine of New
Englander Ira Gay, patented in 1836, had, instead of a cylindrical planing cutter, a beveled planing disk (or flattened cone) (see Figure 11). In
November 1838, "Gay's New and Valuable Patent Planing Machine" was
on display and for sale at a fair held by the Franklin Institute. Builder Jacob
Swimley promptly ordered one from Gay's assignee, Samuel Shepherd. It
arrived in March 1840, accompanied by mechanic Thomas Hill, who came
from New Hampshire to set it up for Swimley and stayed on in Philadelphia (thus augmenting by one the local population of mechanics). It had
a tonguing-and-grooving feature that Samuel Shepherd and David Baldwin had patented in 1837. Swimley paid $10,500 for both patent rights
to construct, sell, and use the Gay-Shepherd-Baldwin machine in Philadelphia City and County and in a twenty-one-square-mile strip across the
river in New Jersey. When he saw it, Mark Richards thought Gay's plan-

34. James M. Patton, affidavit, Oct. 9, 1846, in *B. C. Smith* v. *Patton,* 1846; George B.
Sloat, bill of complaint, Oct. 27, 1846, in *Sloat* v. *Patton,* 1847; justices Grier and Kane,
decree, Apr. 14, 1847, in *Sloat* v. *Patton,* 1847; George B. Sloat, complaints, May 10, 1851,
Jan. 28, 1853, in *Sloat* v. *Patton,* 1851.

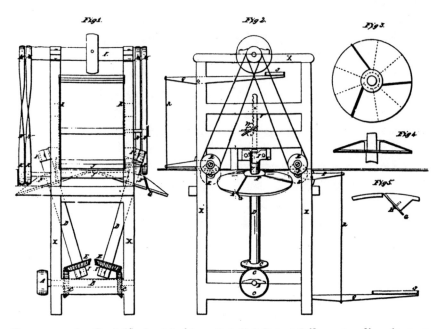

FIGURE 11. Ira Gay's Planing Machine, 1836. *U.S. Patent Office, microfilm of patent drawing, redrawn and photolithographed after 1870.* Ira Gay, machine tool–maker in Dunstable, N.H., received a patent for this planing machine June 25, 1836. Straight plane irons were set into the beveled face of the disk at the angles shown in *Fig. 3*, and one-half of the face was presented parallel to the board, as indicated in *Fig. 1*.

ing disk did not infringe upon the Woodworth planing cylinder, but that Shepherd's thick circular saws for tonguing and grooving did infringe on the Woodworth "duckbill" cutters for tonguing and grooving. Edgar H. Richards and John Hemphill sued Swimley in 1840–1841. They asked for an injunction but apparently withdrew the complaint before a hearing was held.[35]

Almost a decade later, the planing machine assignees were different

35. Handbill dated Nov. 7, 1838, in records of *Edgar H. Richards and John Hemphill v. Jacob Swimley,* U.S. Circuit Court, Eastern District, Pennsylvania, April sess. 1840, NA, Philadelphia. Jacob Swimley, affidavit, Nov. 21, 1840, in *Richards v. Swimley,* 1841; *Transfers of Patent Rights,* Liber G, 30; *Digests of Patent Assignments,* G-1, 12; John K. Kane, *Opinions of the Hon. John K. Kane, Delivered in the Circuit Court of the United States for the Eastern District of Pennsylvania, in Equity, October Sessions, 1846, in the Cases of Smith and Sloat against Mercer and Another, Plympton and Others, and Patton and Another, upon Questions Arising under an Alleged Invasion of the Woodworth Patent Planing Machine* (Philadelphia, 1846), 11.

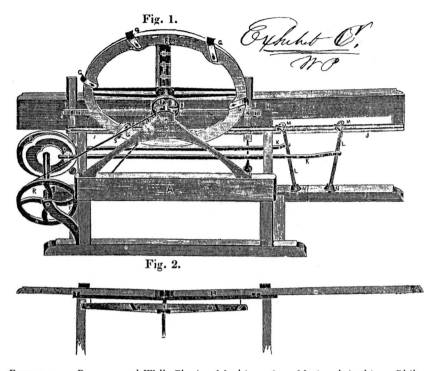

FIGURE 12. Barnum and Wells Planing Machine, 1849. *National Archives, Philadelphia.* Daniel Barnum and Thomas J. Wells of New York City received U.S. Patent 6185 for this vertical-disk planing machine, Mar. 13, 1849, as an improvement on the Bramah machine in planing boards (*Fig.* 2 shows the fence *H* that curved the board, so that it was not mangled). This printed copy of the patent drawing was Exhibit C in *Wilson* v. *Barnum* in 1849.

and lawsuits even more frequent. New Yorker Daniel Barnum's planing machine, patented in 1849, did not have the Woodworth/Emmons pressure rollers to constrain the board, but had gently curved iron plates that slightly bent the board as it was progressing past the whirling cutters on its planing disk (see Figure 12). The slight and fleeting curvature of the board was sufficient to keep it from being drawn into the revolving cutters and chewed up, which defect was frequent—some said inevitable—with the Bramah-type planer. The complainants argued that this feature served the same function as Woodworth's pressure rollers and therefore was "the same," that is, a mechanical equivalent.[36]

Were cutter disks significantly different from Woodworth's cutter cylin-

36. *Wilson* v. *Barnum*, 1849.

ders? *Were* Woodworth's pressure rollers significantly different from curved iron plates? *Were* thick circular saws different from Woodworth's "duck-bill" tonguing-and-grooving cutters? The evidence brought forth for witnesses' inspection in these trials was very frequently and necessarily nonverbal or "material" evidence.[37] Besides two-dimensional drawings (Figures 7–12), the witnesses studied three-dimensional models that embodied actual or hypothetical variations on the machines in question— small planers with and without pressure rollers, planing cylinders or disks, and so forth. Sometimes the material evidence was, not a machine, but its product: witnesses examined boards planed by one machine or another for distinctive tool marks left by the different types of cutters, such as the faint parallel lines across the board from "dubbing" by the Woodworth cylindrically rotating knives. Such boards and even shavings were brought to court and exhibited.[38]

Testimony was taken in distant cities, so that material evidence, along with the questions to be put to the witnesses, was frequently bodily transported for these out-of-state sessions. In addition, the witnesses themselves frequently traveled considerable distances to testify. For example, John Gibson and Charles Gould of Albany, New York, who had been familiar with Daniel Barnum's machine in operation there, came and looked at Barnum's machine in Philadelphia and testified it was "the same" as the one in Albany, except that Barnum had separated it into two machines—one for tonguing and grooving and one for planing. Since Barnum's Albany machine had been placed under perpetual injunction in a previous lawsuit, the implication was that he had moved his operation to Philadelphia in order to escape the injunction.[39]

In the long run, this experience affected the categories in which inventors and other relevant mechanicians recognized similarities and differences in the machines they were inventing, using, and disputing, so that their socially constructed definition of a certain patented machine could

37. For the importance of nonverbal thinking and communication in invention and engineering, see Brooke Hindle, *Emulation and Invention* (New York, 1981), esp. chap. 6, "The Contriving Mind"; Hindle, "How Much Is a Piece of the True Cross Worth?" in Ian M. G. Quimby, ed., *Material Culture and the Study of American Life* (New York, 1978); and Eugene S. Ferguson, "The Mind's Eye: Nonverbal Thought in Technology," *Science*, CXCVII (1977), 827–836.

38. Isaac Brock, U.S. Navy Lieutenant Wood, depositions in *Wilson v. Barnum*, 1849.

39. John Gibson and Charles D. Gould, affidavits in *Wilson v. Barnum*, 1849.

change to include features of the contending machines that initially appeared to be quite distinct. Daniel Barnum said in his own defense that the difference of his from Woodworth's machine was like the difference of parallel from perpendicular, a difference he called "immutable." Thomas Hill said simply that "the two planing wheels are substantially different—the one is a cylinder, the other a disk." [40] But the assignees, the patent lawyers, and the eyewitnesses called on both sides of these disputes had to take seriously the possibility that the Woodworth planing patent could legally expand to include Gay's planing discs and Barnum's iron plates, which were nowhere mentioned in the Woodworth patent specification.

So did subsequent would-be inventors and builders of planing machines have to take note of the way the evidence was "construed" in a lawsuit. So did the patent examiners in Washington, scrutinizing patent applications for originality, notice what was said in the courts about mechanical equivalence. To the extent that the Woodworth assignees won their cases—which they predominantly did, sometimes after appeal on up to the Supreme Court—against the users of planing machines with a variety of different physical features, the Woodworth planer underwent gradual social construction by the persons interacting during management of patents.

Judges and juries were also important participants in this social construction of the patented machines. In every case, they had to decide what were significant and insignificant mechanical differences. Thus, for example, Judge John K. Kane pondered the actions of machines in three Woodworth cases in 1846, including one against Patton, and posed an analytical spectrum of geometric shapes for cutters, ranging from a (Woodworth) cylinder, through cones of various degrees (including Gay's flattened one), to a (Bramah) disk. He pointed out:

> This deviation from the strict form of the Woodworth machine towards that of the Bramah, or from the Bramah towards the Woodworth, may go on increasing, till the appropriate action of the original machine effectively disappears; the cylinder, by a series of progressive changes, having lost itself in the disc, or the disc in the cylinder. It is impossible to define, for the practical objects of a judicial decree, that angle or degree of deviation at which one of these geometric forms shall be said to pass into the other.

40. Affidavits by Barnum and Hill in *Wilson v. Barnum*, 1849.

He then noted that the Bramah machine was "unprotected by a patent in this country" (while Woodworth's was so protected). Thus, since the law had to solve a practical problem rather than a geometric one, he pronounced: "Where is the line of separation? Obviously, it is at the point of the first deviation from the free machine to that of which the use is prohibited." In Kane's view, then, the penumbra of protection around the Woodworth patent extended all the way along that spectrum of variations from the cylinder up to, and stopped short only at, the flat-sided disk. Thus, since the beveled disk at issue (Gay's) deviated from "the simple disc of Bramah," its action was too much like that of "the Woodworth rotary cutter," and Kane was "therefore of opinion that the planing machine of the defendants is an infringement of the complainant's patent-right." [41] He issued special injunctions against the defendants in all three cases. Kane's judicial equation of cone with cylinder contributed to the social reconstruction of Woodworth's machine, that is, its inclusion, conceptually, of noncylindrical cutting wheels.

By this time, the Woodworth patent, instead of expiring in 1842, had been in quick succession renewed for seven years, reissued with somewhat changed specification, and renewed again for an additional seven years, so that it was now to be in force until the end of 1856, a total of twenty-eight years instead of the normal fourteen. Patent regulations allowed a patentee who decided—perhaps in the course of a lawsuit—that the wording of his specification was defective to withdraw his patent, reword the specification, and apply to get the patent *reissued* for the remainder of the original patent term. The Patent Act of 1836 also allowed for a seven-year *extension* of the term of a patent if the patentee showed that he had failed to reap "a reasonable remuneration for the time, ingenuity and expense bestowed upon [his invention], and the introduction thereof into use." William W. Woodworth, the inventor's son, obtained an extension by this procedure in 1842, a reissue in 1845, and the second extension by special act of Congress in 1845, that is, long before the end of the first extension. A patented invention was supposed to remain the same in an extension or a reissue; in fact, however, these provided unintended occasions for making changes—hence for social construction—in the scope of a patent. After experience with the Woodworth and other patents, in 1861 Congress lengthened the patent term from fourteen to seventeen years and no longer allowed the Patent Office to grant extensions.

41. John K. Kane, *Opinions*, 14, 15.

Woodworth's assignees in the 1840s met increasing suspicion and dismay among not only would-be planing machine inventors and users but also the broader community of mechanically knowledgeable persons and even the general public. They became known as "the Woodworth patent monopoly," who were accused of reaping great profit from fixing a high price for planed boards and of having worked some skulduggery in getting the patent both extended so long and reissued with broader coverage. When William W. Woodworth, the inventor's son, petitioned Congress for an act to extend it for a further full term of fourteen years, that is, to 1870 and a bill for this was put before the House of Representatives in March 1850, a nationwide campaign got underway to stop it.

Public Protest Movement

On a rainy evening, April 3, 1850, a "mass meeting" assembled in the county courthouse in Philadelphia to protest the Woodworth patent and draw up a resolution sent to the Pennsylvania legislature in Harrisburg and to Congress in Washington. Although not identified as such in the resolution or newspaper accounts, six of the nine officers named at the meeting were lumber merchants, and one was a sashmaker and glazier. One of the lumber merchants was the "wealthy citizen" John Naglee, who was considered "sometimes very active in politics" and "very respectable" although "not always as consistent as zealous." The Pennsylvania General Assembly did later pass a resolution, as did several other state legislatures, "that our Senators be instructed and our Representatives in Congress requested to use all honorable means to prevent any further extension" of the Woodworth patent.[42] Although unsuccessful in 1850, in 1852 and 1854 the Woodworth assignees attempted again to get the extension bill through Congress, and reams of anti-Woodworth patent petitions arrived at the Capitol in Washington. So did pro-Woodworth patent petitions, albeit in smaller numbers.

42. Records of the U.S. Senate Committee on Patents and the Patent Office, Legislative Records Division, SEN 31A–H13, Record Group 46, NA, Washington, D.C.; *Pennsylvania Inquirer,* Apr. 6, 1850; entries in *McElroy's Philadelphia City Directory* for 1851; *Boyd's Philadelphia City Business Directory* for 1859–1860, 340; Freedley, *Philadelphia Manufactures,* 565; Member of Bar, *Wealthy Citizens of Philadelphia,* 17; resolution approved Apr. 5, 1852, in records of the Committee on Patents and the Patent Office, SEN 32A-H14.2, RG 46, NA, Washington, D.C.

The *Scientific American* printed a suggested wording for an anti-Woodworth "remonstrance" for the convenience of local anti-Woodworth protesters, but not all the petitions that arrived at Congress had the same wording; some were earnestly reasonable in tone while some bristled with terms like "fraud," "oppressive and extortionate," "Shylock," "sham defenses," "ruinous suits," "bloated monopoly," "merciless and unscrupulous."[43] They were usually printed rather than handwritten, a trait that indicates some forethought and organization on the local level rather than spontaneous bursts of outrage. Who were the local organizers in Philadelphia, and what kind of persons signed the anti- and the pro-Woodworth petitions?

Most of the petitions arriving in Washington on this issue referred simply to "citizens of" a particular town, county, state, or even the whole United States and did not otherwise identify their signers.[44] A number of the anti-Woodworth petitions from Philadelphia, however, were explicitly from members of an "interested" group. On February 29, 1852, 108 "Builders and Lumber Dealers" signed their petition. Two weeks later, "Joseph R. Atkins, Stephen Wale, and 195 other carpenters of the City and County of Philadelphia" sent another protest.[45] Clearly, both lumber merchants and carpenters disliked paying the prices that the Woodworth planing mills charged and for that reason were economically interested in preventing the further extension of the patent. Preferably, as the 1850 "mass meeting" had urged, they wanted the patent immediately "repealed."

Concerning the carpenters' motivations, there is no opinion expressed in any of the petitions, nor in lawsuit testimony, that the Philadelphia journeymen carpenters regarded the planer as a threat to their livelihood by mechanizing their former hand labor.[46] They presumably had plenty of

43. Two 1850 petitions from citizens of Philadelphia and of Pennsylvania, signed first by John Naglee and Son and by Wm. Dilworth, respectively, in records of Committee on Patents and the Patent Office, SEN 31A-H13, RG 46, NA, Washington, D.C.

44. Suggestive of geographically widespread coordination of effort, many printed petitions left a blank for this information to be filled in.

45. Records of the Committee on Patents and the Patent Office, SEN 32A-H14.2, RG 46, NA, Washington, D.C.

46. Cf. Charles R. Tompkins, *A History of the Planing-Mill, with Practical Suggestions for the Construction, Care, and Management of Wood-working Machinery* (New York, 1889), 10. Clancey, "Cylinder Planing Machine," 205, also disputes Tompkins, finding no antimachine motive per se for opposition to Woodworth's patent assignees in New England.

work otherwise in filling the demand for new housing. The grounds for complaint were instead the height of the prices fixed by the planing monopoly and its aggressiveness in suing infringers, even users of machines that were initially perceived as technically different from the Woodworth planer.

By 1852 the civic leadership of Philadelphia City and County was in accord with the lumber merchants and carpenters that the Woodworth patent should not be extended. An 1852 petition from Philadelphia, containing fifty-eight signatures, is labeled "Petition of Select and Common Councils, Board of Trade, Wardens of the Port, Aldermen and Recorder," plus "Hon. Henry Horn" and "Hon. James Harper." Two others contain signatures of fourteen commissioners of the Northern Liberties district and fifteen of the commissioners of Spring Garden district, with which is scrawled the note that four who were "interested" would not sign.[47]

The obnoxious litigious behavior of the assignees, especially James G. Wilson, had become well enough known that persons who were not directly interested economically agreed with the anti-Woodworth argument. They believed that the assignees, who were not after all the inventor, did not deserve to continue collecting an indirect tax from all house owners in the country. In 1852 *Scientific American* reported that Wilson had received $2,131,752 in twenty-four years as assignee of the Woodworth patent and that "the gross earning of the patent amount to $15,000,000 per annum."[48]

The pro-Woodworth petitions from Philadelphia tend to be briefer, genteelly handwritten, and expressing anxiety "that the inventive genius of the country should be adequately protected" or belief "that every principle of well founded American policy will be maintained by securing to the heirs of so great a benefactor a just reward for his labor and genius." To the extent that they can be identified from the city directory, pro-Woodworth signers seemed not to have signed in groups, as the anti-Woodworth petitioners sometimes did. Predictably, of more than 100 signatures on three pro-Woodworth petitions, only one lumber merchant, one planing miller, and two carpenters were among the identifiable signers. But of four anti-Woodworth petitions *not* explicitly identified with interest groups, con-

47. Records of Committee on Patents and the Patent Office, SEN 32A-H14.2, RG 46, NA, Washington, D.C.

48. "The Profits of Patents," *Scientific American,* VII (1852), 306; *Scientific American,* VII (1852), 317.

taining 225 signatures over all, two turn out to harbor clusters of nine lumber merchants in one and ten carpenters in the other.[49]

With these exceptions, most of the identifiable signers on both sides of the issue were scattered over a wide range of occupations, suggesting that the organizers of neither side suborned factory workers nor (excepting the lumber merchants and carpenters) rallied occupational organizations for their signatures. A few higher-status occupations—a lawyer, an editor, a physician, a treasurer—appear in the pro-Woodworth ranks and are missing from the anti-Woodworth side, but there is no really pronounced difference in socioeconomic character between the signers in the two lists. This suggests that both sides of this controversy succeeded in soliciting support from a wider range of people than just the narrow groups themselves that were economically interested in the outcome.

From all parts of the country where the Woodworth patent monopoly held sway over the planing of boards (and the cutting of moldings for sash and blind work), similar petitions flooded the files of the House and Senate patent committees in Washington in 1850, 1852, and 1854. Pro-Woodworth congressmen reintroduced the extension bill every session, but the patent finally expired—after twice the normal patent period—on December 28, 1856.

We now turn to the patent transformation in Philadelphia of another woodworking machine, the irregular turning lathe.

The Blanchard Lathe

Since prehistory, regular lathes have been able to shape a wide variety of wooden objects with circular cross sections, such as dishes or chair rungs, but not with irregular cross sections, such as ax handles or hat blocks. Hence, before Thomas Blanchard's machine transformed the shaping of irregular wooden objects, lastmakers, wheelwrights, and gunsmiths used hand tools to make shoe lasts, wheel spokes, and gunstocks. A good lastmaker could make three to six pairs of lasts a day with hand tools. But even the earliest Blanchard lathe could rough shape or fine shape a last

---

49. Pro-Woodworth petitions of Mar. 15, 22, Apr. 12, 1852, Records of the Committee on Patents and the Patent Office, SEN 32A-H14.2; anti-Woodworth petitions of Apr. 16, 1850, SEN 31A-H13, Feb. 25, 1852, SEN 32A-H14.2, Feb. 20, 1854, SEN 33A-H14–H14.1, RG 46, NA, Washington, D.C.

in less than ten minutes (see Figure 13), so two lastmakers with a helper could run a Blanchard lathe and among them make an average of sixty pairs of lasts a day, including the remaining handwork.[50]

Blanchard had invented his machine in Massachusetts in 1819 and put it to work at the Springfield Armory to make gunstocks for military muskets. But he also quickly obtained incorporation for the Blanchard Gunstock Turning Factory in order to sell his patented intellectual property in various territories to makers of shoe lasts, hat blocks, tool handles, wheel spokes, and so on (see Figure 14). Almost immediately, as with the Woodworth patent, someone else claimed he had already invented it. The assignees of one Azariah Woolworth, who had recently invented but not yet patented a lastmaking machine in Waterbury, Connecticut, threatened to sue Blanchard. Woolworth's machine and Blanchard's differed in several important respects. Where Woolworth's machine used a reciprocating cutter and rotated the workpiece only in between passes of the cutter, Blanchard's machine used a rotary cutter and rotated the workpiece continuously (see Figure 15). These different motions left parallel toolmarks on Woolworth lasts, a spiral toolmark on Blanchard lasts. Hence Blanchard explicitly stated in his patent specification that his machine "is different from the last-making machine made and used in Waterbury in Connecticut."[51] To resolve the dispute out of court, however, the Blanchard backers bought out all rights to the Woolworth machine, which was patented in June 1820, half a year after Blanchard's patent. When both patents expired in 1834, only Blanchard's was renewed, by act of Congress, for fourteen years.

Lastmakers in New England adopted Blanchard's machine at scattered waterpower sites in the 1820s, sometimes incorporating one or another feature of the Woolworth machine into their own particular "Blanchard lathe." Since a lastmaker could make at least three times more lasts a day with a machine than by hand, it wasn't long before handmade lasts were rare in New England. Boston shoemaker Nathaniel Faxon, who in many years sold more than six thousand lasts annually in addition to the ones

50. James March and Leonard Smith, depositions quoted in Simon Greenleaf, master's report in *Thomas Blanchard* v. *Chandler Sprague,* U.S. Circuit Court, Massachusetts District, 1838, 22–23, RG 21, NA, Boston Branch.

51. Thomas Blanchard patent specification, "An Engine for turning or cutting irregular forms...," Jan. 20, 1820, Records of the Patent and Trademark Office, *Restored Patents,* IV, 1817–1822, 363, NA, Washington, D.C.

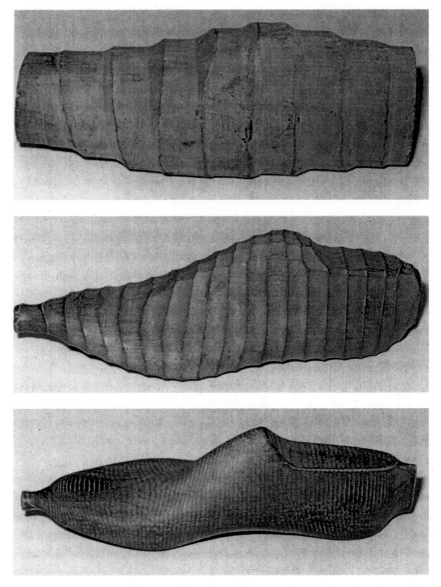

FIGURE 13. Stages in Process of Turning a Last by Blanchard Lathe. *Springfield Armory Museum, Springfield, Mass.* These sample lasts in process all show the spiral path of the cutting in a Blanchard lathe. The cutting wheel (*Fig. 14*) rotated very rapidly while the workpiece rotated slowly.

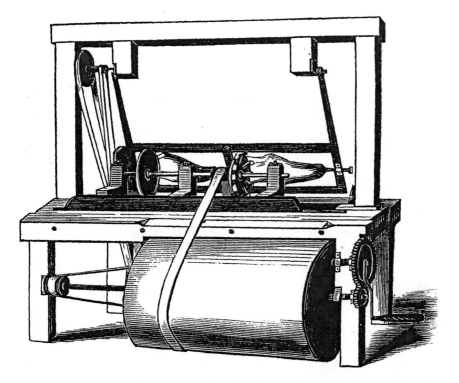

FIGURE 14. Blanchard Lathe for Making Shoe Lasts, 1820. *Henry Howe,* Memoirs of the Most Eminent American Mechanics . . . *(New York, 1840).* Thomas Blanchard received a patent for this "self-acting" machine on Jan. 20, 1820. Driven by the belt around the large drum, the cutting wheel on the right rotated rapidly while the tracing wheel on the left pressed against a model last. The model and the workpiece rotated at the same slow rate while the tracer and cutter moved past left to right, the tracer following the curves of the model. The cutter, swinging with the tracer in their frame, duplicated its motions, cutting a spiral around the workpiece (see Figure 13).

he used himself, said in 1838 that fewer than a hundred in the last seven years were handmade.[52]

In contrast to the prompt interest of Philadelphians Inslee *et al.* in the recently patented Woodworth planer in 1830, Blanchard lathes or other lastmaking machines apparently took nearly a decade to reach Philadel-

---

52. Samuel Cox, deposition, Dec. 6, 1837, in *Blancard v. Sprague,* 1838; Carolyn C. Cooper, *Shaping Invention: Thomas Blanchard's Machinery and Patent Management in Nineteenth-Century America* (New York, 1991), 172; Nathaniel Faxon, testimony quoted in Simon Greenleaf's master's report, *Blanchard v. Sprague,* 1838, 40–43.

COMPARISON OF FEATURES

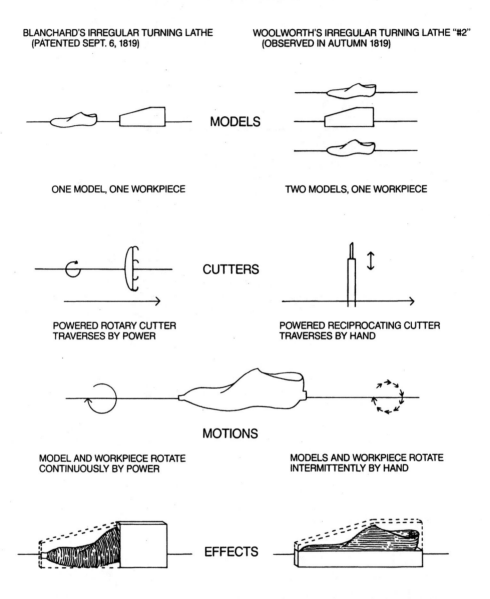

BLANCHARD'S IRREGULAR TURNING LATHE
(PATENTED SEPT. 6, 1819)

WOOLWORTH'S IRREGULAR TURNING LATHE "#2"
(OBSERVED IN AUTUMN 1819)

MODELS

ONE MODEL, ONE WORKPIECE

TWO MODELS, ONE WORKPIECE

CUTTERS

POWERED ROTARY CUTTER
TRAVERSES BY POWER

POWERED RECIPROCATING CUTTER
TRAVERSES BY HAND

MOTIONS

MODEL AND WORKPIECE ROTATE
CONTINUOUSLY BY POWER

MODELS AND WORKPIECE ROTATE
INTERMITTENTLY BY HAND

EFFECTS

CUTS SPIRAL PATH AROUND WORKPIECE

CUTS LINEAR PATH ALONG WORKPIECE

F I G U R E 15. Comparison of Blanchard's Irregular Turning Lathe with Azariah Woolworth's Waterbury Lastmaking Machine, 1819. *Drawn by Lyn Malone.* Instead of
rotating *while* cutting took place, the models and workpiece in Woolworth's machine
stood still and were rotated *between* the successive lateral sweeps of the cutter along
the length of the workpiece. The result was separate linear cuts of the cutter along the
workpiece.

phia from New England, and without benefit of formal patent assignment or license. In other words, mechanization of lastmaking in Philadelphia took place by actions that Blanchard regarded as patent infringement during the second term of his patent. As in the case of the planing machine, patent litigation records show the migration of individual mechanics and machines to Philadelphia from locations in New England. In about 1839 lastmaker Erastus Underwood bought "an old machine" in Boston for making lasts and used it until 1842. He had a new one built in 1843 by Sylvester S. Chase, of New Haven County in Connecticut, who moved to Philadelphia and set up shop as machinist and then moved across the river to Camden, New Jersey. Another mechanic in Philadelphia, Lea Pusey, learned from Chase how to make such a machine, which he did in the winter of 1844 for Joseph Brown, another Philadelphia lastmaker.[53]

In Philadelphia, where high-quality hats and shoes were made, one might expect the larger-scale hatmakers or shoemakers to take an earlier interest in introducing the Blanchard lathe to make shoe lasts or hat blocks, which were important components in their respective manufacturing processes. But Amos K. Carter, Blanchard's assignee from 1843 for Pennsylvania and three or four other mid-Atlantic states, was not a Philadelphian and had nothing to do with hats or shoes. He was a coachmaker and wheel-spoke turner in Newark, New Jersey, who reportedly paid Blanchard twelve thousand dollars for his patent rights in four states but seemed to be more interested in collecting license fees in Philadelphia than in using the lathe himself.[54]

When Carter began showing up at the workshops of lastmakers in Philadelphia, he found that Brown, Underwood, and others had recently started using machines. Carter declared them to be Blanchard lathes and threatened lawsuits if they didn't start paying royalties. Some lastmakers agreed to pay Carter his fee of two cents per last for the Blanchard patent right, but others, like Underwood and Brown, refused, saying their machines

---

53. Erastus Underwood, deposition, Erastus Underwood, Sylvester S. Chase, affidavits, June 4, 1844, Lea Pusey, affidavit, June 5, 1844, in *Blanchard Gunstock Turning Factory v. A. Bond and Erastus Underwood,* U.S. Circuit Court, Eastern District, Pennsylvania, April sess. 1844, NA, Philadelphia; Philadelphia and Camden city directories, 1845–1848; Franklin, N.J., 1850 manuscript census returns; Salem City, N.J., 1860 manuscript census returns.

54. Sylvester S. Chase, affidavit, June 5, 1844, in *Blanchard Gunstock Turning Factory v. Bond and Underwood,* April sess. 1844.

were different from Blanchard's. Machinists Chase and Pusey each denied ever seeing a Blanchard machine or his patent before building their machines for Underwood and Brown, respectively. Pusey said he had taken care "not to interfere [with any patent] . . . and used nothing that had not been used ever since I can remember."[55] Carter, as assignee of the Blanchard Gunstock Turning Factory, brought suit against Brown and against Underwood and his associate Amos Bond.

Bond had already owned part of a lastmaking operation in New Hampshire that was sued by Blanchard and, like Barnum and his planing machine, apparently moved to Philadelphia in order to resume business in a new location, with Underwood. Relocation to Philadelphia was apparently no huge effort. A lastmaker with a machine would be able to rent space in a woodworking steam mill (see Figure 16). Bedstead maker Biddle Reeves, for instance, rented out extra shop space and power in his factory on St. John Street to lastmaker Isaac Eldridge and subsequently found himself sued, along with Eldridge, by the Blanchard Gunstock Turning Factory. Isaac Eldridge had bought Underwood's former machine third- or fourthhand and rebuilt it.[56] In the lawsuits that Blanchard and his incorporated Gunstock Turning Factory conducted against various Philadelphia lastmakers from 1844 to 1850, both sides tried to demonstrate their points by exhibiting machine models, sample lasts, and drawings and by testimony of geographically scattered practical mechanics and engineers. Blanchard's patent was again renewed by act of Congress for fourteen years from 1848.

Like the Woodworth planer, the Blanchard lathe underwent socially constructed expansion through such lawsuits, so that its patent was construed to cover mechanical variations that looked quite different initially. Isaac Eldridge provided a particularly dramatic demonstration of this phenomenon. He not only attended his own trials in Philadelphia from 1848 to 1850; he also traveled to Boston to attend other ongoing Blanchard trials. From the model machines he saw and testimony he heard—including from the aging Woolworth himself, who came to Philadelphia to testify— Eldridge apparently learned about the features of Woolworth's 1819 lastmaking machine, which Blanchard's patent specification had explicitly

<hr/>

55. Chase, June 4, 1844, Pusey, June 5, 1844, *ibid.*

56. James Whitehead, affidavit, Jan. 17, 1848, in *Blanchard Gunstock Turning Factory v. Eldridge et al.*, U.S. Circuit Court, Eastern District, Pennsylvania, April sess. 1848, RG 21, NA, Philadelphia.

W. B. SAURMAN & CO.'S
STEAM PATTERN AND MODEL WORKS,

Southeast corner of Broad and Wallace Streets.

STEAM SAWING,
PLANING, TURNING, AND WOOD SCREW CUTTING,
OF EVERY DESCRIPTION,

PROMPTLY DONE TO ORDER,

AT THE

Lowest Cash Prices.

W. B. S. and Co. would beg leave also, to inform Builders, Carpenters, and others not having Steam Power in their own establishments, that they can have the use of Wood-working Steam tools at the above Works, for a reasonable charge per day or hour.

FIGURE 16. Woodworking Steam Power for Hire, 1852. *Advertisement in* Philadelphia as It Is in 1852 *(Philadelphia, 1852), 368.* The availability of extra space and power for rent in steam mills, such as those of Biddle Reeves or W. B. Saurman and Company, made it easier for small woodworking operations to get started in Philadelphia and to move from place to place.

said was different from the Blanchard lathe. Eldridge reasoned he would be safe from suit with a Woolworth machine, even though it was slower and less productive than the Blanchard lathe. When Eldridge's first machine was placed under injunction, he deliberately made himself a series of machines like that obsolete 1819 Woolworth machine in one feature after another: for instance, a reciprocating cutter instead of Blanchard's rotary cutter, intermittent rotation instead of Blanchard's continuous rotation. Eldridge showed off sample lasts to impress prospective customers and to prove he had a non-Blanchard machine. Witnesses reported seeing the lasts, and from their appearance some doubted the machine that had produced them was really non-Blanchard. Lasts were shown in court, as nonverbal evidence. Each time, the court sent a commission to inspect the machine and report back; each time, the judge enjoined Eldridge from further use of the changed machine; each time, Eldridge remodeled it.

The court appointed Philadelphia engineer W. W. Hubbell to report on the features of Eldridge's second machine. During Eldridge's demonstration for Hubbell, the building shook, the power transmission belt broke three times, and a bearing caught on fire, but Hubbell saw enough to decide that Eldridge's machine differed from a Blanchard lathe only in having a double-edged reciprocating cutter instead of a rotary one and a "friction column" as a tracer instead of a friction wheel. This was insufficient difference to satisfy Judge Kane. As he issued a new injunction, the judge complimented Eldridge for his mechanical ingenuity but scolded him for devising something "less useful and more costly than that which was known before."[57]

But Eldridge rebuilt his machine again and again resumed making lasts. The court next sent a commission of three well-respected and active members of the Franklin Institute—John C. Cresson, John F. Frazer, and Charles B. Trego—to inspect Eldridge's third machine. They reported that it had a cutter wheel and a tracer that passed "rapidly from one end of the model to the other and backward . . . and at each end of the motion the model and rough material receive a small and equal motion around their longer axes." So Eldridge's machine now resembled Woolworth's Waterbury lastmaking machine, not in the cutter, but in the intermittent rotation

57. W. W. Hubbell, master's report, Feb. 27, 1849, in *Thomas Blanchard* v. *Isaac Eldridge et al.,* in Equity, U.S. Circuit Court, Eastern District, Pennsylvania, October sess. 1848, RG 21, NA, Philadelphia. For Kane's remarks, see 3 *Federal Cases,* 622–623, or *Scientific American,* IV (1849), 27.

of the workpiece. The commissioners pointed out that these were different kinematics from those set forth in Blanchard's specification: "The only method proposed by Mr. Blanchard is that in which the friction wheel or tracer describes a spiral over the whole surface of the model and causes the cutters to act in a similar direction." They unanimously concluded that Eldridge's method "is not a mere colourable and unimportant change from the method described in Mr. Blanchard's patent, but that it is essentially different" and that therefore "the machine of Mr. Eldridge is different in its principle and mode of operation from that described in Mr. Blanchard's patent." After their report, the court refused Blanchard's motion for attachment of Eldridge's property.[58]

But when Blanchard took Eldridge to trial yet again to seek an injunction against his machine, he not only marshaled additional expert witnesses to testify to the mechanical equivalence of the continuous and intermittent motions. He also brought into court a model lastmaking lathe that was capable of either motion, by the shifting in or out of a simple mechanical linkage. It could perform in the Blanchard-specified continuous, spiral-path fashion *or* in the Woolworth mode, with intermittent rotation of the workpiece between lateral lengthwise passes of the cutter.

Justice Robert C. Grier, the judge in this case, complained that he would "have had much less difficulty in arriving at a conclusion . . . in the present case but for the opposite opinions expressed by gentlemen of the highest reputation for learning, judgment, and practical skill in mechanics." He finally based his opinion on an analysis of the spiral motion into the separate lateral and circular motions of the cutter and workpiece, respectively, as Blanchard's courtroom model had demonstrated could be done. He concluded that "such a change . . . affecting the motions of the model and guide only in the figure of their path or the relative lines of their movements in no case changes the principle . . . of the machine."[59] He granted the injunction against Eldridge's Woolworth-like machine.

58. John F. Frazer, John C. Cresson, and Charles B. Trego, "Report of the Commissioners," Dec. 1, 1849, in *Thomas Blanchard v. Isaac Eldridge et al.,* October sess. 1848 (for credentials of these commissioners, see Sinclair, *Philosopher Mechanics,* esp. 254–258); *Equity Docket Books,* III (1848–1858), U.S. Circuit Court, Eastern District, Pennsylvania, October sess. 1848, 25, RG 21, NA, Philadelphia.

59. Justice Grier's opinion of Sept. 19, 1850, in *Thomas Blanchard v. Biddle Reeves, Isaac Eldridge, et al.,* U.S. Circuit Court, Eastern District, Pennsylvania, April sess. 1850, in 3 *Federal Cases,* 639, 640.

Thus, as a piece of intellectual property whose boundaries were socially drawn and redrawn by patent management, Blanchard's invention had expanded in concept by 1850 to include all the features of a machine that Blanchard had explicitly excluded from his patent specification in 1820. This had come about, in part, by the exercise over the years of nonverbal, artifactual demonstrations in the courtroom, by and for "expert witnesses," of mechanical equivalence in the artifacts. What had not seemed equivalent to Blanchard thirty years earlier had become so to him and to his increasingly mechanically knowledgeable peers, whose training took place not only in school and on the job but also in the patent courts. Their various perceptions were of course not immune to their changing economic and other "interests," but they also reflected a shaping of consensus on what were or were not "purely technical" distinctions. The influence ran in both directions.

Consequences of Patent Management

Woodworth's twenty-eight-year-old planer patent finally expired in 1856 after the vociferous protest campaign spearheaded by the magazine *Scientific American* and organized, in Philadelphia at least, by carpenters and lumber merchants. People resented paying royalties to the owners of a long-dead inventor's patent. Thomas Blanchard, however, stayed alive until 1864; and although he had sold off much of his patent rights, he continued to reap some direct income from royalties until the patent expired in 1862, after forty-two years as private but socially expanded intellectual property. Despite some criticism of his own and his assignees' behavior, his patent did not provoke a mass protest.[60] In 1861 Congress passed revisions to the patent system that eventually eliminated patent extension by Patent Office procedures and greatly diminished reissue, from which both Blanchard and Woodworth had benefited. Both machines continued to be manufactured and used after expiration of their much-prolonged patents.

What may we learn from these stories of the adoption of two major patented woodworking machines in pre-1850 Philadelphia? We may first

---

60. For instances of criticism, see *Scientific American*, III (1848), 245, 270, 349, VIII (1852), 253. For a discussion of this magazine's role in the Woodworth patent controversy, see Michael Borut, "The 'Scientific American' in Nineteenth-Century America" (Ph.D. diss., New York University, 1977), 127–143.

note that farseeing self-interest does not seem to have been at work. The earliest innovators of these woodworking machines in Philadelphia did not belong to those industries that were the beneficiaries of the very large productivity gains that the machines brought about in house construction and shoemaking, respectively. Instead, lastmaking and planing became separate businesses established by innovators who were primarily patent speculators. As new industries, they disrupted established woodworking arrangements, causing discomfort and resistance. Yet they met opposition, not by craftsmen who mistrusted machinery and the changes it brought about, but by competitors who wished to use different machines and by customers who wanted lower prices. Far from rejecting the new technology, they wanted to reap its benefit faster than the patent managers allowed them to do. These stories also show geographic mobility by mechanics from New England and New York who brought particular machines to Philadelphia and then settled there and were years later called on as witnesses in patent suits. Such mobility was encouraged by patent litigation itself, which induced travel and exchange of information among the newly emerging population of mechanics and engineers. They also exchanged information with a larger group of participants in patent management, including judges, juries, and lawyers. The stories also make clear that much of the information exchanged during lawsuits was nonverbal, consisting of exhibited artifacts or pictures.

Most of all, these stories point to the importance of the patent system as the mediating institution that drew together the inventors in one part of the country and the local innovators in another. As managers of their patents, inventors and innovators who became litigants may well have regarded it as an exasperating and costly way to proceed. Both detractors and supporters of the patent system have been shrill from time to time ever since its inception two hundred years ago. But if we look past the surface to see how people actually interacted within that system, we will find unexpected insights into its effects on technology, such as the mobility of mechanics, the transmission of nonverbal information, and the social construction of technical knowledge. As these interactions concerning patented woodworking machines took place, Philadelphians helped transform the machines and the patent system, even as the machines helped transform Philadelphia.

Judith A. McGaw

"So Much Depends upon a Red Wheelbarrow"

Agricultural Tool Ownership in

the Eighteenth-Century Mid-Atlantic

My title derives from one of the best-known Imagist poems, by William Carlos Williams. The poem goes:

> so much depends
> upon
>
> a red wheel
> barrow
>
> glazed with rain
> water

The author thanks members of the Philadelphia Center for Early American Studies and the Transformation of Philadelphia Project for helping her formulate and develop the research from which this article derives. She is especially indebted to Stephanie Wolf, Jean Soderlund, Marianne Wokeck, Billy Smith, and Lois Carr. Research was funded by the National Endowment for the Humanities and the National Science Foundation. For comments on the final manuscript she especially thanks Mary Kelley, Ruth Schwartz Cowan, Gary Kulick, and Fredrika Teute.

> beside the white
> chickens.[1]

I begin by invoking the Imagist spirit because their mission—replacing the generalizations and abstractions of Victorian poetry with what they called "direct treatment of the 'thing'"—is an approach historians studying early American agricultural technology might profitably emulate.[2] Unfortunately, a salient characteristic of scholarship treating early American farming is nicely captured by a *Peanuts* cartoon a graduate student gave me some years ago. The strip features Sally standing in front of her class and holding a piece of paper. "This is my report on Mr. John Deere," she says. "In 1837, Mr. Deere invented the self-polishing steel plow which was a great help to farmers . . ." In the next panel she has been interrupted and replies: "Plow? No, Ma'am I've never seen a plow . . ." After a pause she adds, "I've never even seen a farmer!"[3] Paraphrasing Williams, this essay argues that, for understanding technological change in America, much depends upon our seeing farmers and plows, red wheelbarrows and white chickens.

The Imagists also set a standard worth emulating by insisting that writers "use the language of common speech, but . . . employ always the *exact* word, not the nearly-exact, nor the merely decorative word," to quote Ezra Pound. Alas, when we read the history of technology, we hear few echoes of common speech, literally or metaphorically speaking. We have mostly studied technology as an expression of leaders—inventors, experimenters, large corporations, or governments; we have mostly ignored common people's technological expression—what tools they chose to own or generally employed, for example.[4]

Nowhere is our ignorance of the mundane more evident than in scholar-

1. William Carlos Williams, *Collected Poems 1909–1939—Vol. I.* Copyright 1938 by New Directions Pub. Corp.

2. F. S. Flint, quoted in William Pratt, ed., *The Imagist Poem: Modern Poetry in Miniature* (New York, 1963), 18.

3. Charles Schultz, *Peanuts* (United Feature Syndicate, Inc., 1981).

4. Pratt, ed., *Imagist Poem*, 22. The comparatively narrow focus of scholarship to date is well characterized in John M. Staudenmaier, *Technology's Storytellers: Reweaving the Human Fabric* (Cambridge, Mass., 1985); and Staudenmaier, "What SHOT Hath Wrought and What SHOT Hath Not: Reflections on Twenty-five Years of the History of Technology," *Technology and Culture*, XXV (1984), 707–730.

ship on early American agricultural technology. With a few noteworthy exceptions, secondary literature tells us little about which tools, practices, and knowledge early farmers and farm wives customarily employed. Instead, the scholarship features famous firsts—inventors' contrivances and agricultural reformers' proposals, with little sense of how these tools and ideas fared after their debut. We have more often looked at the machine on the drawing board than at the wheelbarrow in the garden.[5]

Despite our inattention to common practice, historians have not been reluctant to characterize early American tool ownership, albeit in highly generalized terms. Indeed, one noteworthy feature of our literature is the virtual absence of clear and specific images of preindustrial technology. Rather than deriving from records that feature farmers, farm wives, and

5. In addition to heavy dependence on the agricultural press and other writings of agricultural reformers, most works in the history of agriculture rely on anecdotal evidence, especially travelers' accounts and data on the practices of elite farmers such as the founding fathers. After more than half a century, the standard works remain Percy Wells Bidwell and John I. Falconer, *History of Agriculture in the Northern United States, 1620–1860* (Washington, D.C., 1925); and Lewis Cecil Gray, *History of Agriculture in the Southern United States to 1860*, 2 vols. (Washington, D.C., 1933). Other important works with more focused coverage include Clarence H. Danhof, *Change in Agriculture: The Northern United States, 1820–1870* (Cambridge, Mass., 1969); Howard S. Russell, *A Long, Deep Furrow: Three Centuries of Farming in New England* (Hanover, N.H., 1976); Hubert G. Schmidt, *Agriculture in New Jersey: A Three-Hundred-Year History* (New Brunswick, N.J., 1973); Stevenson Whitcomb Fletcher, *Pennsylvania Agriculture and Country Life, 1640–1840* (Harrisburg, Pa., 1950); Wayne Caldwell Neely, *The Agricultural Fair* (New York, 1935); Albert Lowther Demaree, *The American Agricultural Press, 1819–1860* (New York, 1941); and Donald B. Marti, *To Improve the Soil and the Mind: Agricultural Societies, Journals, and Schools in the Northeastern States, 1791–1865* (Ann Arbor, Mich., 1979).

Notable exceptions to these generalizations are the more recent works of economic and social historians, many of which survey general patterns of farm practice by exploiting sources such as probate inventories, tax lists, and census data. In general, however, these studies have concerned themselves principally with economic issues such as change in the standard of living or in agricultural productivity. In consequence, they tend to offer only highly generalized pictures of tool ownership. Outstanding overviews of this literature for the colonial era are John J. McCusker and Russell R. Menard, *The Economy of British America, 1607–1789*, rev. ed. (Chapel Hill, N.C., 1991), esp. 277–308; and Richard B. Sheridan, "The Domestic Economy," in Jack P. Greene and J. R. Pole, eds., *Colonial British America: Essays in the New History of the Early Modern Era* (Baltimore, 1984), 43–85. For the later period, see Jeremy Atack and Fred Bateman, *To Their Own Soil: Agriculture in the Antebellum North* (Ames, Iowa, 1987).

their tools, most scholarship rests on several sorts of unsubstantiated generalizations. One approach has relied heavily on what agricultural reformers and European travelers wrote. It reiterates their claims that common farmers resisted innovation and their assessments of early farm technology as hopelessly primitive, especially on the frontier.[6]

By contrast, the other principal approach portrays the early yeoman as a technological virtuoso. It assumes, albeit implicitly, that tool ownership and tool-wielding skill were common in the preindustrial era. That assumption underlies, for example, the Marxist contention that industrialization degraded work when it shifted tool ownership from tool users to the employers of tool users. It also undergirds revisionist scholarship that depicts industrialization as the demise of an earlier, subsistent, communitarian economy in favor of an increasingly impersonal market-oriented one. Likewise, accounts of women and industrialization contrast restrictive nineteenth-century mill or domestic work with a colonial role presumed to entail an enormous array of agricultural processing tasks, such as spinning, weaving, candlemaking, churning, cheese production, and pickling.[7]

Scholars who assess industrialization more favorably also assume widespread preindustrial tool ownership. For example, one explanation of America's relatively swift industrialization has been that frontier living

6. See note 5, above.

7. The Marxist literature accords virtually no attention to farmers, focusing instead on the small fraction of early industrial workers who had been craftsmen. The absence of significant scholarship on tool ownership among early American craftsmen makes this literature especially problematic. Classic statements in the debate over community self-sufficiency include James A. Henretta, "Families and Farms: *Mentalité* in Pre-Industrial America," *William and Mary Quarterly*, 3d Ser., XXXV (1978), 3–32; Christopher Clark, "Household Economy, Market Exchange, and the Rise of Capitalism in the Connecticut Valley, 1800–1860," *Journal of Social History*, XIII (1979–1980), 169–189; Carole Shammas, "How Self-Sufficient Was Early America?" *Journal of Interdisciplinary History*, XIII (1982–1983), 247–272. More generally, see Allan Kulikoff, "The Transition to Capitalism in Rural America," *WMQ*, 3d Ser., XLVI (1989), 120–143; Gary B. Nash, "Social Development," and James A. Henretta, "Wealth and Social Structure," in Greene and Pole, eds., *Colonial British America*, 233–289. For an outstanding recent summary of the historiography on colonial women's work, see Jeanne Boydston, *Home and Work: Housework, Wages, and the Ideology of Labor in the Early Republic* (New York, 1990), xii–xvii. See also, Judith A. McGaw, "Women and the History of American Technology," *Signs*, VII (1981–1982), 798–828.

nurtured technological creativity. Frontiersmen, it argues, had to be jacks-of-all-trades—to know how to use all of the tools that specialists wielded in more settled communities. The unspoken corollary is, of course, that frontiersmen owned all those tools. Nor is this mythic jack-of-all-trades confined to scholarly claims about the frontier. Often, eighteenth- and nineteenth-century northern rural communities are presumed to have nurtured a youthful variant of the type—the Yankee whittling boy. This handy youth figures prominently in the literature celebrating inventors, much of it popular, but some of it scholarly.[8]

My data on tool ownership in the eighteenth-century mid-Atlantic paint a very different picture. I find, for example, that only a little more than half of farmers or yeomen probably owned plows and that, among farm women, about 20 percent made do without either a pot or a kettle, those huge iron or brass caldrons that colonial restorations invariably hang over the fire. The artifact we most often envision in early American hands—the gun—actually existed in only about half of households. And frontiersmen were only slightly more likely to own firearms: about 60 percent versus about 50 percent for inhabitants of longer-settled regions. Nonetheless, early Americans were far more likely to own guns than to possess that other icon of early American life—the Bible—although, surprisingly, frontier households came closest to owning Bibles as often as guns.[9]

These data are arresting. If many, even most, colonial Americans lacked items we have believed common, even essential, our image of America's traditional technology must be quite distorted: a composite of colonial revival stereotypes and an uncritical acceptance of surviving artifacts as

8. See, for example, Dirk J. Struik, *Yankee Science in the Making* (Boston, 1948). A more sophisticated restatement of this argument is Thomas C. Cochran, *Frontiers of Change: Early Industrialism in America* (New York, 1981), 9–13.

9. These and other data derive from computer-assisted analysis of a 15-year-interval sample of probate inventories, described in greater detail hereinafter. I compiled data for Burlington and Hunterdon counties, N.J., at the New Jersey State Archives, Trenton; data for York County, Pa., at the York County Courthouse, York; and data for Westmoreland County, Pa., at the Westmoreland County Courthouse, Greensburg. I am indebted to the staff of each of these institutions for providing work space, access to materials, and assistance in using them.

Limitations of probate inventory data are discussed later; see especially notes 20, 21, below.

representative. We must do better if we would understand what techno-logical experiences and fingertip knowledge the offspring of American farmers brought to the early factories or assess how the shift to industrial production altered people's relationships to their tools.

THIS essay offers one strategy for doing better. It begins by enunciating why American historians and historians of technology need real knowl-edge of how early Americans farmed. After articulating the large ques-tions that motivated my data collection and shaped my data selection and analysis, I outline a feasible and promising alternative to the uncritical, anecdotal approach that has predominated in the field. By summarizing my research strategy—describing the sources I exploit and the methods I employ—I hope to evoke emulation: to persuade additional scholars to cultivate the abundant household-level evidence that awaits those willing to venture into new terrain and to challenge yet more colleagues to seek out other promising scholarly resources. The methodological discussion also provides a context for assessing and appreciating the preliminary report of findings on colonial tool ownership to which the essay turns next. As a first step away from the unsubstantiated generalizations to which we have grown accustomed, I present concrete evidence drawn from my research in progress and propose some large conclusions that may be drawn from the presence or absence of winnowing fans, dung hoes, candle molds, and dough troughs in eighteenth-century households. Finally, I conclude with a few brief observations provoked by this look at small things—"red wheel barrows" and "white chickens"; I invite my various colleagues to question our accustomed academic approach to early American technology.

My study of early American agricultural tool ownership derives ulti-mately from one big question: What accounts for America's sudden, rapid, and comparatively successful early nineteenth-century industrialization? It has seemed to me that most accounts of the American Industrial Revo-lution, because they begin in 1790, miss a good part of the answer to that question. Likewise, by focusing on manufactures and treating agriculture merely as a belated beneficiary or victim of industry's mechanical cre-ativity, scholarship on technological change in the American Industrial Revolution has begged some critical questions, such as: How did a declin-ing proportion of farm households become able to feed a growing nonfarm population, supply a burgeoning international market, and provide the raw materials most early industry processed? or, What sorts of technical expertise did the sons and daughters of farmers bring to industrial work?

or, How did changes in the goods farm households purchased help to create a market for early industrial commodities? or, What toolmaking skills had become common before industrialization?[10]

In contrast to the narrow focus and abbreviated chronology of American scholarship, studies of British industrialization have historically paid at least some attention to prior and simultaneous agricultural innovation. More recently, a number of scholars have given husbandmen a leading role in the British Industrial Revolution. They have shown that most agricultural innovation, both technical and organizational, occurred in the seventeenth and early eighteenth centuries, much earlier than previously believed. Although historians of industry have simultaneously pushed the origins of the manufacturing revolution well back into the eighteenth century, the new "chronology of improvement now makes a strong case for the close interdependence of agriculture and manufacturing, with the springs of much manufacturing improvement to be found in the early dynamism of the agricultural sector."[11]

At the very least, then, understanding the American Industrial Revolution requires that we examine American industry's colonial agricultural roots. But we also need to look at how farm practice changed during the era of the American Revolution and into the nineteenth century if we are to link agricultural innovation to industrial development. Nor can we limit our attention to the farm activities of the late-colonial era. The new scholarship on Britain certainly raises the possibility that the initial British settlers of any given region engaged in agricultural innovation from the outset. Many of their activities paralleled, even when they did not precisely reenact, those of progressive agriculturalists who remained at home. But the innovativeness of colonists who undertook such activities as clearing forests and planting maize has generally escaped notice because farmers

---

10. For a sample of some of the best modern scholarship depicting rural areas as affected by or reacting to earlier industrialization and urbanization, see Steven Hahn and Jonathan Prude, eds., *The Countryside in the Age of Capitalist Transformation: Essays in the Social History of Rural America* (Chapel Hill, N.C., 1985).

11. E. L. Jones, *Agriculture and the Industrial Revolution* (New York, 1974), esp. 86; Maxine Berg, *The Age of Manufactures: Industry, Innovation, and Work in Britain, 1700–1820* (New York, 1986), 93–94. See especially H. P. R. Finberg and Joan Thirsk, eds., *The Agrarian History of England and Wales*, V, part 1, *1640–1750: Regional Farming Systems* (Cambridge, 1984), V, part 2, *1640–1750: Agrarian Change* (Cambridge, 1985); Joan Thirsk, *The Rural Economy of England: Collected Essays* (London, 1984).

undertook these novel tasks from the start of colonial agricultural history, making the new practices appear either "natural" or "necessary." They were neither. The relative ease with which colonists adopted new practices simply reflects that these novel tasks entailed relatively straightforward translations of new British agricultural strategies to the colonial situation.[12]

In the absence of significant scholarly attention to common agricultural practice, I concluded that, before I could begin to document change in early American agricultural technology, analyze its origins, or link it to manufacturing, I needed to find out what American farmers and farm wives had done and how they had done it from the beginnings of settlement to the mid-nineteenth century. In other words, I had somehow to catch a glimpse of all those red wheelbarrows and white chickens. Understanding anything else about early American technology ultimately depended upon seeing those.

After reaching such a disconcerting conclusion, I made several strategic decisions that converted an impossible task into a manageable one. First, I acknowledged that there is no such thing as a representative colonial farm, farm community, or farm region. Indeed, I will argue later that attempting to find a typical farm family and a standard array of farm tools misstates the problem in a way that inevitably misleads us.

I focused on the mid-Atlantic region because it was the only major region where industry and agriculture flourished side by side and because the mid-Atlantic's emphasis on mixed livestock and grain farming set the pattern for much later American agriculture. The Middle Colonies qualify as the quintessential American region in other respects as well. As Frederick Jackson Turner noted nearly a century ago, the region "mediated between New England and the South, and the East and West," and "it had a wide mixture of nationalities, a varied society, the mixed town and county system of local government, a varied economic life, many religious sects."[13] Insofar as these factors influenced agricultural technique, the mid-Atlantic region represents the best microcosm of early American farm practice.

12. This argument is spelled out in Judith A. McGaw, "Agriculture in the Industrial Revolution: Some Anglo-American Comparisons," in British Society for the History of Science and the History of Science Society, "Program, Papers, and Abstracts for the Joint Conference" (Manchester, England, July 1988), 65–75.

13. Turner quoted in Sheridan, "The Domestic Economy," in Greene and Pole, eds., *Colonial British America*, 59.

Yet, the mid-Atlantic has received far less scholarly attention than has either New England or the South.

Given the rich diversity of the region, five counties warranted close examination. Each county both shares features with and differs from the others. Thus, comparing data from the various counties should indicate which of several relevant factors—length of settlement, ethnic composition, access to markets, relative affluence, and natural endowment, for example—offer the best explanation of particular agricultural patterns. Examining several counties simultaneously should also provide a built-in reminder that there were many "right" answers to the question of how best to farm, a useful corrective to the biases of both secondary literature and the writings of early American agricultural reformers.[14] Limiting my choice of counties to New Jersey and Pennsylvania also kept manageable the number of different currencies and the various legal parameters to be considered. At the same time, my choices were likely to disclose technological differences associated with the westward course of settlement, an aspect of colonial agriculture that has received far less attention than have North-South differences.

Burlington County, New Jersey, offers a case of very early settlement, beginning roughly in the last quarter of the seventeenth century. Close to Philadelphia and to good water transportation, it represents nearly ideal access to a rapidly growing urban market. It also exemplifies a heavily British population with strong Quaker influence. By the nineteenth century, Burlington should illustrate farming practice in a region that rusticated, for its growth slowed early.[15] By contrast, Hunterdon County,

14. I eliminated from consideration counties whose soil and topography rendered farming especially difficult. Numerous studies of New England agriculture have already made abundantly clear the characteristics of agricultural practice under adverse environmental conditions. The random sample of counties featured in Atack and Bateman, *To Their Own Soil,* also includes a number of poorly endowed counties. Indeed, their work inadvertently illustrates that, especially with respect to natural endowment, counties are far from homogeneous units, making random sampling a highly questionable strategy.

My selection of counties also sought to minimize the number of boundary changes sample counties underwent over the period studied.

15. My selection of New Jersey counties was guided by Peter O. Wacker, *Land and People: A Cultural Geography of Preindustrial New Jersey: Origins and Settlement Patterns* (New Brunswick, N.J., 1975); Schmidt, *Agriculture in New Jersey;* John P. Snyder, *The Story of New Jersey's Civil Boundaries, 1606–1968* (Trenton, N.J., 1969); John E. Pomfret, *Colonial New Jersey: A History* (New York, 1973); Pomfret, *The Province of West*

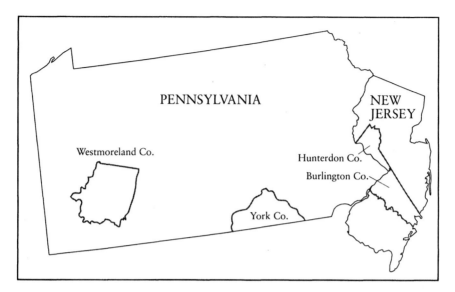

BURLINGTON, HUNTERDON, YORK, AND WESTMORELAND COUNTIES

New Jersey, the county just north of Burlington in West Jersey, shows a rapid transition from frontier to settled farming region. Initial settlement lagged about two generations behind that of Burlington, but during the second quarter of the eighteenth century Hunterdon's population came to

New Jersey, 1609–1702: A History of the Origins of an American Colony (Princeton, N.J., 1956); Richard P. McCormick, *New Jersey from Colony to State, 1609–1789* (Newark, N.J., 1981). On Burlington, see also Jean R. Soderlund, *Quakers and Slavery: A Divided Spirit* (Princeton, N.J., 1985); E. M. Woodward, *History of Burlington County, New Jersey, with Biographical Sketches of Many of Its Prominent Men* (Philadelphia, 1883); Carl Raymond Woodward, *Ploughs and Politicks: Charles Read of New Jersey and His Notes on Agriculture, 1715–1774* (Philadelphia, 1974). I deliberately selected New Jersey counties in the Philadelphia region so as to minimize the number of diverse transportation and marketing factors to be considered.

   Burlington possesses many similarities to and should offer useful comparisons with Chester County, Pa., the one county in the region of which numerous studies already exist. See, for example, Lucy Simler, "Tenancy in Colonial Pennsylvania: The Case of Chester County," *WMQ*, 3d Ser., XLIII (1986), 542–569; Paul G. E. Clemens and Simler, "Rural Labor and the Farm Household in Chester County, Pennsylvania, 1750–1820," in Stephen Innes, ed., *Work and Labor in Early America* (Chapel Hill, N.C., 1988), 106–143; Joan M. Jensen, *Loosening the Bonds: Mid-Atlantic Farm Women, 1750–1850* (New Haven, Conn., 1986); Mary M. Schweitzer, *Custom and Contract: Household, Government, and the Economy in Colonial Pennsylvania* (New York, 1987).

equal and then to surpass that of its neighbor to the south. Hunterdon also differed from Burlington in its significant Dutch and substantial German population, although British and New England influences were also prominent. Like Burlington, the county sent crops by water to Philadelphia, but its northern reaches also traded with New York.[16]

Moving west, York and Adams counties in south central Pennsylvania exemplify a mid-eighteenth-century frontier, a rich natural endowment, and a highly mobile population that went on to shape the southern backcountry as well as the near Midwest. Adams, where Gettysburg is situated, was formed out of York in 1800. Settlement of the combined eighteenth-century county began about a generation after that of Hunterdon. Like Hunterdon, it attracted a large German population. Unlike Hunterdon, it drew an almost equally large Scotch-Irish contingent, concentrated especially in the newer, western portions of York—those that became Adams County. York and Adams counties also differed from counties to the east in their relatively poor market connections. Navigation on the lower Susquehanna was so poor as to render the river an obstacle rather than a thoroughfare. Instead, as the Baltimore market developed in the late eighteenth century, York citizens petitioned for more roads to nearby Maryland. Overland travel to Philadelphia took far longer, although Harrisburg and Lancaster merchants served as convenient middlemen and attracted substantial York commerce into Philadelphia's commercial orbit before the Revolution.[17]

Located beyond the Appalachian Mountain barrier, Westmoreland

16. See the New Jersey histories cited in note 15, above. My choice of Hunterdon was also influenced by the availability of Hubert G. Schmidt's *Rural Hunterdon: An Agricultural History* (New Brunswick, N.J., 1946), and by the remarkable collection of manuscript materials housed at the Hunterdon County Historical Society, upon which Schmidt's work is based.

17. Selection of Pennsylvania counties was abetted by David J. Cuff *et al.*, eds., *The Atlas of Pennsylvania* (Philadelphia, 1989); Philip S. Klein and Ari Hoogenboom, *A History of Pennsylvania* (University Park, Pa., 1973); Joseph E. Illick, *Colonial Pennsylvania: A History* (New York, 1976); James Weston Livingood, *The Philadelphia-Baltimore Trade Rivalry, 1780–1860* (Harrisburg, Pa., 1947); John Flexer Walzer, "Transportation in the Philadelphia Trading Area, 1740–1775" (Ph.D. diss., University of Wisconsin, 1968); Diane Lindstrom, *Economic Development in the Philadelphia Region, 1810–1850* (New York, 1978). The selection of York and Adams counties also complements James T. Lemon, *The Best Poor Man's Country: A Geographical Study of Early Southeastern Pennsylvania* (New York, 1976), which concentrates on Chester and Lancaster counties.

County, Pennsylvania, was created on the eve of the Revolution and offers a chance to examine a late-eighteenth-century frontier. After the Revolution, Westmoreland farmers clearly perceived their situation as different from that of farmers to the east, at least judging from their participation in the Whiskey Rebellion. In contrast to York and Adams, Scotch-Irish settlers were a predominant early presence in Westmoreland, making study of the county an opportunity to examine closely generalizations contrasting their slovenly farming methods with those of their German-American contemporaries. Westmoreland also offers a case of dependence on western markets and on Ohio River–borne commerce.[18]

Having selected five sample counties, my first task was to see whether their residents owned "red wheel barrows" and "white chickens." Without a clear image of what tools people commonly owned, I could not deduce what skills farm family members customarily possessed or determine whether agricultural practices were changing. Indeed, with roughly 80–90 percent of the colonial population engaged in agriculture, understanding the cultural and economic significance of any early American technology will entail viewing it within the context of agricultural technology.[19] Since so much depends upon seeing those red wheelbarrows, I wanted to see as many of them as I could—to find a record that exists for enough individuals to capture the mundane technology of ordinary folk.

The record that forms the backbone of this study is the probate inventory, a document created, as the name suggests, when someone died. Probate inventories were not, as the name might suggest, limited to those with wills to probate. They exist for intestate as well as testate decedents. Briefly, inventories were, and are, intended to protect an estate's assets by providing a legal record of personal property ownership. (Real property was not usually listed in Pennsylvania and West Jersey probate inventories.) In some colonies and circumstances they were required whether or not the decedent had much personal property. For example, sometimes

18. In addition to works in note 17, see A Gentleman of the Bar, *Early History of Western Pennsylvania, and of Western Expeditions and Campaigns, from MDCCLIV to MDCCCXXXIII* (Pittsburgh, Pa., 1847); William Findley, *History of the Insurrection in the Four Western Counties of Pennsylvania* . . . (Philadelphia, 1796); Thomas P. Slaughter, *The Whiskey Rebellion: Frontier Epilogue to the American Revolution* (New York, 1986). On presumed ethnic differences in farming, see Lemon, *Best Poor Man's Country,* xiv.

19. Sheridan, "Domestic Economy," in Greene and Pole, eds., *Colonial British America,* 43.

the law required an inventory whenever the decedent left minor children. Courts also ordered inventories when a potential heir or other interested party called for one. Those with no assets to protect clearly needed no inventories, so these documents miss the very poor. They do, however, report the assets of many men and women of modest means, and they are clearly more inclusive than other social history sources such as wills, travelers' accounts, account books, or diaries.[20]

They are also far richer in technological detail. Indeed, the level of specificity often astonishes. Inventories distinguish weeding hoes from grubbing hoes, itemize goods of negligible value such as wooden trenchers, and list parts—plowshares, harrow teeth, and wagon covers—as well as plows, harrows, and wagons. All told, then, inventories offer an unparalleled and essentially unutilized resource for historians of technology, both in the attentiveness to tools and in the coverage of the tool-owning population. They also exist continuously from early settlement through the mid-nineteenth century, unlike many social history records that either end or commence with the political transformations of late-eighteenth-century America.[21]

20. Many scholars have made creative use of probate inventory data, especially to address standard of living questions. Work on the Chesapeake and New England has been particularly rich. See, for example, the collection of articles by Lorena S. Walsh, Gloria L. Main, and Lois Green Carr, gathered under the general title "Forum: Toward a History of the Standard of Living in British North America," with comments by Jackson Turner Main, Billy G. Smith, and John J. McCusker, *WMQ*, 3d Ser., XLV (1988), 116–170; Lois Green Carr, "Diversification in the Colonial Chesapeake: Somerset County, Maryland, in Comparative Perspective," in Carr, Philip D. Morgan, and Jean B. Russo, eds., *Colonial Chesapeake Society* (Chapel Hill, N.C., 1988); Lois Green Carr, "The Development of the Maryland Orphan's Court, 1654–1715," in Aubrey C. Land *et al.*, eds., *Law, Society, and Politics in Early Maryland* (Baltimore, 1977); Peter Benes, ed., *Early American Probate Inventories* (Boston, 1989); Carole Shammas, "The Domestic Environment in Early Modern England and America," *Jour. Soc. Hist.*, XIV (1980–1981), 3–24; Sarah F. McMahon, "A Comfortable Subsistence: The Changing Composition of Diet in Rural New England, 1620–1840," *WMQ*, 3d Ser., XLII (1985), 26–65. Alice Hanson Jones, *American Colonial Wealth: Documents and Methods*, 3 vols. (New York, 1977), offers a large selection of inventories drawn from all of the colonies together with a discussion of her methodology in using inventories to assess colonial wealth.

21. On the limitations of inventories, see the works cited in note 20, above. Ultimately in order to extend my discussion from the tool-owning population to the population in

Despite their many assets, like other historical documents, inventories will lead us astray unless we keep in mind why they were created. Otherwise, as one reads through probate inventories, the sense they convey of walking through the house and around the barnyard with the inventory takers can lull one into unwarranted confidence in a document's completeness. Inventories, as noted, were designed to protect an estate's personal property. But inventory takers certainly knew that not all assets needed equal protection. Obviously, listing assets such as cash or silverware afforded insurance against such valuables' disappearance into someone's pocket before settlement.

What is less obvious, until one begins reading wills, inventories, and administrators' accounts, is the need to protect much of the estate's personal property from disappearing into someone's stomach or into the woods. Joseph Wills, administrator of the estate of Daniel Wills of Burlington County, communicated this situation graphically when his 1729 accounts claimed credit for "two hogs appraised in the Inventory at 10/ [shillings] Each[,] which either Strayd away with strange hogs or were Destroyd by Wolves or Dogs so that they never came to this Accomptants use"![22]

By contrast, other items were evidently deemed so secure that their absence from inventories does not necessarily spell rarity. Furniture attached to the walls of the house—built-in bedsteads or benches, for example—understandably seemed safe from depredations. In consequence, where

---

general, I will need to use wills and tax records to correct for age and wealth biases. As discussed below, I have taken care to avoid inclusion of those for whom old age meant the end of active farming—the principal distortion that results from decedents' likelihood of being older than the general population. Since the wealth bias of inventory data tends to overstate tool ownership, my data tend to be biased against one of my principal conclusions: that tool ownership was much more limited than we have generally assumed. On most other questions the fact that inventories are limited principally to those owning tools introduces no evident bias.

Inventory coverage for the regions I have surveyed does change over time, however. By the second quarter of the 19th century in established communities such as Burlington, inventories had become far less specific, often limiting themselves to noting "Sundries in the Kitchen," "Sundries in the Dining Room," "Sundries in the Bedroom," and so forth. These changes appear to reflect two related trends: the increasing use of specialists to perform inventories and the increasing standardization of rooms' contents.

22. Administrator's Accounts, Daniel Wills Estate, Dec. 12, 1729, file no. 1931-37, New Jersey State Archives, Trenton.

featherbeds and other bedding are enumerated but bedsteads not listed, we cannot safely conclude that people were reduced to sleeping on the floor. Nor, of course, can we conclude that they were not.[23]

More important, like all historical documents, inventories were shaped by prevailing gender assumptions. Legally, widows were entitled to a certain share of the property. Sometimes an inventory designates items as "belonging to the wito," to borrow a phrase from one York County inventory. Comparing such inventories with others is important in assessing whether local inventory takers tended to omit goods they assumed to be the widow's. Likewise, certain items were enumerated so rarely as to suggest that some property was commonly perceived as belonging to women, despite the formal legal assumption that a married woman's property belonged to her husband. For example, women's clothing appears almost exclusively in women's inventories; it is rarely listed in the estates of a woman's husband or father.[24] Alternatively, some items associated with women may have been considered by male enumerators to be natural extensions of the woman or deemed so trivial because associated with women that they were not found worthy of enumeration in women's or men's estates. For example, I have found only eight needles and three thimbles in the more than 350 inventories I have examined closely thus far. Other items generally associated with women's work—poultry, for example—were also uncommon, if judged by inventories. Yet we know from other evidence, notably faunal remains in archaeological excava-

23. These observations reflect my reading of works cited above in note 20 as well as my work with inventories. Discussions with Stephanie Wolf helped me identify many of these limitations.

24. Probate Inventory, Joseph Klipfer Estate, Jan. 6, 1774, York County Courthouse, York, Pa. My analysis of such comparisons to date suggests the likelihood of local variations in this practice, so that testing for the presence of the widow's portion needs to be done for each sample county and for each temporal sample.

The inventory of John McNeil is the exception to excluding women's goods that proves the rule. It was compiled in a remote part of York (later Adams) County on Aug. 30, 1762, a lag of several years after McNeil's will was proved at the courthouse. The enumerators, one of whom signed with an X, naively included several gowns, "peticoats," and shifts. At the end of the inventory, however, someone noted in a different hand: "deduct the Woman's Cloaths w.ch ought not have been valued." Inventory, John McNeil Estate, Aug. 30, 1762, York County Courthouse, York, Pa.

tions, that, even in frontier Westmoreland, domestic fowl contributed substantially to the diet.[25]

To generalize, then, our common practice of treating some documents as literary evidence, in which case they receive close reading and attention to nuance and to social construction, and of treating other documents as quantitative evidence, in which case we code them up and crunch out the numbers, entails a false dichotomy between literary and social scientific sources. Certainly inventories are both. They can, for example, be used to tell us that before the American Revolution a significant proportion of Burlington and Hunterdon County households—about 20–25 percent—owned slaves. But they can also be used to reveal something of slavery's meaning in the Quaker mid-Atlantic, if we note that Burlington inventory takers rather consistently listed slaves between the farm tools and the livestock.

Although part of my intention in quoting Williams's poem was to underscore the literary character of the historical enterprise, I do not minimize the importance of the quantitative. For example, I needed to devise a sampling strategy that would avoid aberrant years: times when events such as wars disrupted record keeping or times when epidemics or warfare made the population of decedents unusually young, for example. I found that I could best assure comparability among years if I used a fifteen-year interval in sampling. Starting with 1774, the last good pre-Revolutionary year, I moved forward in fifteen-year increments to 1849, a date chosen so as to conclude my study with results from the first full federal agricultural census, that of 1850. I got as close as possible to initial settlement by moving backward from 1774 in fifteen-year increments, collecting inventories as far back as records permitted.

This strategy has already generated a daunting array of information. My earliest sample comes from Burlington in 1714. Hunterdon supplies inventories beginning in 1729. York, founded in 1749, offers its earliest sample in 1759, and Westmoreland, organized in 1773, provides a small sample the following year. In order to ensure ample representation of in-

---

25. James B. Richardson III and Kirke C. Wilson, "Hannas Town and Charles Foreman: The Historical and Archaeological Record, 1770–1806," *Western Pennsylvania Historical Magazine*, LIX (1976), 180–181. Without exception, those who signed inventories as enumerators were male, so we know that women's items, if present, were viewed—or ignored—through male eyes.

ventories compiled in each season of the year and of assets left by various sorts of persons, I chose not to sample within years but to collect all inventories filed in each sample year. I have also chosen to retain virtually all of the original data.[26]

The discussion that follows rests on a data base containing all of the eighteenth-century inventories for Burlington and Hunterdon, New Jersey, and for Westmoreland, Pennsylvania, and all of the York/Adams County inventories prior to 1789, about 350 inventories. That translates into an inventory data base of roughly twenty-six thousand records—individual inventory items, that is. The data I present here are drawn from a subset of 250 inventories chosen to include only individuals who were farmers and who owned at least some household goods.[27]

Before exploring what these data suggest about early American agricultural technology, one final general question remains to be answered:

26. I had the benefit of creating my data bases using far more flexible and powerful software than was available to those who created most of the important early inventory data bases. Using *Paradox 3.0*, I was able to avoid any coding and to retain variant spelling, phrasing, or nomenclature of items. This capacity is especially crucial for studying technology, because farm tools appear under an extraordinary variety of names and because we know too little about what actual differences these nominal distinctions may imply to make their accurate reduction to fewer categories possible.

My inventory data base also includes various adjective fields that should ultimately permit me to associate characteristics with monetary values and to deduce characteristics of items enumerated only in generic terms—to determine, for example, whether a "spinning wheel" is a wool or flax wheel, based on its monetary value.

27. I chose to include as farmers all those who listed themselves or were listed as yeomen, husbandmen, planters, or farmers in any of the documents such as wills, administrations, and accounts that accompany inventories, as well as those for whom no occupation was listed. I also included as farmers those who gave two occupations, one of them among those mentioned above. Inventories of those for whom only a nonfarm occupation or status (widow and gentleman were also designations) was listed make clear that virtually everyone was engaged in agriculture to some degree. Ultimately, my study is intended to wrestle with the issue of what it meant to be engaged in agriculture in this period and to relate inventory differences to differences in occupational designation. For the present study, however, the criteria I have set for inventories' inclusion are intended to assure that my focus on the diversity of farm practice does not simply result from including individuals whose principal arena of activity lay elsewhere.

I found that including only those individuals whose inventory listed a bed assured the presence of household goods while not eliminating those whose array of goods was modest.

What was early American agricultural technology? To date, restrictive definitions of technology have impeded the study of early farming technology. Accounts of colonial farming in particular regularly dismiss its technology as "primitive" and technological change as "absent." Either characterization reflects an unfortunate tendency "to limit the definition of technology to those things which characterize the technology of our own time, such as machinery and prime movers," a definition that makes the nineteenth-century reaper seem to herald the dawn of American agricultural technology.[28]

Fundamentally, historians' definitions of technology reflect the fact that we have written mostly about nineteenth- and twentieth-century technology and have given little thought to early modern technology or to farm technology generally. It suffices to say here that agricultural technology includes far more than machines, implements, and the knowledge of how to use them. At minimum it must also include the plants and animals that humans have developed, together with knowledge of plant and animal behavior; the methods of identifying land suited to particular purposes and of modifying and organizing that land; the construction methods, structures, and procedures devised for storing crops and housing livestock; and the knowledge of how to modify crops, land use patterns, and storage techniques to adapt to various climates. Under the household organization of labor that characterized early modern agriculture, the tools, skills, and knowledge employed in processing food, fiber, and other farm products— most of them wielded by women—must also qualify as agricultural technology. Assuredly, the agricultural enterprise could not have functioned successfully without them.[29] For purposes of this discussion, then, agricultural tools will include items such as pots, churns, stills, spinning wheels, maize, turnips, cattle, and sheep as well as implements such as plows, wagons, hoes, and axes.

What do probate records of such items allow us to conclude about

28. Melvin Kranzberg and Carroll W. Pursell, Jr., eds., *Technology in Western Civilization*, I, *The Emergence of Modern Industrial Society: Earliest Times to 1900* (New York, 1967), 4–6.

29. Judith A. McGaw, "No Passive Victims, No Separate Spheres: A Feminist Perspective on Technology's History," in Stephen H. Cutcliffe and Robert C. Post, eds., *In Context: History and the History of Technology: Essays in Honor of Melvin Kranzberg* (Bethlehem, Pa., 1989), 172–191, argues similarly for a scholarship that transcends narrow and gender-laden definitions of manufacturing.

tool ownership on early American farms? As suggested already, one very clear message is that there was no standard array of household and farm utensils that virtually all colonists owned or needed to own. That many individuals made do without plows, kettles, firearms, or Bibles reminds us, as we need always be reminded when we leave the era of modern technology, that the belief in "one best way" to perform a given task is an artifact of industrialized production. Indeed, the ability of modern households to own a standard array of tools reflects the existence of systems of transportation, communication, and manufacturing inconceivable in early modern societies. I stress the point because the history of technology has focused so narrowly on the recent period that aberrant features of industrial technology easily get read back into discussions of earlier eras.

Another conclusion warranted by these data is that eighteenth-century Americans were not technologically self-sufficient. Patterns of tool ownership imply extensive participation in the market. For example, only five inventories listed candle molds, and only about 5 percent enumerated tallow. A far larger proportion—24–40 percent—included candles or candleholders, implying the acquisition of candles elsewhere. Likewise, although yard goods appear in most inventories, weaving equipment shows up in fewer than 5 percent of Burlington farm households. Even in more recently settled Hunterdon and York, where poorer transportation limited access to imported textiles and urban weavers, only about 15 percent of farm inventories generally listed looms. Where the prospect of acquiring textiles from Britain or from local population centers was extremely limited, as in the very remote settlements of late-eighteenth-century Westmoreland County or of York in 1759, the proportion of farm households prepared to weave still reached only 25 percent—hardly an impressive tally.[30]

30. The conclusion that early Americans were not self-sufficient is hardly new. Important earlier articles on the issue are cited in notes 7 and 20, above. Unfortunately, the work of these and other scholars has had relatively little currency in or influence on the history of technology. Moreover, early Americanists have tended to emphasize the implications of this finding for economic development and the standard of living while neglecting its implications for technological skill and occupational specialization.

On the issue of domestic, craft, and mill production of textiles, see Gail Barbara Fowler, "Rhode Island Handloom Weavers and the Effects of Technological Change, 1780–1840" (Ph.D. diss., University of Pennsylvania, 1984); and Adrienne Dora Hood, "Organization and Extent of Textile Manufacture in Eighteenth-Century, Rural Pennsylvania: A Case Study of Chester County" (Ph.D. diss., University of California–San Diego, 1988), notable recent exceptions to the general neglect of household and craft technologies.

Eighteenth-century farmers not only purchased essential commodities such as cloth but also depended on mills or specialists to process much of what they grew. There is, for example, no evidence that any farmer owned tools to grind his own grain; and, except in frontier York with its exceptionally poor transportation facilities, most people who grew flax lacked both flax brakes to prepare the fiber, and many lacked hackles to comb it. More households owned spinning wheels, although only in York and in early Hunterdon did substantially more than half do so, a pattern probably indicative of the frontier population's relative youthfulness rather than of its relationship to the market.[31] If eighteenth-century agricultural regions had been technologically self-sufficient, we might expect, at the very least, to find a region's ownership of spinning wheels to parallel its fiber production, but, except in York, the proportion of households owning sheep rather consistently outran the proportion prepared to spin wool. Indeed, in many years more Burlington decedents owned herds of sheep (twenty or more) than owned spinning wheels of any sort. Woolen wheel ownership exceeded the number of substantial sheep herds only in York and Westmoreland, where, given the persistence of wolves, large herds of sheep remained uncommon.

Similarly, tool ownership patterns clearly imply extensive wood-processing specialization. Remarkably, even common tasks such as firewood preparation entailed exchanges with outsiders. Throughout the eighteenth century three-quarters or more of Hunterdon and Burlington households lacked mauls and wedges for splitting wood. Saws of various sorts and froes, for cleaving shingles, were even less common. As do the turning lathes listed in craftsmen's inventories, these data remind us that specialized production and employment long antedated American settle-

31. Spinning wheels signify families early enough in their lifecycles to have young daughters at home or proximity to neighbors with young daughters. If anything, these data overstate the extent to which households engaged in spinning, because at this stage in my work I have focused simply on the presence or absence of a tool. In fact, many tools listed in inventories reflect an earlier stage in the household's history and may not have been in current use. Close reading of inventories makes this especially clear in the case of spinning wheels; a number of them are grouped with broken or discarded goods in areas clearly set aside for storage rather than for implements currently employed.

I have taken care to construct my data base so as to retain all data in their original order and, where several items were grouped on the same line, its original groupings. I intend to explore more fully what can be learned from the organization of inventory items and to add implied location data to the data base as a result of this analysis.

ment. It is hardly surprising, then, that colonists presumed self-sufficiency to be unnecessary and, probably, undesirable. Certainly they quickly made it uncommon. The most dramatic case in point is that of frontier farmers, who generally distilled their grain into whiskey, a commodity whose high value relative to its bulk enabled it profitably to be transported long distances to market. Despite the relative cheapness of stills, most farmers relied on others to process their grain. Similarly, in eastern counties, where apple trees had had a chance to mature, most farmers depended on a small minority who owned cider mills.

Two additional aspects of tool ownership signal the market's mounting importance over the eighteenth century. First, early in the century Burlington farmers rarely (less than 15 percent of the time) owned wagons. Instead, between three-fifths and one-half of households had carts—relatively small, cheap, two-wheeled vehicles better designed for hauling small loads around the farm than for carrying crops to market. After midcentury, this pattern was reversed. About three-fifths to one-half of Burlington farmers came to own wagons, while cart ownership declined dramatically. At the same time, roughly two-thirds of Hunterdon farmers owned wagons, the greater proportion reflecting longer average overland distances to market. Even in York, where wagon trips to market were less frequent, nearly two-fifths of households owned them. Only in remote Westmoreland did many farmers have to make do with pack saddles. During the same years, substantial proportions (from one-quarter to one-half) of farm households in each county came to own steelyards, signifying more frequent occasions to weigh goods for exchange.[32]

Diverse patterns of tool ownership also reveal eighteenth-century mid-Atlantic farmers to have been a distinctly innovative lot. And, far from deriving from the romanticized jack-of-all-trades conjured up as part of the frontier subsistence myth, early American readiness to try new technology had several more mundane sources. First, it built on the common experience of the many colonists who came from a place—Britain—in

---

32. In fact, the data on wagon ownership in York is understated because of the inclusion of farmers from relatively unsettled areas on the periphery of what became Adams County. Like Westmoreland farmers, these men also frequently owned horses and saddles but no wagons.

Steelyards are relatively light, simple balance scales on which the commodity to be weighed is suspended from the end of the short arm of the balance beam and a counterbalancing weight is suspended from the long arm.

the throes of an agricultural revolution. If we take differences in British and colonial circumstances into account, we find that land reorganization and reclamation, experimentation with and adoption of new crops, and increased attention to livestock husbandry engaged colonial yeomen as well as their British counterparts. Whereas the British brought new land into cultivation by adopting crop rotations suited to light, upland soil or by draining swamps or by irrigating meadows, early British Americans achieved the same outcome by acquiring the utterly alien skills needed to clear the continent's dense woodlands. And, like Englishmen at home who acquired their land reclamation technologies from their Dutch neighbors, mid-Atlantic English colonists developed forest destruction to a fine art by borrowing the technology of their Native American and Scandinavian predecessors and of their German contemporaries.[33]

Among their tools, farmers' frequent ownership of axes best documents their openness to new land-making technology. Until relatively late in the eighteenth century, for example, Burlington and Hunterdon County farmers were at least as likely to own axes as plows, and in York and Westmoreland they were far more likely to. Furthermore, despite the near certainty that most arrived from England without axes or tree-felling experience, a substantial proportion (one-quarter to one-third) of Burlington and Hunterdon decedents owned more than one axe.

Mid-Atlantic farmers also paralleled British agricultural revolutionaries in readily adopting new crops, such as Indian corn, listed rather consistently in 50–60 percent of inventories. The extent to which this formerly novel grain had achieved acceptance is best suggested by the evidence that almost all inventories after midcentury employ the words "wheat" or "rye" or "oats" to enumerate European grains; they reserve the word "corn" for maize. Indeed, so early and so thoroughly was Indian corn integrated into the mixed-farming regime brought by English colonists that one of the supreme ironies of eighteenth- and nineteenth-century agricultural history is apparent. The same reformers who praised English

33. J. D. Chambers and G. E. Mingay, *The Agricultural Revolution, 1750–1880* (London, 1966); Peter Mathias, *The First Industrial Nation: An Economic History of Britain, 1700–1914* (London, 1969), 66–70; Jones, *Agriculture and the Industrial Revolution,* esp. 69, 75; Berg, *Age of Manufactures,* 95; John R. Stilgoe, *Common Landscape of America, 1580 to 1845* (New Haven, Conn., 1982), 170–182, 184–186; Nicholas P. Hardeman, *Shucks, Shocks, and Hominy Blocks: Corn as a Way of Life in Pioneer America* (Baton Rouge, La., 1981), 51–58.

husbandmen for their willingness to experiment with new crops ignored the far more general adoption of maize by American farmers and, instead, derided them for their reluctance to grow turnips.[34]

Also like their British contemporaries, mid-Atlantic farmers accorded particular attention to livestock husbandry. In every sample county, horses and cattle were far more common than plows or even axes for most of the century, and hay crops grew in frequency and value. Churn ownership suggests that farm women's work also shifted toward livestock husbandry. Burlington churn ownership increased to nearly one-fifth of estates by the late eighteenth century while in Hunterdon and York more than one-half and about two-thirds, respectively, of all inventories listed churns.[35]

Repeated evidence that York County decedents more often employed progressive technology reflects the prominence of Germans in its population. But the case of German farmers' apparent progressiveness well illustrates that technological progress is always in the eye of the beholder. In fact, for German settlers the techniques that British farmers had recently adopted with so much fanfare represented established practice.[36] So, for

34. Lemon, *Best Poor Man's Country*, 183. Given that Indian corn was unlikely to appear in estate inventories during the months when livestock could be turned out to forage, 50%–60% is an extremely high proportion.

Among other things, the convergence of maize adoption with the techniques of the English agricultural revolution is suggested by the development of various techniques to preserve corn leaves and stalks as green winter fodder, producing a succulent feed fully the equal of the more celebrated European turnip, swede, and mangel-wurzel. Likewise, although Englishmen borrowed from Native Americans their checkerboard cornfield arrangement and crisscross cultivation system, the persistence of this system probably owed much to its similarity to one advocated by Jethro Tull in the early 18th century. See Hardeman, *Shucks, Shocks, and Hominy Blocks*, 89–90, 96–98.

35. The early and continued evidence in mid-Atlantic inventories of livestock's importance complements the preserved physical evidence of mid-Atlantic innovation in barn construction (Stilgoe, *Common Landscape*, 152–154, 158). Discussions of barns and other outbuildings have rested principally on anecdotal evidence. The next phase of my work calls for analysis of both residences and outbuildings in my sample counties through a data base derived from records of the United States Direct Tax of 1798.

36. Schmidt, *Agriculture in New Jersey*, 49–50, 79–80, 123; Fletcher, *Pennsylvania Agriculture*, 48–51, 125–126, 137, 172, 387–388. Again, most discussions of German farming practices rest on anecdotal evidence. Surprisingly, given the prominence of Hanoverians in English society in the 18th century, there has been virtually no attention to German influence on British farming, although it is worth noting that Dutch influence sparked many

example, York farmers often owned tools for manuring: dung forks, dung hooks, and dung shovels appear in two-thirds of their inventories by 1774. By contrast, almost no Burlington farmers' inventories record any such implements. York farmers were also far more likely to wield cradles at harvesttime (20 percent of York farmer decendents versus, at most, 5–10 percent of Burlington farmers). And they generally owned hoes, indicative of more intensive cultivation (about 70–80 percent of York farmers' inventories versus about 30 percent of Burlington farmers' inventories during the same years). York's settlers often used those tools to cultivate potatoes and turnips (about 40 percent and 25 percent, respectively)—"the new root crops" from an English perspective. Farmers in York also employed far more winnowing fans (nearly one-third of York inventories listed them, versus only one Burlington inventory).

However commonplace German farmers may have found these technologies, viewed through British eyes they were innovative. It is all the more noteworthy, then, that by 1774 only about half of York farmers owning dung tools were German and even fewer potato growers were (about one-third). By contrast, fifteen years earlier, Germans had owned most dung tools (about three-fourths of them). In other words, tool ownership patterns show German agricultural technology diffusing earlier and more readily than most secondary scholarship has surmised.[37]

Hunterdon County also had a substantial German population, and one derived from essentially the same sources as supplied York. But whereas York Germans arrived early in that county's history and generally pre-

---

reforms and that during this period the Germans were also generally referred to as Dutch. See Joan Thirsk, ed., *Agricultural Change: Policy and Practice, 1500–1750* (Cambridge, 1990), 305; Chambers and Mingay, *Agricultural Revolution,* 59–60.

Lemon, *Best Poor Man's Country,* suggests that ethnic differences in farming were less evident in 18th-century Pennsylvania than most scholars have assumed.

37. As noted above, Lemon interprets anecdotal evidence of similarities between ethnic groups as meaning that differences were minimal to begin with. As befits a geographer, however, his emphasis is on space rather than on time. By supplying evidence of change over time in ethnic tool ownership patterns, my data suggest colonial interaction rather than similar heritages as the source of much 18th-century similarity. Nonetheless, the broad European commonalities to which Lemon points must have made technological diffusion occur more readily.

Attribution of ethnicity rests on surname analysis. I am grateful to Marianne Wokeck, who lent her considerable expertise to this task.

dominated numerically, Hunterdon Germans arrived after their county's other principal ethnic groups and made up only a quarter of the population.[38] As a result, techniques associated with German farming diffused far more selectively in Hunterdon. Where a technology proved well suited to established practice and could be introduced as a solitary innovation, German techniques fared well there. So, for example, winnowing fans became even more common there than in York (they appeared in about 35 percent of Hunterdon farm inventories). This adoption is hardly surprising, for they were clearly well suited to Hunterdon's late-eighteenth-century emphasis on wheat, a crop listed in 65–72 percent of farm inventories.[39] Similarly, although at midcentury Hunterdon farmers, like their Burlington neighbors, rarely harvested with cradles, their increasing reliance on hay reinforced the increased presence of German exemplars to make cradle use more common in Hunterdon than in York. By 1789 two-fifths of Hunterdon farm inventories listed cradles, and only half of these are identifiable as German or Dutch. On the other hand, dung implements appear almost exclusively in German and Dutch inventories in Hunterdon, reflecting both the absence of evident soil depletion and the greater array of new behaviors dunging entailed.

Comparison of York and Hunterdon can also help establish how much of York's apparent innovativeness derived from German influence and how much from the influence of the frontier. Again, data on tool ownership serve as a useful corrective to the belief, pervasive in American popular culture, that frontier living meant being reduced to older, more primitive ways of doing things—a belief we readily incorporate into historical scholarship where evidence is absent. Tool ownership patterns indicate that, far from reverting to obsolete technology, frontiersmen generally brought the latest technology with them. For example, large, heavy, iron-reinforced, covered market wagons (from which the famed Conestoga was derived) showed up far more often in late-eighteenth-century York than in Hunterdon inventories of the same era. And most kettle-owning farm women in Westmoreland and York employed modern iron kettles, whereas traditional brass kettles continued longer in Hunterdon and, especially, Burlington.

Similarly, comparison of Burlington, Hunterdon, York, and Westmore-

---

38. Lemon, *Best Poor Man's Country,* 43–50; Schmidt, *Rural Hunterdon,* 33–35, 42.

39. This is an extremely high proportion, since grain was one of many items whose probability of being enumerated varied seasonally.

land inventories indicates that the technological commitments made by a region's pioneers could be hard to break. For example, over the eighteenth century, mid-Atlantic farmers followed a general European trend away from the use of oxen and toward the use of horses as draft animals. Thus, oxen were relatively common in early Burlington, showing up in more than one-quarter of inventories before 1730. By contrast, oxen appear in virtually no inventories in York, where most settlement occurred after midcentury. They are entirely absent from Westmoreland, settled even later. In Burlington, however, oxen represented established practice, and they remained relatively common even late in the century. Hunterdon, settled only a generation earlier than York, also relied heavily on horses from the outset, but a persistent minority used oxen late in the century—apparently Hunterdon immigrants from New England, an early and persistent ox-using region.[40]

Frontier conditions also encouraged technological innovation more directly. Not surprisingly, all of the various rough woodworking tools appeared far more often in York than in Burlington or Hunterdon inventories. Likewise, specialized hoes were a feature of frontier agriculture. Most commonly, inventories distinguished between grubbing hoes and weeding hoes. Such distinctions were especially common in York, ten times more common than in Burlington inventories. Widespread ownership of both weeding and grubbing hoes certainly made sense under frontier conditions. Corn, an ideal crop on new land, generally received hoe cultivation—hence the weeding hoe—and new land also required farmers to use hoes to grub out roots, stones, and other debris.[41]

Frontier and German influence intersected somewhat differently to shape food-processing technology. One of the more striking technological differences between German and British settlers was that Germans were accustomed to cook and heat with stoves, whereas the British relied on open hearth cookery and heating. Like many German technologies, stoves eventually became the British American's modern technology, but there is less evidence of their colonial diffusion. In English Burlington, for example, despite ready access to stoves in nearby Philadelphia, only one stove was enumerated in any eighteenth-century inventory. Rather, food preparation methods appear to have been highly resistant to change, a pattern anthropologists have often ascribed to the deep-seated cultural

---

40. Bidwell and Falconer, *Agriculture in the Northern United States*, 111–113.
41. Hardeman, *Shucks, Shocks, and Hominy Blocks*, 55–58, 83–86.

conditioning of dietary preferences. A look at the technology involved suggests instead that the domestic craftswomen who prepared most meals understandably resisted adopting new tools that threatened their hard-won proficiencies, altered the quality of the work experience, and would have required them to learn a new repertoire of skills.[42]

Domestic tool ownership patterns in York support this interpretation. As a frontier community with a very poor transportation system, York was a difficult place to obtain a stove. Nonetheless, nearly half of York's German decedents had managed to acquire stoves by the 1770s (about 20 percent of all inventories). Even more striking is the suggestion implicit in York's domestic tool ownership pattern that German farm wives who lacked stoves merely tolerated fireplace cooking as a temporary expedient. Trammels, elaborate contrivances for hanging and adjusting pots over the fire, are absent from York inventories although they appear often in Burlington ones. Instead, York mistresses relied on simple pothooks. For varying the heat they applied, they placed their pots on pot racks designed to elevate pots and frying pans over coals on the hearth. Not only did this cooking method come closer to replicating stove cooking, but also pot racks served as complements to legless cooking vessels—vessels one would prefer to retain or purchase if one expected ultimately to acquire a stove.

By contrast, Hunterdon County domestic tool ownership patterns suggest how farm women came to adopt new technology. Although more favorably situated to acquire stoves, few Hunterdon decedents owned them (about 10 percent of inventories). Nor did Hunterdon Germans show a preference for pot racks (fewer than 10 percent of inventories enumerated pot racks, even fewer than in Burlington); county farm wives of all ethnic groups employed trammels. Essentially, this very different behavior reflects differences in the timing of German settlement. As relative latecomers, Hunterdon Germans generally moved into existing houses—houses that embodied a preference for fireplace cooking.[43] Moreover,

42. Schmidt, *Rural Hunterdon*, 268–269; Fletcher, *Pennsylvania Agriculture*, 387–388, 399. I am indebted to Stephanie Wolf for suggesting the interpretation of the resistance of craftswomen.

43. Schmidt, *Rural Hunterdon*, 93, notes the relative uniformity of houses, despite the presence of various ethnic groups. He also finds general adoption of the separate "Dutch cellar" for storing fruit and vegetables. His observations are, however, based on surviving structures.

Hunterdon Germans tended to be poorer than those who could afford the longer trek and greater farm-making investment required in York. Most probably arrived without pots and pans and acquired kitchen utensils from craftsmen accustomed to supplying Hunterdon's established British settlers.

Ethnic differences in food processing are also manifest in the distribution of specialized tubs between York and Burlington. With one exception all of the powdering tubs—tubs used to preserve meat by salting—appear in Burlington inventories. By contrast, all of the pickling tubs show up in York inventories, evidence of a divergent meat preservation tradition. Cabbages and implements for processing cabbage are, likewise, limited almost exclusively to York and Hunterdon German inventories. In sum, the distribution of food processing implements in eighteenth-century inventories hints at a larger, historically invisible array of technologies: the skills and knowledge essential to transforming crops into food.

A L T H O U G H we can glimpse only obliquely the skills and knowledge that constituted most of colonial agricultural technology, inventories offer a rather direct look at the material component of that technology: the tools, livestock, and plants that early Americans commonly employed. This preliminary account of several hundred such documents demonstrates how productive of new and revised understandings of early agricultural practice the actual evidence of tool ownership can be. Looking at farmers' possessions reveals that there existed no standard array of implements farmers and farm wives owned or needed to own; it shows the belief in "one best way" to perform a task to be a historical artifact, an intellectual by-product of industrial society that we inadvertently project onto the preindustrial past.[44]

Certainly, colonial farm household tools show a diversity that we deem uncommon in the world we inhabit. In part this diversity reflected the considerable and growing involvement of farmers and farm wives in networks of exchange—with one another and with the local representatives of distant producers. In part the diversity reveals selective adoption from

44. Indeed, given the dearth of studies of the actual technological practice of 20th-century workers, farmers, or housewives, it is unclear whether the notion of "one best way" is much more descriptive of the reality of 20th-century life than it is of life in earlier eras. So long as we continue to write principally the supply-side history of technology, it is likely to remain unclear.

the varied menu of innovations that American farmers employed virtually from the start of settlement: land-clearing and woodworking tools, new field and garden crops, and the various creatures, structures, and implements that embodied a growing commitment to livestock husbandry. And, whereas the ethnic diversity of the colonial population may have introduced much of this technological diversity, close examination of late-eighteenth-century tool ownership discloses extensive cross-cultural borrowing, especially in regions where ethnic diversity began early. Inventories also reveal diversity emerging from the diverse times at which counties were settled. Technological patterns established early evidently continued to influence local farmers' and farm wives' technological options. Given the continuous westward course of settlement, then, temporal diversity gradually assumed a geographic manifestation.

These preliminary observations about early American tool ownership carry obvious implications for the large question with which I began: What accounts for America's sudden, rapid, and comparatively successful early nineteenth-century industrialization? They suggest that the wellsprings of American willingness and ability to innovate were mundane rather than mythic, endemic rather than heroic. Judging from these data, the sources of early American technological innovation—and, by extension, America's early industrialization—should sound familiar to American historians: the rise of a market economy, the selective transfer of European culture, the social consequences of ethnic diversity, and the significance of the frontier. These are themes to which students of the American past perennially recur.

If these themes also best elucidate our early technological history, then "so much depends" on closely scrutinizing the small technologies found on early American farms—much more than a new portrait of early tool use. If, as these preliminary findings suggest, traditional historical themes will play a predominant role in explaining America's technological history, historians of technology will have to question the wisdom of their increasing specialization and separation from the larger historical profession. More than most historians, we should recognize and avoid the high cost of unnecessarily reinventing the wheel. Likewise, once early American historians recognize their perennial concerns as having great relevance to understanding the problem of technological innovation, they may wonder at their former willingness to delegate study of this crucial issue to historians of technology. Like all contemporary Americans, early American historians should recognize and avoid the high potential costs of leaving technology solely to the technology experts. Finally, if early American technological diversity helped precipitate America's remarkable nineteenth-century technological innovativeness, the findings discussed above hold a moral for us all, early American historians and historians of technology alike. Studies of early American technology need all of our various approaches; scholarly innovation thrives best when a field's cultivators know there is no "one best way" to farm.

NINA E. LERMAN

Books on Early American Technology, 1966–1991

Introduction

The following bibliography is a composite of many ideas and approaches. It embodies, of necessity, my own view of what constitutes history of technology, and it incorporates the outlooks of other scholars on the works they find relevant in their particular fields. This compilation is offered, therefore, not as a summary, but as a starting point, a preliminary navigational chart to encourage further exploration. It is intended as an update to Brooke Hindle's earlier bibliography on the same subject, and it shares with that work the "ultimate objective" Hindle articulated in his introductory essay: "to raise technology to its proper place within the context of early American history."[1]

Some twenty-five years have passed since Hindle compiled his bibliog-

This bibliography benefited immeasurably from the suggestions and advice of the following scholars: Edward C. Carter II, Carolyn Cooper, Donald C. Jackson, Larry D. Lankton, Steven Lubar, Patrick Malone, Judith McGaw, Glenn Porter, Robert C. Post, Alex Roland, Philip Scranton, Darwin Stapleton, Stephanie Grauman Wolf. The author also thanks Richard Brodhead, Jennifer Gunn, Christian Gelzer, Joyce Roselle, R. Srinivasan, and Keith Wailoo for behind-the-scenes logistics and support.

1. Brooke Hindle, *Technology in Early America: Needs and Opportunities for Study* (Chapel Hill, N.C., 1966); quote from "The Exhilaration of Early American Technology: An Essay," 28 (or above, 67).

raphy, years in which the history of technology in general has become a far more established field of study. Technology has received recognition for its roles in phenomena otherwise considered political, social, intellectual, or aesthetic, and the impact of these realms on the nature of technological activity has been increasingly recognized. Yet somehow, in the case of *early* American history, technology has not achieved the central position Hindle suggested for it in 1966. The body of literature published since then that can be strictly classified as "history of early American technology," while important, remains relatively small. Early American historians have neglected technology, and historians of technology have neglected early America.

Nonetheless, a vast number of books have been published that treat early American technology at least in part. The works collected here are divided into categories based on the ones Hindle used, but modified to reflect the changing currents of historical scholarship, the shape of the field as I see it now. In explanation of topics I have included under a rubric of "technology," I offer an analogy. I do so not only for the sake of historians who have not much considered the question of how to study technology but also for those who have—the question has no single definitive answer even among those who ponder it often. Consider a historical document, such as the Bill of Rights, but one about which we know little. It engenders many questions, ranging from the immediate and obvious (what does it say? what does it not say?) to the more distant (how has it been used? who has used it? what kind of meanings has it had to its users?). We may also want to know who wrote it, and why; who approved of it, and who dissented; and what changes, if any, were made between its initial and final forms. In broader perspective, we may ask on what models it may have been based and whether it, in turn, became a model for other documents. In short, we can ask for concrete details and for general context; we can ask about origins, and we can ask about subsequent uses and effects.

Written documents and technological artifacts are both the result of human activity, and these same questions may be asked about a technology.[2] In the case of technology, we must understand the specific content, the hardware and its workings. We must ask what came before it, why it

---

2. For a more extended exploration of documents and artifacts, see Stephanie Grauman Wolf, "Documentary Sources for the Study of the Craftsman," in Ian M. G. Quimby, ed., *The Craftsman in Early America* (New York, 1984).

came into being, who brought it into being. We must explore the ways in which it was used, intended and unintended, and the intended and unintended effects such uses produced. We must understand that technological change is the product of human choice and endeavor to understand the choices made about technology—by inventors, producers, and users. And when we ask about choice, we must ask both who chooses, in a given instance, and who does not. Finally, technologies are as interrelated as are (for example) constitutional provisions, laws, and judicial precedent—different sizes of lens are appropriate for examining different aspects of such systems. A hammer, a technique of construction, and an entire city can all be studied as technology.

The realm of technology is indeed huge. Hindle divided it into categories based partly on the purpose of the particular technology in question and partly on what he called the "special relationships" of technology as he saw them in early America. Any such categories are interrelated, and while they often suggest interesting juxtapositions, they also create artificial divisions—readers must be aware of the linkages relevant to their particular topics of interest. For example, "Power" is a category separate from both "Manufacturing and Labor" and "Civil Engineering and Transportation"; it includes both waterwheels and steam engines. The fuel burned for steam power is treated elsewhere: lumber is included with "Agriculture," and coal falls logically into "Mining and Metals." "Manufacturing and Labor" includes manual and machine-based production, but the inventors of the machines are discussed under "Patents and Invention." Transportation is grouped with "Civil Engineering and Transportation," a category that includes not only roads and ships but also building and architecture.

I have retained Hindle's overall structure for several reasons. First, Hindle had a clear understanding of the terrain, and his distinctions make sense. Second, the works Hindle cited—particularly those published before 1850—remain extremely useful. The existence of the present bibliography should by no means preclude use of the original, and a parallel structure may help to reinforce this continuity. On the other hand, several modifications have seemed appropriate to the changes in the historical literature in the past twenty-five years and to the scope of the present endeavor. For the benefit of readers using both bibliographies, these modifications are noted in the outline below. The present bibliography includes only book-length publications that consider American technology before 1850 and were published since Hindle's bibliography (1966).

1. General Surveys and Bibliographies. This category includes larger works, international in scope, useful for background information, and published bibliographies relevant to early American technology. Hindle's extensive discussion of sources, which he treats in a separate section, combines bibliographic works with published materials from before 1850. Recent publications of primary source materials appear here throughout the topical sections.

2. American Surveys. General works and recent publications of journals, papers, and documents treating multiple aspects of American technology in this period are grouped together here. Any book whose subject fitted into a more specific category was placed there; for example, a book about the American system of manufactures would be found in Section 10, "Manufacturing and Labor." Works focusing on cities, which Hindle included here, are now in Section 8, "Settlement Patterns and Urban Growth."

3. Agriculture. This category includes material on lumber and furs in addition to farming. It includes books on foodways when they are relevant to agricultural practice, but most material on food processing (which Hindle included here) is now found in Section 4, "Household Technology." Processing of nonfood agricultural products may also be found in Section 10, "Manufacturing and Labor." Whaling, which Hindle included with lumber, may be found with other maritime pursuits in Section 7, "Civil Engineering and Transportation."

4. Household Technology. One of the most significant changes in early American scholarship is the recognition that the household provides as viable and interesting a locus for study as the field, the factory, or the construction site. Neglect of this topic, which focuses primarily on women's activities, has only begun to be rectified—tasks such as food processing and clothing manufacture remain sources of unexplored technological riches. This category is entirely new; much of the material could not have been easily categorized in the existing structure.

5. Mining and Metals. This section includes works on the raw materials and the early stages of their processing; forges are treated here, but metalworking is treated in Section 10, "Manufacturing and Labor."

6. Military Technology. The technologies of the military include not only firearms but also organization, strategy, and the day-to-day accoutrements of armies and navies. Note that more general works on ships and shipbuilding are included in Section 7, "Civil Engineer-

ing and Transportation," and techniques of firearm manufacture are included in Section 10, "Manufacturing and Labor."

7. Civil Engineering and Transportation. This broad category comprises a range of topics: engineers; surveying and cartography; ships, ship-building, and maritime pursuits; steamboats and riverboats; canals; railroads; roads; and building and architecture. The subsection on ships includes whaling (which Hindle included under agriculture) and shipping, both oceangoing enterprises. The one on building and architecture is construed narrowly; broader works on urban development may be found in Section 8, "Settlement Patterns and Urban Growth."

8. Settlement Patterns and Urban Growth. Much of the change in the physical environment of early America was neither designed nor engineered. Nonetheless, such changes were often technological, or they were the effects of technological change, or they in turn inspired a technological response. Demographic change, too, is closely related to technologies: the relocation of people, whether to the city or to the frontier, meant the relocation of workers with varying technical skills and provoked technological responses to new resources and new problems. This section is new.

9. Power. This category includes material on water and steam power and on power transmission—for example, in mill construction.

10. Manufacturing and Labor. As Hindle noted in 1966, the term "manufacture" can refer to both hand and machine production. Both are included under this heading. Note that, while paper and printing are included here, circulation of printed matter is treated in Section 12, "Communication." Household manufactures, such as clothing, are treated in Section 4, "Household Technology."

—. Heat, Light, and Electricity. The topics Hindle treated in this category have not received much attention in recent years. The scant material that might have been placed here seemed most logically grouped elsewhere, either in Section 4, "Household Technology," Section 9, "Power," Section 10, "Manufacturing and Labor," or Section 11, "Education, Organization, and Science."

11. Education, Organization, and Science. This section comprises not only formal education and apprenticeship, technological institutions, and technology-related science but also works on the spread of literacy and numeracy in early America.

12. Communication. Communication is important on two levels: not

only did it become increasingly dependent on technologies such as printing and transportation in this period, but its increase also facilitated various means of technology transfer. This is a new category.

13. America and Technology Transfer. In Hindle's bibliography this section was titled "America and Europe." Recent scholarship, however, has recognized the contributions of all Americans, immigrant and native, free and unfree, to all of the technologies included in this bibliography. Works emphasizing the transfer of skills, techniques, and hardware from one group of people to another are included in this section.

14. Patents and Invention. This section focuses on the creation of technologies recognized as new, the process of such creation, and the legal mechanism designed to protect and encourage such work.

15. Technology in Business and Economic History. Hindle's broad category "Technology in American History" has been narrowed here. Books related to any one of the categories above have been listed there; more general works are listed in Section 2, "American Surveys."

1. GENERAL SURVEYS AND BIBLIOGRAPHIES

Although the present bibliography focuses on America, the important general survey by Melvin Kranzberg and Carroll Pursell, Jr., should be noted. The two-volume *Technology in Western Civilization* (1967) adds an American synthesis, incorporating a more contextual perspective, to the British and French entries noted by Hindle.[3] Its first volume treats technology before 1900. Kranzberg has also produced another survey, this one written with Joseph Gies, entitled *By the Sweat of Thy Brow: Work in the Western World* (1975). It ranges from the Stone Age through the twentieth century and so encompasses the transition from craft-based to mechanized production, among other issues.

Also worldwide in scope, Eugene S. Ferguson's comprehensive 1968 *Bibliography of the History of Technology* provides a solid foundation for updates such as the annual bibliographies in *Technology and Culture*. The American Studies Information Guide Series includes a volume *Technology and Values in American Civilization* (1980), by Stephen H. Cutcliffe, Judith A. Mistichelli, and Christine M. Roysdon. Ewald Rink has com-

3. Charles Singer *et al.*, eds., *History of Technology* (London, 1954–1958); and Maurice Daumas, ed., *Histoire générale des techniques* (Paris: 1962–).

piled a bibliography of early works, *Technical Americana: A Checklist of Technical Publications Printed before 1831* (1981). Several sections of the topically arranged *Books about Early America: 2001 Titles* (1989), compiled by David L. Ammerman and Philip D. Morgan, contain relevant material. Of some use may also be Martha Jane Soltow and Mary K. Wery's *American Women and the Labor Movement, 1825–1974: An Annotated Bibliography* (1976). More specific bibliographies, also mentioned in the appropriate topical section, include Joel Schor and Cecil Harvey, *A List of References for the History of Black Americans in Agriculture, 1619–1974* (1975); Peter M. Molloy, *History of Metal Mining and Metallurgy: An Annotated Bibliography* (1986); and Darwin H. Stapleton, *The History of Civil Engineering since 1600: An Annotated Bibliography* (1986).

2. AMERICAN SURVEYS

Recent surveys of American technology before 1850 have tended to borrow the standard political periodization based on the upheaval of the Revolution. For the earlier period, John J. McCusker and Russell R. Menard's *Economy of British America, 1607–1789* (1991) treats technology in a broad economic context. Several general works on industrialization focus on the later side of the divide. Brooke Hindle and Steven Lubar's *Engines of Change: The American Industrial Revolution, 1790–1860* (1986) provides a synthesis of case studies in manufacturing, agriculture, and transportation based on the existing literature and stresses the European sources and American adaptations of the technologies it examines. Thomas C. Cochran's *Frontiers of Change: Early Industrialism in America* (1981) treats three periods, 1785–1825, 1825–1840, and 1840–1855, and explores the American commitment to industrialization, changes in the business system and the organization of production (as opposed to mechanization), and the central roles of the mid-Atlantic region and manufactures other than textiles, in contrast to the more traditional emphasis on the New England textile industry.

The essay collection *Material Culture of the Wooden Age* (1981), edited by Hindle, includes material on both sides of the Revolutionary divide and explores links between the often academically separated fields of history of technology and material culture. Its essays are organized topically, treating "Home and Farm," "Transportation," and "Production." In *Material Life in America, 1600–1860* (1988), editor Robert Blair St. George has collected a wide sampling of recent work in material culture, on topics ranging from foodways to architecture. The more traditional approach to material culture, such as may be found in Ivor Noël Hume's *Guide to Artifacts of Colonial America* (1970), often provides useful reference materials for historians of technology but tends not to explore historical context. More recently, catalogs of museum exhibits have bridged this gap; see, for example, the three-volume *New England Begins: the Seventeenth Century* (1982), compiled by Jonathan L. Fairbanks and Robert F. Trent for the Department of American Decorative Arts and Sculpture at the Museum of Fine Arts, Boston.

In contrast, labor historians, who have recently begun to explore work and workers in the nonunion context of early America, provide historical context but tend to gloss over technological detail. Their efforts can provide the alert historian of technology with guideposts to interesting questions about production technologies. A useful example is Stephen Innes's edited collection *Work and Labor in Early America* (1988). See also Rosalyn Baxandall, Linda Gordon, and Susan Reverby's *America's Working Women: A Documentary History* (1976), and the first volume, *The Black Worker to 1869,* of Philip S. Foner and Ronald L. Lewis's *The Black Worker: A Documentary History from Colonial Times to the Present* (1979).

Explorations of technology and cultural values have proliferated, producing broad works such as the collection *Technology and Social Change in America* (1973), edited by Edwin T. Layton, Jr., or Elting E. Morison's *From Know-How to Nowhere: The Development of American Technology* (1974). An edited collection of source documents by Carroll W. Pursell, Jr., *Readings in Technology and American Life* (1969), also emphasizes the social context of technology. On early America and industrialization specifically, John F. Kasson's *Civilizing the Machine: Technology and Republican Values in America, 1776–1900* (1976) explores positive and negative attitudes toward technology. Neil Longley York, in his *Mechanical Metamorphosis: Technological Change in Revolutionary America* (1985), finds a shifting outlook and increasing commitment to indigenous American technology in the period 1760–1790.

Published editions of letters and travel journals also provide insight into aspects of American technology. Documents such as *Detailed Reports on the Salzburger Emigrants Who Settled in America* (1968), edited by George Fenwick Jones *et al.;* Timothy Dwight's *Travels in New England and New York* (1969), edited by Barbara Miller Solomon; Francis Baily, *Journal of a Tour in Unsettled Parts of North America in 1796 and 1797* (1969), edited by Jack D. L. Holmes; C. L. Fleischmann's 1852 *Trade,*

Manufacture, and Commerce in the United States of America (1970); or the expedition journals of Meriwether Lewis and William Clark supply a wealth of varied detail. A broad range of relevant topics is also treated in the nine volumes thus far of *Letters of the Delegates to the Continental Congress, 1774–1789*, edited by Paul H. Smith (1976–). The papers of Benjamin Franklin, Thomas Jefferson, Alexander Hamilton, Henry Laurens, and Frederick Law Olmsted are also related to a range of technological topics.

3. AGRICULTURE

The technologies of agriculture, including less tangible techniques as well as the tools and machines involved in producing either subsistence or surplus crop yield, remain understudied. General histories of agriculture in America include *Eighteenth-Century Agriculture: A Symposium,* a special issue of the journal *Agricultural History* (XLIII, no. 1 [1969]), edited by John T. Schlebecker, devoted to eighteenth-century agriculture; Darwin P. Kelsey's edited collection, *Farming in the New Nation: Interpreting American Agriculture, 1790–1840* (1972), which includes a section "Science and Technology in Agriculture"; and the early portions of John T. Schlebecker's 365-year overview *Whereby We Thrive: A History of American Farming, 1607–1972* (1975). R. Douglas Hurt's book *American Farm Tools: From Hand Power to Steam Power* (1982) is unusual in its focus on the development of the hardware of eighteenth- and nineteenth-century agricultural technology. Two edited collections of documents, Wayne D. Rasmussen's four-volume *Agriculture in the United States: A Documentary History* (1975) and David B. Greenberg's *Land That Our Fathers Plowed: The Settlement of Our Country as Told by the Pioneers Themselves and Their Contemporaries* (1969), should also be useful. Books on more specific aspects of agriculture include regional studies, studies focusing on particular crops or systems of labor organization, uncultivated harvests such as lumber and fur, and the relationships of foodways with agriculture.

The proliferation of regional and case studies in recent historical scholarship provides a wealth of specific examples of agricultural practice and organization. Some, such as James C. Bonner's *History of Georgia Agriculture, 1732–1860* (1964), Clarence H. Danhof's *Change in Agriculture: The Northern United States, 1820–1870* (1969), Hubert G. Schmidt's *Agriculture in New Jersey: A Three-hundred-Year History*

(1973), or Robert Leslie Jones's *History of Agriculture in Ohio to 1880* (1983), focus on agricultural practices and markets. Others, social histories of agricultural communities, reveal not only specific activities in the fields but also the larger social and economic organizations supporting a particular agricultural system. For the Northeast, Howard Russell describes the daily work and community interactions of farm men and women in *A Long Deep Furrow: Three Centuries of Farming in New England* (1976). Darrett B. Rutman's study *Husbandmen of Plymouth: Farms and Villages in the Old Colony, 1620–1692* (1967) uses wills and inventories to study day-to-day husbandry in the seventeenth century. Benjamin Labaree's broader *Colonial Massachusetts: A History* (1979) treats agriculture in the context of a general history. Moving south, Robert D. Mitchell's *Commercialism and Frontier: Perspectives on the Early Shenandoah Valley* (1977) traces the shift from agriculture toward manufacture over the course of the eighteenth century. Gregory A. Stiverson's *Poverty in a Land of Plenty: Tenancy in Eighteenth-Century Maryland* (1977), Carville V. Earle's *Evolution of a Tidewater Settlement System: All Hallow's Parish, Maryland, 1650–1783* (1975), and Lois Green Carr, Russell R. Menard, and Lorena S. Walsh's *Robert Cole's World: Agriculture and Society in Early Maryland* (1991) document the intertwining of settlement patterns, social structures, land and resource use, and agricultural practice. The collection *Colonial Chesapeake Society* (1988), edited by Carr, Philip D. Morgan, and Jean B. Russo, is useful on agriculture and also craft work. Converse D. Clowse's book *Economic Beginnings in Colonial South Carolina, 1670–1730* (1971) studies crops, livestock, lumbering, the production of naval stores such as tar and pitch, and the Indian and fur trades. Similarly, Robin F. A. Fabel's *Economy of British West Florida, 1763–1783* (1988) treats plantation agriculture, tar and pitch production, and maritime trade. The reference biography of Thomas Jefferson, edited by Merrill D. Peterson (1986), includes essays on Jefferson's views on agriculture and slavery, among other topics; Edwin Morris Betts has edited *Thomas Jefferson's Farm Book, with Commentary and Relevant Extracts from Other Writings* (1976). And, lest we forget that the Europeans were not the only Americans who had ideas about how to practice agriculture, we have books such as R. Douglas Hurt's *Indian Agriculture in America: Prehistory to the Present* (1987) and also Von Del Chamberlain's *When Stars Came Down to Earth: Cosmology of the Skidi Pawnee Indians of North America* (1982). The contributions of African agricultural practices and agricultural labor were also crucial; in addition to works noted below, see Joel Schor and Cecil Harvey's *List of References for the History of Black Americans in Agriculture, 1619–1974* (1975).

Several studies focus on particular crops. Rice culture was heavily dependent on slavery, as demonstrated in Daniel C. Littlefield's *Rice and Slaves: Ethnicity and the Slave Trade in Colonial South Carolina* (1981) and Julia Floyd Smith's *Slavery and Rice Culture in Low Country Georgia, 1750–1860* (1985). Peter A. Coclanis provides an economic history of South Carolina rice culture in *The Shadow of a Dream: Economic Life and Death in the South Carolina Low Country, 1670–1920* (1989). Midwestern grain agriculture is a nineteenth-century story, but the early sections of *The Grain Trade in the Old Northwest* (1966), by John G. Clark, and *Threshing in the Midwest, 1820–1940: A Study of Traditional Culture and Technological Change* (1988), by J. Sanford Rikoon, treat conditions before 1850.

Plantation-based cotton and tobacco growing, because so much a part of an economic and political system, have received extensive treatment by historians. Because slavery was a critical element in the technological systems of cotton and tobacco production, the enormous body of literature on slavery is relevant to the study of plantation agriculture. Unfortunately, only the most crop-related works can be listed here. Other books treating slaves and slavery are mentioned in Sections 4, "Household Technology"; 5, "Mining and Metals"; 7, "Civil Engineering and Transportation"; and 10, "Manufacturing and Labor." On cotton, see Harold D. Woodman, *King Cotton and His Retainers: Financing and Marketing the Cotton Crop of the South, 1800–1925* (1967); Gavin Wright, *The Political Economy of the Cotton South: Households, Markets, and Wealth in the Nineteenth Century* (1978); and John Hebron Moore, *The Emergence of the Cotton Kingdom in the Old Southwest: Mississippi, 1770–1860* (1988). On tobacco, T. H. Breen describes the importance of management techniques to producing a good crop in *Tobacco Culture: The Mentality of the Great Tidewater Planters on the Eve of the Revolution* (1985). Paul G. E. Clemens puts regional tobacco growing in an international context in *The Atlantic Economy and Colonial Maryland's Eastern Shore: From Tobacco to Grain* (1980). Gloria L. Main focuses on the economy and material culture of Maryland in *Tobacco Colony: Life in Early Maryland, 1650–1720* (1982). Allan Kulikoff uses quantitative methods to treat structural issues of political economy and society in *Tobacco and Slaves: The Development of Southern Cultures in the Chesapeake, 1680–1800*

(1986). Melvin Herndon's *William Tatham and the Culture of Tobacco* (1969) includes a facsimile of Tatham's 1800 essay on tobacco culture and commerce as well as Herndon's comments on post-1800 developments. Two edited collections of source materials on slavery, Willie Lee Rose's *Documentary History of Slavery in North America* (1976) and Michael Mullin's *American Negro Slavery: A Documentary History* (1976), may also be useful.

Aspects of labor organization related to agriculture are described by William Van Deburg in *The Slave Drivers: Black Agricultural Labor Supervisors in the Antebellum South* (1979), by David E. Schob in *Hired Hands and Plowboys: Farm Labor in the Midwest, 1815–60* (1975), by David W. Galenson in *White Servitude in Colonial America: An Economic Analysis* (1981), and by A. Roger Ekirch in *Bound for America: The Transportation of British Convicts to the Colonies, 1718–1775* (1987).

Several books on lumbering and timber economies have appeared since Hindle's compilation. A useful survey, which includes treatments of technologies and resource use, is provided in the first two parts of *This Well-Wooded Land: Americans and Their Forests from Colonial Times to the Present* (1985), by Thomas R. Cox, Robert S. Maxwell, Phillip Drennon Thomas, and Joseph J. Malone. Charles F. Carroll describes seventeenth-century timber use in *The Timber Economy of Puritan New England* (1973). Philip L. White documents the combination of subsistence farming with logging in *Beekmantown, New York: Forest Frontier to Farm Community* (1979). And logging, transportation, and log milling beginning in the eighteenth century are described in John Hebron Moore's *Andrew Brown and Cypress Lumbering in the Old Southwest* (1967).

The early American fur trade, like the timber economy, falls into a category of uncultivated harvests. Besides providing insight on resource use, it also took place at the intersection of two cultures' technologies, and illuminates the complex interactions of Europeans with Native Americans. Work on the fur trade has been mostly anthropological or ethnohistorical in approach, but is often based heavily on artifactual evidence. For examples, see Carolyn Gilman, *Where Two Worlds Meet: The Great Lakes Fur Trade* (1982); Calvin Martin, *Keepers of the Game: Indian-Animal Relationships and the Fur Trade* (1978); and the collections edited by Shepard Krech III, *Indians, Animals, and the Fur Trade: A Critique of "Keepers of the Game"* (1981) and *The Subarctic Fur Trade: Native Social and Economic Adaptations* (1984).

Another field of research that can shed light on agricultural practice is the folklife study of foodways. Peter Benes has edited a collection of essays, *Foodways in the Northeast* (1984), which treats patterns in food production and consumption ranging from farmhouse design to the distribution of brick ovens to zooarchaeology (the study of materials such as animal bones to determine what people ate, how it was butchered, and so forth). A specific case may be found in *French Subsistence at Fort Michilimackinac, 1715–1781: The Clergy and the Traders* (1985), by Elizabeth M. Scott. The early portions of *The Lifeline of America: Development of the Food Industry* (1964), by Edward C. Hampe, Jr., and Merle Wittenberg, may also be of use. Explorations of spice and flavoring use straddle the categories of agriculture and household technology; on the agricultural side James E. Landing's book *American Essence: A History of the Peppermint and Spearmint Industry in the United States* (1969) is a useful reminder that crops other than grain, cotton, and tobacco had a place in the American economy from at least the late eighteenth century on.

4. HOUSEHOLD TECHNOLOGY

If there is one entirely new category that has been "discovered" since Brooke Hindle's bibliography, it is that of household technology—technologies used primarily by women and that industrialized late or not at all. Ruth Schwartz Cowan's pathbreaking work, including her overview *More Work for Mother: The Ironies of Household Technology, from the Open Hearth to the Microwave* (1983), has led the way, but few have yet followed.[4] Most of the books mentioned in this section are social and labor histories or studies in material culture, and most of them do not explicitly explore technological choice. They do not ask how technologies came to be shaped as they did, and their illuminations of how technologies were used and how they affected the lives of those who used them tend not to reflect the options these users faced. One exception, because of its focus on demand as well as distribution patterns, is Carole Shammas's *Pre-Industrial Consumer in England and America* (1990). Jeanne Boydston's *Home and Work: Housework, Wages, and the Ideology of Labor in the Early Republic* (1990), a comprehensive history of housework, explores the changing content of women's work and stresses the importance of domestic labor to the process of industrialization. A useful artifactually based social history is Barbara Clark Smith's *After the Revolution: The Smithsonian*

---

4. Much of the best recent scholarship, as is often the case with relatively new topics, remains in article form and must unfortunately be excluded from this compilation.

History of Everyday Life in the Eighteenth Century (1985). Books such as those by Mary Beth Norton, *Liberty's Daughters: The Revolutionary Experience of American Women, 1750–1800* (1980), or Joy Day Buel and Richard Buel, Jr., *Way of Duty: A Woman and Her Family in Revolutionary America* (1984), are helpful in discussing women's work and women's roles, but are not concerned explicitly with technology. Similarly, books about servants, including Faye E. Dudden's *Serving Women: Household Service in Nineteenth-Century America* (1983), may reveal details about technology as they describe the lives of domestic servants. A wealth of artifactual details and examples may be found in books such as Henry Glassie's *Pattern in the Material Folk Culture of the Eastern United States* (1968) and Gertrude Z. Thomas's *Richer than Spices: How a Royal Bride's Dowry Introduced Cane, Lacquer, Cottons, Tea, and Porcelain to England, and So Revolutionized Taste, Manners, Craftsmanship, and History in both England and America* (1965). Women's diaries can be invaluable sources when they survive; a useful published example is *The Diary of Elizabeth Drinker* (1991), edited by Elaine Forman Crane. Details from a sampling of late-eighteenth-century probate inventories are provided by Alice Hanson Jones's *American Colonial Wealth: Documents and Methods* (1977).

Cooking and sewing are as much craft skills as any traditionally male tasks such as metalwork and can be examined in the same terms. In fact, certain kinds of cooking and sewing, such as baking and tailoring, were performed by men outside the household and are usually included under the rubrics of crafts or artisanship. Women's tasks are included here, rather than in the section on crafts, because they were performed within the household. Their neglect is part of the favoritism shown paid labor taking place outside the home, and their exploration is still restitutive. When household skills and technologies are better understood, comparisons of work and technological change across gender divisions should prove as enlightening as those by region, trade, or class. Perhaps some future bibliography of early American technology will consider the craft skills of the entire population together.

Among the few works on cooking is Jane Carson's *Colonial Virginia Cookery* (1966, 1985), which describes methods of cooking and serving, different cuisines, and cooking equipment. In *Loosening the Bonds: Mid-Atlantic Farm Women, 1750–1850* (1986) Joan M. Jensen explores the rise of dairying activities in her social history of changing family relationships on the mid-Atlantic farm. Exhibit catalogs such as *The Larder Invaded:*

Reflections on Three Centuries of Philadelphia Food and Drink (1987), by Mary Anne Hines, Gordon Marshall, and William Woys Weaver, provide a forum for artifact-based studies whose usefulness often extends beyond the lives of the exhibits. Another exhibit catalog is among the best works on sewing: Betty Ring's *"Let Virtue Be a Guide to Thee": Needlework in the Education of Rhode Island Women, 1730–1830* (1983) uses school-girl samplers to study women's education, which in this period invariably included needlework. Elsewhere, the focus has been on the nineteenth century and later, when the production of clothing moved outside the home: see Claudia Kidwell and Margaret Christman, *Suiting Everyone: The Democratization of Clothing in America* (1979). Two books on quilting in Pennsylvania, Jeannette Lasansky's *Pieced by Mother: Over One Hundred Years of Quiltmaking Traditions* (1987) and the collection she edited, *In the Heart of Pennsylvania: Symposium Papers* (1986), provide detail on the skills needed for that technology and offer some insight on the early period. Less representative groups such as the Shakers have received attention, as in Beverly Gordon's book *Shaker Textile Arts* (1980). In addition to cooking and sewing, many skills later dominated by male doctors were mastered mostly by women; see Laurel Thatcher Ulrich's book *A Midwife's Tale: The Life of Martha Ballard, Based on Her Diary, 1785–1812* (1990).

Labor organization in the household, as in other workplaces, varied widely. Household technology, therefore, may be assumed also to have varied. For a survey of structure and room use, see *Common Places: Readings in American Vernacular Architecture* (1986), edited by Dell Upton and John Michael Vlach. Many social histories explore family life or women's roles, and from them we can glean clues; technology is rarely addressed explicitly. But behind "household moveables" and the entries in probate records lie technologies, if we choose to think of them that way. On New England, both John Demos in *A Little Commonwealth: Family Life in Plymouth Colony* (1970) and Toby L. Ditz in *Property and Kinship: Inheritance in Early Connecticut, 1750–1820* (1986) use evidence of who-had-what to construct their studies. Nancy F. Cott, in *The Bonds of Womanhood: "Woman's Sphere" in New England, 1780–1835* (1977), uses diaries to discuss not only female friendships but also the decline of middle-class women's involvement in production. Laurel Thatcher Ulrich's *Good Wives: Image and Reality in the Lives of Women in Northern New England, 1650–1750* (1982), Claudia L. Bushman's *"Good Poor Man's Wife"* (1981), and Katherine Kish Sklar's *Catharine*

Beecher: A Study in American Domesticity (1973) discuss household activities of women, although Sklar's is primarily a psychological biography. Moving southward, Sharon V. Salinger focuses on the household as an economic unit in *"To Serve Well and Faithfully": Labor and Indentured Servants in Pennsylvania, 1682–1800* (1987). Christine Stansell provides a later and more urban view in *City of Women: Sex and Class in New York, 1789–1860* (1986).

In the South, social histories such as Catherine Clinton's *Plantation Mistress: Woman's World in the Old South* (1982) and Suzanne Lebsock's *Free Women of Petersburg: Status and Culture in a Southern Town, 1784–1860* (1984) include discussions of women's work. The side-by-side cultures of owners and slaves are both of interest, as slaves were household workers but also struggled to maintain households of their own; see Elizabeth Fox-Genovese's *Within the Plantation Household: Black and White Women of the Old South* (1988). Both Eugene Genovese's *Roll, Jordan, Roll: The World the Slaves Made* (1975) and Mechal Sobel's *The World They Made Together: Black and White Values in Eighteenth-Century Virginia* (1987) may be of use. Several archaeological studies describe building layouts and artifacts; see William M. Kelso, *Kingsmill Plantations, 1619–1800: Archaeology of Country Life in Colonial Virginia* (1984), and Theodore R. Reinhart, ed., *The Archaeology of Shirley Plantation* (1984).

Household technologies of the frontier were often different from those in settled areas, but settlers also chose which items to carry with them and which to leave behind. Books such as Douglas Edward Leach's *Northern Colonial Frontier, 1607–1763* (1966) include descriptions and details of daily life. Log cabin building has received more explicit attention; see the subsection on building in Section 7, "Civil Engineering and Transportation."

Finally, the household technologies of the people who preceded both the Europeans and the Africans on this continent should be explored. American children are often told Thanksgiving stories about how the Good Indians saved the starving Pilgrims in their first winter in America; such winter survival is a matter of household technology. Again, recent scholarship has not focused explicitly on technology, but provides many useful details. On New England, see *Indian New England before the Mayflower* (1980), by Howard S. Russell. Traveling south along the East Coast, C. A. Weslager focuses on the period before 1815 in *The Delaware Indians: A History* (1972), and Helen C. Rountree explores life in Virginia before it was so named in *The Powhatan Indians of Virginia: Their Traditional Cul-*

ture (1989). James P. Ronda describes both white and Native American views farther west in *Lewis and Clark among the Indians* (1984).

5. MINING AND METALS

North America was a continent of many resources, some more accessible, and of more immediate commercial value, than others. The ubiquitous plentitude of timber allowed a reliance on wood for construction and fuel that had become unimaginable in deforested Britain, and made the need to mine American coal and iron reserves less pressing. Early American ventures to exploit the precious and nonferrous metals were extremely modest in scope. Nonetheless, metals were mined, and iron was puddled, forged, and smelted (the last operation in charcoal-fired blast furnaces). As the nineteenth century progressed, wrought iron, cast iron, and eventually steel supplanted wood in many construction and production applications, and anthracite, bituminous coal, and coke replaced wood and charcoal as industrial fuels. Historians of mining and metals have focused attention on developments after 1850, and histories of the early American period are not numerous; see Peter M. Molloy's *History of Metal Mining and Metallurgy: An Annotated Bibliography* (1986).

A useful overview of colonial mining and use of gold, silver, copper, lead, and iron is provided by James A. Mulholland in *A History of Metals in Colonial America* (1981). *Wealth Inexhaustible: A History of America's Mineral Industries to 1850* (1985), by M. H. Hazen and R. M. Hazen, is less cohesive but also surveys the early American period. Several works focus on the developing American iron industry. W. David Lewis, in *Iron and Steel in America* (1976), provides a general overview of colonial ironmaking. Paul F. Paskoff's *Industrial Evolution: Organization, Structure, and Growth of the Pennsylvania Iron Industry, 1750–1860* (1983) and Peter Temin's *Iron and Steel in Nineteenth-Century America: An Economic Inquiry* (1964) focus on economics and how economics influenced the rate of technological change in the industry. Studies of particular ironworks include James M. Ransom, *Vanishing Ironworks of the Ramapos: The Story of the Forges, Furnaces, and Mines of the New Jersey–New York Border Area* (1966), and W. David Lewis and Walter Hugins, *Hopewell Furnace* (1983). Note that metalworking is treated in Section 10, "Manufacturing and Labor."

Pennsylvania coal, as well as iron, has received attention: Frederick Moore Binder's 1955 dissertation has been published as *Coal Age Empire: Pennsylvania Coal and Its Utilization to 1860* (1974), John N. Hoffman explores a specific case in *Anthracite in the Lehigh Regions of Pennsylvania, 1820–45* (1968), and the marketing of coal in the early nineteenth century is explored in *Philadelphia's First Fuel Crisis: Jacob Cist and the Developing Market for Anthracite Coal,* by H. Benjamin Powell (1978). In *St. Clair: A Nineteenth-Century Coal Town's Experience with a Disaster-prone Industry* (1987), Anthony F. C. Wallace takes a more anthropological approach to the workings of a mining community. In the South, the iron and coal industries depended on slave labor. Ronald L. Lewis, in his book *Coal, Iron, and Slaves: Industrial Slavery in Maryland and Virginia, 1715–1865* (1979), finds a shift from ownership to hiring of slaves by industrialists. Lewis also offers a national view in *Black Coal Miners in America: Race, Class, and Community Conflict, 1780–1980* (1987). Robert S. Starobin treats mining among other cases in *Industrial Slavery in the Old South* (1970). Katherine A. Harvey's edition of *The Lonaconing Journals: The Founding of a Coal and Iron Community, 1837–1840* (1977) provides a wealth of detail on mining in western Maryland. Some of the environmental costs of coal (and metal) mining are documented in Duane Smith's *Mining in America: The Industry and the Environment, 1800–1980* (1987).

Other resources besides coal and iron were mined in early America but have received little attention. Limerock was mined for plaster and mortar in Maine as early as 1733; see Roger L. Grindle's *Quarry and Kiln: The Story of Maine's Lime Industry* (1971). Early copper mining and milling, this time in New Jersey, is treated in the first parts of *Copper for America: The Hendricks Family and a National Industry, 1755–1939,* by Maxwell Whiteman (1971). The upper Michigan copper district, which opened in the 1840s, is treated by Donald Chaput in *The Cliff: America's First Great Copper Mine* (1971) and by Larry Lankton's study of traditional and modern mining technologies, *Cradle to Grave: Life, Work, and Death at the Lake Superior Copper Mines* (1991).

6. MILITARY TECHNOLOGY

Military technology includes, as does other technology, both hardware and organization. On hardware, *Firearms in Colonial America: The Impact on History and Technology, 1492–1792,* by M. L. Brown (1980), discusses the European roots of colonial firearms and views technological development in a social context. Because weapons tend to survive as artifacts and are often found in museums and private collections, many books focus on taxonomy and identification rather than context and use.

For details and photographs, see Warren Moore, *Weapons of the American Revolution and Accoutrements* (1967), and George C. Neumann, *The History of Weapons of the American Revolution* (1967). Giles Cromwell, *The Virginia Manufactory of Arms* (1975), may also provide useful details. Another less-considered element of military hardware is the fort; see the early parts of Willard B. Robinson's survey *American Forts: Architectural Form and Function* (1977) for illustrations and architectural informa-·tion, and the chapter on fortifications in Neil Longley York's *Mechanical Metamorphosis: Technological Change in Revolutionary America* (1985).

Most works on the organizational aspects of the military focus on the Revolution and on the early formation of national military policy. The exception is Patrick M. Malone's *Skulking Way of War: Technology and Tactics among the Indians of New England* (1991), which describes the seventeenth-century adoption of firearms by Native Americans and of forest warfare techniques and strategies by European Americans. On the Revolution, the essay collection edited by Ronald Hoffman and Peter J. Albert, *Arms and Independence: The Military Character of the American Revolution* (1984), discusses political and economic influences on military organization. The logistics of army supply distribution are discussed by Erna Risch in *Supplying Washington's Army* (1981) and by E. Wayne Carp in *To Starve the Army at Pleasure: Continental Army Administration and American Political Culture, 1775–1783* (1984). On the British side, the difficulties of long-distance military control are treated by David Syrett in *Shipping and the American War, 1775–1783: A Study of British Transport Organization* (1970) and by R. Arthur Bowler in *Logistics and the Failure of the British Army in America, 1775–1783* (1975). Barbara Graymont's study *The Iroquois in the American Revolution* (1972) uses archaeological and anthropological evidence and treats military technologies explicitly. Two collections of documents treat the Revolution: Paul K. Walker's *Engineers of Independence: A Documentary History of the Army Engineers in the American Revolution, 1775–1783* (1982) and John Dann's *Revolution Remembered: Eyewitness Accounts of the War for Independence* (1980). The papers of General Nathanael Greene and of Henry Bouquet have also been collected and are being published. The roots of national peacetime military policy are explored in Richard H. Kohn, *Eagle and Sword: The Federalists and the Creation of the Military Establishment in America, 1783–1802* (1975). The next chapter, the Republican version of the military establishment, is provided by Theodore J. Crackel's book *Mr. Jefferson's Army: Political and Social Reform of the Military Establishment, 1801–1809* (1987). The early essays in Merritt Roe Smith's edited collection *Military Enterprise and Technological Change: Perspectives on the American Experience* (1985) are also useful. Bottom-up views of the military are the focus of Harold L. Peterson's *Book of the Continental Soldier* (1968) and Steven Rosswurm's *Arms, Country, and Class: The Philadelphia Militia and "Lower Sort" during the American Revolution, 1775–1783* (1987).

On the water, the Revolutionary navy has received attention from William M. Fowler, Jr., in *Rebels under Sail: The American Navy during the Revolution* (1976) and Nathan Miller in *Sea of Glory: The Continental Navy Fights for Independence, 1775–1783* (1974). The contributions of officers to military tradition are the focus of the essay collection *Command under Sail: Makers of the American Naval Tradition, 1775–1850* edited by James C. Bradford (1985). An alternative view of the military system is highlighted in Jerome R. Garitee, *The Republic's Private Navy: The American Privateering Business as Practiced by Baltimore during the War of 1812* (1977). The social side of military policy is explored by James E. Valle in his *Rocks and Shoals: Order and Discipline in the Old Navy, 1800–1861* (1980). And before his steamboat adventures, Robert Fulton's career focused on military technology; see Wallace Hutcheon, *Robert Fulton: Pioneer of Undersea Warfare* (1981). Two document collections provide access to sources in naval history: *Naval Documents of the American Revolution* (1964–), edited by William Bell Clark, and *The Naval War of 1812: A Documentary History,* edited by William S. Dudley (1985–).

7. CIVIL ENGINEERING AND TRANSPORTATION

This broad section can be divided into many interrelated categories, as the compound nature even of its title may suggest. Material in "Engineers" and in "Surveying and Cartography" is followed by categories for major types of transportation. Included with the material on each mode of transportation are studies of those who worked with and used it; while these books do not always treat technology explicitly, they provide a context often missing in earlier works. Shipping (although primarily commercial) and whaling (although arguably not "engineering") are included with ships and shipbuilding, so that oceangoing technologies are grouped together. Naval policy and organization remain in the section on the Military. "Building and Architecture," the final category here, is defined relatively narrowly; broader works may be found below in Section 8, "Settlement Patterns and Urban Growth."

Engineers

Although engineers are often the heroes of histories of technology, in this period theirs was but a fledgling profession. Engineers were trained by apprenticeship, and all who were not military engineers were civil engineers. Darwin H. Stapleton's comprehensive *History of Civil Engineering since 1600: An Annotated Bibliography* (1986) is a valuable reference. The early parts of Monte A. Calvert's book, *The Mechanical Engineer in America, 1830–1910: Professional Cultures in Conflict* (1967), focus more specifically on early America. The shift toward professionalization among engineers was a late-nineteenth-century phenomenon, as is demonstrated by the focus on individuals rather than engineering societies in the literature on the early period. For examples, see Daniel Larkin, *John B. Jervis, an American Engineering Pioneer* (1990), and *The Reminiscences of John B. Jervis, Engineer of the Old Croton* (1971), edited by Neal Fitzsimons; James P. Baughman, *Charles Morgan and the Development of Southern Transportation* (1968); Gene D. Lewis, *Charles Ellet, Jr.: The Engineer as Individualist, 1810–1862* (1968); Robert Hunter and Edwin Dooley, Jr., *Claudius Crozet: French Engineer in America, 1790–1864* (1989); and K. H. Vignoles, *Charles Blacker Vignoles: Romantic Engineer* (1982). Of particular interest on topics from waterworks to steamboats are the newly published papers, journals, and drawings of Benjamin Henry Latrobe, which include extensive and important editorial essays.

Surveying and Cartography

Surveying, obviously a critical element in activities such as road building and land apportionment, remains largely understudied. The one study focusing on surveying techniques is Sarah S. Hughes's *Surveyors and Statesmen: Land Measuring in Colonial Virginia* (1979). Hughes explains the role of the surveyor and changes in seventeenth-century surveying technique and explores the interaction of surveying and politics. Some information can also be found in the engineering biographies listed above: for example, Charles Blacker Vignoles spent the years 1817–1823 surveying and mapping Florida. The other source of information, often indirect in nature, is the literature on cartography: maps are technological artifacts and their makers skilled craftspeople. One useful reference is

Walter W. Ristow's *American Maps and Mapmakers: Commercial Cartography in the Nineteenth Century* (1985), which includes many reproductions. *The Hammond-Harwood House Atlas of Historical Maps of Maryland, 1608–1908,* compiled by Edward C. Papenfuse and Joseph M. Coale III, presents more than one hundred maps and other illustrations, with historical notes. Books on particular mapmakers can provide details on technique: see Hubert G. Schmidt's new edition of *George Washington's Mapmaker: A Biography of Robert Erskine* (1928, 1966) by Albert H. Heusser, and Edmund and Dorothy Smith Berkeley, *Dr. John Mitchell: The Man Who Made the Map of North America* (1974). On an important eighteenth-century mapmaker and surveyor, see Walter Klinefelter, *Lewis Evans and His Maps* (1971). On the late eighteenth and early nineteenth century, see the *Atlas of the Lewis and Clark Expedition* (1983), edited by Gary E. Moulton.

General Transportation

Historians in recent years have failed to produce synthetic work on transportation. The exception, a collection compiled by the Indiana Historical Society, *Transportation and the Early Nation* (1982), focuses on the effects of transportation technologies in the Midwest.

Ships, Shipbuilding, and Other Maritime Pursuits

As prevalent and important as shipbuilding was, few have studied the subject. Pioneering work has been done by Joseph A. Goldenberg in his technically detailed overview *Shipbuilding in Colonial America* (1976). Also useful may be Alexander Laing, *American Ships* (1971), and Paul Johnstone, *Steam and the Sea* (1983). On shipping, particular trades, and economic context, we find more. Several of the essays in *Ships, Seafaring, and Society: Essays in Maritime History* (1987), edited by Timothy J. Runyan, treat early American subjects. James F. Shepherd and Gary M. Walton explore changes in economic organization and regional specialization in *Shipping, Maritime Trade, and the Economic Development of Colonial North America* (1972). John W. Tyler uses insurance records to document the extent of smuggling among Revolutionary merchants in *Smugglers and Patriots: Boston Merchants and the Advent of the American Revolution* (1986). Specialized or exotic trading may not be typical but has received attention. On the China trade, see Jonathan Goldstein, *Philadelphia and the China Trade, 1682–1846: Commercial, Cultural, and Attitudinal Effects* (1978); and the museum exhibit catalogs by Margaret C. S. Christman, *Adventurous Pursuits: Americans and the China Trade, 1784–1844* (1984), and Jean Gordon Lee, with an essay by Philip Chadwick Foster Smith, *Philadelphians and the China Trade, 1784–1844* (1984). On the slave trade, see Philip D. Curtin, *The Atlantic Slave Trade: A Census* (1969), Walter Minchinton *et al.*, eds., *Virginia Slave-Trade Statistics, 1698–1775* (1984), and Jay Coughtry, *The Notorious Triangle: Rhode Island and the African Slave Trade, 1700–1807* (1981). Ports themselves also influenced the nature of trade; see James H. Levitt, *For Want of Trade: Shipping and the New Jersey Ports, 1680–1783* (1981), and Eugene R. Slaski, *Poorly Marked and Worse Lighted: Being a History of the Port Wardens of Philadelphia, 1766–1907* [1980].

Another important use of ships is, of course, found in fishing and whaling. These were often community-based activities, as illustrated by Christine Leigh Heyrman's study of Marblehead and Gloucester, *Commerce and Culture: The Maritime Communities of Colonial Massachusetts, 1690–1750* (1984). Robert Owen Decker's book *Whaling Industry of New London* (1973) provides an overview of Connecticut whaling, including some details on shipboard life. On the hardware of whaling, see Thomas G. Lytle, *Harpoons and Other Whalecraft* (1984). Edouard A. Stackpole offers a transatlantic perspective in his study *Whales and Destiny: The Rivalry between America, France, and Britain for Control of the Southern Whale Fishery, 1785–1825* (1972).

The sailors themselves provide a useful focus, as their work involves the technology in question, and their skills are part of it. Details of sailors' lives are placed in context in Marcus Rediker's analysis *Between the Devil and the Deep Blue Sea: Merchant Seamen, Pirates, and the Anglo-American Maritime World, 1700–1750* (1987). The journals of whalemen and merchant seamen form the basis of Margaret Creighton's study *Dogwatch and Liberty Days: Seafaring Life in the Nineteenth Century* (1983).

Steamboats and Riverboats

River transportation other than steamboats has received almost no attention in recent years. An exception is Erik F. Haites, James Mak, and Gary M. Walton's *Western River Transportation: The Era of Early Internal Development, 1810–1860* (1975), which treats flatboats and keelboats as well as steamboats, augmenting Louis Hunter's classic *Steamboats on the Western Rivers* of 1949. Its focus is on the Louisville–New Orleans trunk route, and its exploration of the interdependence of new and older technologies provides useful insights on technological change. On the early development of the steamboat, see Edith McCall's *Conquering the Rivers: Henry Miller Shreve and the Navigation of America's Inland Waterways* (1984), *The Autobiography of John Fitch* (1976), edited by Frank D. Prager, and also Brooke Hindle's discussion of Robert Fulton in *Emulation and Invention* (1981).

Canals

Although a synthetic treatment of American canal technology remains to be written, several studies of specific canals have been published. *The National Waterway: A History of the Chesapeake and Delaware Canal, 1769–1965* (1967), by Ralph D. Gray, chronicles a canal first planned in the 1650s but not finally completed until 1829. *The Dismal Swamp Canal* (1970), by Alexander Crosby Brown, describes a canal dug by hand in the 1790s. Harry N. Scheiber's *Ohio Canal Era: A Case Study of Government and the Economy, 1820–1861* (1969) explores the origins, construction, and financing of a state-owned canal and its eventual eclipse by that rival means of transport, the railroad. Stories of other canals may be found in Mary Stetson Clark, *The Old Middlesex Canal* (1974), and Robert McCullough and Walter Leuba, *The Pennsylvania Main Line Canal* (1973). For illustrations (with caption-style text) of Ohio canals, see Frank Wilcox's posthumous work, edited by William A. McGill, *The Ohio Canals* (1969). Robert M. Vogel explores another aspect of canal technology in *Roebling's Delaware and Hudson Canal Aqueducts* (1971).

Railroads

The railroad, like the canal, combined technological and economic developments. John H. White, Jr., details the technological side of one component of the system in *American Locomotives: An Engineering History, 1830–1860* (1968) and *The John Bull: 150 Years a Locomotive* (1981). Focusing more on the business side of railroad development, John F. Stover's *History of the Baltimore and Ohio Railroad* (1987) describes how competition with nearby Philadelphia sparked action by Baltimore's civic leadership. Stephen Salsbury's *The State, the Investor, and the Railroad: The Boston and Albany, 1825–1867* (1967) treats technological innovation in various contexts and explores the roles of government and private investment in railroad development. Robert J. Parks's *Democracy's Railroads: Public Enterprise in Jacksonian Michigan* (1972) provides a case study of the finances, successes, and failures of state-legislated improvements in one state. From the workers' perspective, the story of railroad development gets a different slant: in *Working for the Railroad: The Organization of Work in the Nineteenth Century* (1983), Walter Licht explores work organization, labor recruitment, and the workers' role in the creation of a large-scale bureaucracy. A glimpse of the passenger's view is offered by Eugene Alvarez in *Travel on Southern Antebellum Railroads, 1828–1860* (1974).

Roads and Bridges

Despite the extensive focus on the railroad, the historian must remember the more mundane but persistent use of ordinary roads for transportation. A useful summary of stagecoach travel is *Stagecoach East: Stagecoach Days in the East from the Colonial Period to the Civil War* (1983), by Oliver W. Holmes and Peter T. Rohrbach. In *The Great Wagon Road from Philadelphia to the South* (1973), Parke Rouse, Jr., explores wagoners and stagecoach travel along a single route. Bridges, replacing ferries and fords, played a critical role in the development of viable and convenient roadways; see Lee Nelson's account of Philadelphia's wood-and-iron arch bridge, *The Colossus of 1812: An American Engineering Superlative* (1990). Henry Pickering Walker puts wagon technology in context in *The Wagonmasters: High Plains Freighting from the Earliest Days of the Santa Fe Trail to 1880* (1966). On urban road transportation, see Graham Russell Hodges's *New York City Cartmen, 1667–1850* (1986).

Building and Architecture

Several overviews set the stage for a host of more specific studies on aspects of early American building and architecture. The first two chapters of Carl W. Condit's *American Building: Materials and Techniques from the First Colonial Settlements to the Present* (1968) deal most specifically with technology. The collection *Building in Early America: Contributions toward the History of a Great Industry* (1976), edited by Charles E. Peterson, focuses on the mid-Atlantic region and is oriented toward historic restoration. See also the early portion of Gwendolyn Wright's *Building the Dream: A Social History of Housing in America* (1981). The first eight chapters of Marcus Whiffen and Frederick Koeper's *American Architecture, 1607–1976* (1981) treat the period before 1860 and include material on the architecture of Native Americans as well as European settlers and on town planning and economic issues. Also useful may be Roger G.

Kennedy's *Architecture, Men, Women, and Money in America, 1600–1860* (1985).

The architecture of specific communities, such as the Shakers in Dolores Hayden's *Seven American Utopias: The Architecture of Communitarian Socialism, 1790–1975* (1976) or groups of Moravian settlers in William J. Murtagh's *Moravian Architecture and Town Planning: Bethlehem, Pennsylvania, and Other Eighteenth-Century Settlements* (1967), provide interesting comparisons with more standard notions of early American architecture. Regional studies also reflect this architectural diversity. On the Northeast, see Thomas Hubka's *Big House, Little House, Backhouse, Barn: The Connected Farm Buildings of New England* (1984); *The New World Dutch Barn* (1968), by John Fitchen; *Early Nantucket and Its Whale Houses* (1966), by Henry Chandlee Foreman; *The New England Meeting Houses of the Seventeenth Century* (1968), by Marian Card Donnelly; and Abbott Lowell Cummings's *Framed Houses of Massachusetts Bay, 1625–1725* (1979). A reprinted document of interest here is *The Rules of Work of the Carpenters' Company of the City and County of Philadelphia, 1786*, edited by Charles E. Peterson (1971). On Virginia and Maryland, see Henry Glassie's *Folk Housing in Middle Virginia: A Structural Analysis of Historic Artifacts* (1975), Dell Upton's *Holy Things and Profane: Anglican Parish Churches in Colonial Virginia* (1986), and H. Chandlee Foreman's *Old Buildings, Gardens, and Furniture in Tidewater Maryland* (1967). Migration to the frontier, of course, produced its own characteristic building styles: see C. A. Weslager, *The Log Cabin in America: From Pioneer Days to the Present* (1969), and Donald Hutslar, *The Architecture of Migration: Log Construction in the Ohio Country, 1750–1850* (1986). The works listed above treat, not surprisingly, the buildings of the European newcomers. They are somewhat balanced by Peter Nabokov and Robert Easton's comprehensive survey *Native American Architecture* (1989), which includes information on regions, building types, and building materials and technologies.

The influence of individuals, when their work is known, has also been studied: Harold Kirker's illustrated *Architecture of Charles Bulfinch* (1969) is a useful reference, and his work with James Kirker, *Bulfinch's Boston, 1787–1817* (1964), puts Bulfinch's work in a larger municipal context. Jack McLaughlin's *Jefferson and Monticello: The Biography of a Builder* (1988) details the lengthy development of a famous American landmark.

Other Engineering Works

The transportation or diversion of water for various purposes posed no small challenge in early America. Early American dams were predominantly constructed to divert water, as for a canal, or to enhance waterpower at a mill seat. For a general history, see Norman Smith's *History of Dams* (1972). Louis C. Hunter's *Waterpower in the Century of the Steam Engine* (1979), volume I of *A History of Industrial Power in the United States, 1780–1930,* is also useful. Related materials may be found in Section 9, "Power."

Fighting fire was one incentive for transporting water, as illustrated in George Anne Daly and John J. Robrecht, *An Illustrated Handbook of Fire Apparatus* (1972), and John M. Peckham, *Fighting Fire with Fire* (1973). Irrigation for agriculture was another; see, for example, the case of technology transfer in *The Old World Background of the Irrigation System of San Antonio, Texas* (1972), by Thomas F. Glick. When water was more plentiful, it could be used for more elaborate gardening, such as that described in the collection edited by Robert P. Maccubbin and Peter Martin, *British and American Gardens in the Eighteenth Century: Eighteen Illustrated Essays on Garden History* (1984).

Urban engineering, including waterworks, has received attention from some urban historians, and historians of public health. John Duffy's *History of Public Health in New York City, 1625–1866* (1968) treats housing, the role of industry, and such developments as street cleaning, water supply, and garbage disposal. *The Private City: Philadelphia in Three Periods of Its Growth* (1968), by Sam Bass Warner, Jr., includes information on the Philadelphia waterworks. The Benjamin Henry Latrobe Papers, noted above, provide more detail on the Philadelphia case.

8. SETTLEMENT PATTERNS AND URBAN GROWTH

Material in this section is related to technology in several ways. First, because technological choices are intimately connected to the environment in which people make them, geographic and demographic treatments of settlement patterns provide a crucial context for understanding technological change. Second, because technological change often reshapes the environment in which it takes place, studies of that environment can illuminate aspects of the technologies in question. Third, past experience influences technological choice: both hardware and know-how migrate with the people who possess them. Finally, the human-made physical environment *is* technology, technology often invented, designed, and engineered collectively and anonymously.

Urban Planning and Growth

The city is a technological system as well as a crucible for the interdependence of technology, population, finance, and politics (to name a few). Contemporaries were aware of the importance of urban centers and often saw in America the chance to plan them rationally, in contrast to the more haphazard models of older European cities. The ideas shaping the planning of Boston, Philadelphia, Savannah, and Williamsburg are examined in *The Urban Idea in Colonial America* (1977), by Sylvia Doughty Fries. A later chapter in urban planning may be found in Thomas Bender's *Toward an Urban Vision: Ideas and Institutions in Nineteenth-Century America* (1975). Specific details of plans and regional distinctions are presented by John W. Reps in *The Making of Urban America: A History of City Planning in the United States* (1965) and by Stanley K. Schultz in *Constructing Urban Culture: American Cities and City Planning, 1800–1920* (1989). Southern town planning is treated in works such as John W. Reps's *Tidewater Towns: City Planning in Colonial Virginia and Maryland* (1972) and Joan Niles Sears's *First One Hundred Years of Town Planning in Georgia* (1979).

But not all American communities have been planned in such conscious ways, and not all aspects of planned communities are controlled by deliberate design. The large and diverse literature of regional and community studies now available illuminates the variety of the early American experience and increasingly makes possible descriptions of regional and cultural variation within early America.[5] This literature has emerged and developed mostly in the last twenty-five years; when Brooke Hindle compiled his bibliography, Sumner Chilton Powell's pathbreaking *Puritan Village: The Formation of a New England Town* (1965) had only just been published, and *The Private City: Philadelphia in Three Periods of Its Growth* (1968), by Sam Bass Warner, Jr., had yet to be written. Usually considered social history, such works provide both background and specific information to students of early American technology. Because these studies

5. See, for example, Gary Nash, "Social Development," in Jack P. Greene and J. R. Pole, eds., *Colonial British America: Essays in the New History of the Early Modern Era* (Baltimore, 1984), 233–261.

are now so numerous, they cannot be detailed here. They are reviewed and cited regularly in journals such as the *William and Mary Quarterly,* the *Journal of Social History,* and the *Journal of Interdisciplinary History.* Many are also listed in the compilation by David L. Ammerman and Philip D. Morgan, *Books about Early America: 2001 Titles* (1989), in the section headed "Local and Urban Studies." Community studies focusing on the lives of laborers and craftspeople are included here in Section 10, "Manufacturing and Labor." Works on building and architecture are in Section 7, "Civil Engineering and Transportation."

Settlement Patterns and the Shaping of the Landscape

The geographer's spatial vision of historical change both incorporates technology and, because of the visual and spatial nature of most technologies, provides particularly useful descriptions of the context of technological change. For an overview of what geographers call "spatial systems" and "cultural landscapes," see volume I, *Atlantic America, 1492–1800* (1986), of D. W. Meinig's *Shaping of America: A Geographical Perspective on Five Hundred Years of History.* John R. Stilgoe uses a wealth of detailed examples, from factories to cowpens, to explore the idea of "landscape" as the cumulative effect of human shaping of the environment in *Common Landscape of America, 1580 to 1845* (1982). In the early national period, land policy played a crucial role; the workings of the General Land Office are detailed in *The Land Office Business: The Settlement and Administration of American Public Lands, 1789–1837* (1986), by Malcolm J. Rohrbough. Focused on more specific regions, James T. Lemon's *Best Poor Man's Country: A Geographical Study of Early Southeastern Pennsylvania* (1972) explores settlers, settlement patterns, and land use, what he calls "the interplay of society and land," in an early plea for attention to the mid-Atlantic region; and Peter Wacker describes settlement forms of new and native populations in *Land and People: A Cultural Geography of Preindustrial New Jersey: Origins and Settlement Patterns* (1975). On New York, Sung Bok Kim treats tenancy in the Hudson Valley in *Landlord and Tenant in Colonial New York: Manorial Society, 1664–1775* (1978). The essays in *New Opportunities in a New Nation: The Development of New York after the Revolution* (1982), edited by Manfred Jonas and Robert Wells, present topics such as land policy, economic development, and transportation. William Wyckoff explores the role of wealthy developers, especially around the turn of the nineteenth century, in *The Developer's Frontier: The Making of the Western New York Landscape*

(1988). J. Ritchie Garrison describes the transformation of a rural New England landscape in *Landscape and Material Life in Franklin County, Massachusetts, 1770–1860* (1991). Albert E. Cowdrey provides an examination of changing southern ecology in *This Land, This South: An Environmental History* (1983). John Mack Faragher's *Sugar Creek: Life on the Illinois Prairie* (1986) offers an account of the early nineteenth-century European-American settling of prairie lands. Most of the story William Cronon tells in *Nature's Metropolis: Chicago and the Great West* (1991) takes place after 1850, but the early parts of the book convey the flavor of the transformation brought about by the growth of Chicago as an urban center.

Environmental and demographic studies of nonwhite groups who often had to fit in to European-dominated landscapes or adapt to European domination are also important. In *Changes in the Land: Indians, Colonists, and the Ecology of New England* (1983), William Cronon examines the disparate attitudes toward land and resources that lay behind New England's seventeenth-century transformation. Modeling his study on Cronon's, Timothy Silver also includes the Africans' contribution to reshaping the southern landscape in *A New Face on the Countryside: Indians, Colonists, and Slaves in South Atlantic Forests, 1500–1800* (1990). Gary C. Goodwin describes the "precontact" environment and changes after European arrival in *Cherokees in Transition: A Study of Changing Culture and Environment Prior to 1775* (1977). J. Leitch Wright, Jr., provides an ethnic interpretation of the Creek Confederacy, including discussion of the built environment, in *Creeks and Seminoles: The Destruction and Regeneration of the Muscogulge People* (1986). The early chapters of Stephen J. Pyne, *Fire in America: A Cultural History of Wildland and Rural Fire* (1982) and portions of J. Donald Hughes, *American Indian Ecology* (1987) may also be useful. Ira Berlin explores the social and economic positions of free African-Americans in different parts of the South in *Slaves without Masters: The Free Negro in the Antebellum South* (1974). Related studies may also be found in Section 10, "Manufacturing and Labor."

9. POWER

Although wind and especially animal power was used in America, historians have focused on water and steam. Waterpower was readily available along America's many creeks and streams and played a significant role in American technological development from the days of small colonial mills through the larger-scale industrialization of the nineteenth century. An international context for American use of waterpower can be found in Terry Reynolds's comprehensive *Stronger than a Hundred Men: A History of the Vertical Waterwheel* (1983) and also in several of the essays in *The World of the Industrial Revolution: Comparative and International Aspects of Industrialization* (1986), edited by Robert Weible. The most thorough synthesis of the later period of American waterpower is Louis C. Hunter's *Waterpower in the Century of the Steam Engine* (1979), the first volume of *A History of Industrial Power in the United States, 1780–1930*. Detailed studies of specific sites provide insight into differing styles of power development; see especially the early part of Patrick M. Malone's *Canals and Industry: Engineering in Lowell, 1821–1880* (1983) and also David Gilbert's *Where Industry Failed: Water-Powered Mills at Harper's Ferry, Virginia* (1984). David J. Jeremy presents aspects of the debate over water versus steam power in *Technology and Power in the Early American Cotton Industry: James Montgomery, the Second Edition of his "Cotton Manufacture" (1840), and the "Justitia" Controversy about Relative Power Costs* (1990). A useful view of millraces, dams, and transmission systems is provided by David Macaulay in *Mill* (1983). Local histories and guides, such as A. T. Jackson, *Mills of Yesteryear* (1971) on Texas, or Philip L. Lord, Jr., with Martha A. Costello, illustrator, *Mills on the Tsatsawassa: A Guide for Local Historians* (1983) on a Hudson Valley mill creek, can also be useful.

Unfortunately, American steam power has not received much focused attention, although British books sometimes include useful information. Louis C. Hunter has provided the single general reference on steam power in his posthumously published *Steam Power* (1985), the second volume of *A History of Industrial Power in the United States, 1780–1930*. Specific aspects of the emergence of steam power had been explored previously, in works that remain valuable: by Carroll W. Pursell, Jr., in *Early Stationary Steam Engines in America: A Study in the Migration of a Technology* (1969) and by Bruce Sinclair in *Early Research at the Franklin Institute: The Investigation into the Causes of Steam Boiler Explosions, 1830–1837* (1966). Steam power was, of course, central to the development of railroads and steamboats and to later industrialization; see Section 5, "Mining and Metals," Section 7, "Civil Engineering and Transportation," Section 10, "Manufacturing and Labor," and Section 13, "America and Technology Transfer."

10. MANUFACTURING AND LABOR

The lion's share of production technologies falls under the rubric "manufacturing," a category that encompasses technologies from craft skills to complex mechanical contrivances. Recent studies focusing on such technologies have highlighted the importance of the workers involved, regardless of their ostensible value as given in such labels as "skilled" or "women's work" (see, for example, Judith McGaw's work on papermaking or Merritt Roe Smith's work on arms manufacture, below). This section, therefore, includes a fair number of works in labor history that focus on workers' lives. Such studies serve as useful pointers to the vast and varied range of technological interactions that were part of early American manufactures, even when the historian interested in technology might wish for more attention to the details of the work itself.

Books in this section have been divided, following Hindle, into the following sections: "General Manufacturing"; "Crafts and Craftspeople"; "Tools, Woodwork, Metalwork, and Machining"; "Clocks, Instruments, and the Technology of Science"; "Glass and Pottery"; "Paper, Printing and Book Production"; "Leather"; "Textiles"; "Chemicals"; and "Other Manufacturing." Note that most women's work took place in the household and is treated in Section 4, "Household Technology." Other closely related sections include 9, "Power"; 13, "America and Technology Transfer"; and 14, "Patents and Invention."

General Manufacturing

One of the issues of interest in early nineteenth-century manufactures has been the so-called American System—the nexus of goals, practices, and attitudes that set American manufacture apart from Europe in the eyes of both Americans and their European observers. The essays in Otto Mayr and Robert C. Post's edited collection *Yankee Enterprise: The Rise of the American System of Manufactures* (1981) present a range of issues and questions central to the topic. The early part of David A. Hounshell's *From the American System to Mass Production, 1800–1932: The Development of Manufacturing Technology in the United States* (1984) is also useful, although the focus of the book lies after 1850. In contrast to Hounshell's emphasis on arms manufacture and government contracts, Donald R. Hoke's *Ingenious Yankees: The Rise of the American System of Manufactures in the Private Sector* (1990) traces the roots of the American System to wooden clock manufacture, ax making, and watch production in the period before 1850. Another issue of interest is governmental encouragement of private manufactures; Jacob E. Cooke's biography, *Tench Coxe and the Early Republic* (1978), provides information on Coxe's role as promoter of American industry, and a collection of Coxe's papers has been issued on microfilm. Michael Brewster Folsom and Steven D. Lubar provide a useful collection of source documents in *The Philosophy of Manufactures: Early Debates over Industrialization in the United States* (1981). The so-called McLane Report of 1833—[Louis McLane], *Documents Relative to the Manufactures in the United States of America*—has been reprinted in three volumes (1969), as have *The Report of the Committee on the Machinery of the United States* (1855) and *The Special Reports of George Wallis and Joseph Whitworth* (1854), both in *The American System of Manufactures* (1969), edited by Nathan Rosenberg. Carl Siracusa examines responses to industrialization in Massachusetts in *A Mechanical People: Perceptions of the Industrial Order in Massachusetts, 1815–1880* (1979). A collection of nineteenth-century pamphlets on "the labor question" is reprinted in *Religion, Reform, and Revolution: Labor Panaceas in the Nineteenth Century* (1969), edited by Leon Stein and Philip Taft.

Crafts and Craftspeople

Material on craft technologies and the people who practiced them can be gleaned from several sources. Studies by historians interested in material culture are one of the most important. *The Craftsman in Early America* (1984), edited by Ian M. G. Quimby, provides an overview. David Pye's study, *The Nature and Art of Workmanship* (1968), approaches questions of how to classify craft skills and highlights some useful distinctions. *The Arts in America: The Colonial Period* (1966), by Louis B. Wright, George B. Tatum, John W. McCoubrey, and Robert C. Smith, emphasizes British heritage and treats crafts along with architecture and painting. Museum catalogs, such as Beatrice B. Garvan and Charles F. Hummel's *Pennsylvania Germans: A Celebration of Their Arts, 1683–1850* (1982), are also useful. Many volumes of the *Winterthur Portfolio* (published annually until 1978 and then quarterly thereafter) contain articles of interest here.

Studies of artisanal involvement in political movements often present brief but useful portraits of a community's craftspeople. In *Artisans for Independence: Philadelphia Mechanics and the American Revolution* (1975), Charles S. Olton focuses on master craftsmen. Howard B. Rock's

Artisans of the New Republic: The Tradesmen of New York City in the Age of Jefferson (1979) relates the economic position of the mechanic to political involvement, as does Charles Steffen in *The Mechanics of Baltimore: Workers and Politics in the Age of Revolution, 1763–1812* (1984). Eric Foner's chapter on Philadelphia artisans in *Tom Paine and Revolutionary America* (1976) is also useful. Ian R. Tyrell's *Sobering Up: From Temperance to Prohibition in Antebellum America, 1800–1860* (1979) includes treatment of artisanal involvement in temperance movements.

Community studies and studies of class formation among workers provide summaries of who worked in what industries as well as important contextual settings for studies of craft technologies. Such studies often focus on mid-Atlantic cities; by contrast, economic and technological studies of particular industries have tended to treat New England (see, for example, "Textiles," below). Increasingly, studies of these types do include material on the work workers do and thus the technologies they use: see Billy G. Smith's detailed portraits of day-to-day experience in *The "Lower Sort": Philadelphia's Laboring People, 1750–1800* (1990), Sean Wilentz's descriptions of workshop conditions in *Chants Democratic: New York City and the Rise of the American Working Class, 1788–1850* (1984), and Steven J. Ross's explorations of work patterns and attitudes in *Workers on the Edge: Work, Leisure, and Politics in Industrializing Cincinnati, 1788–1890* (1985). For material on indentured servants' craft skills, see David W. Galenson, *White Servitude in Colonial America: An Economic Analysis* (1981). More culturally or politically oriented studies, which do not focus on the specifics of technological change, include Gary B. Nash, *Forging Freedom: The Formation of Philadelphia's Black Community, 1720–1840* (1988); Bruce Laurie, *Working People of Philadelphia, 1800–1850* (1980); Susan E. Hirsch, *Roots of the American Working Class: The Industrialization of Crafts in Newark, 1800–1860* (1978); Robert E. Purdue, *Black Laborers and Black Professionals in Early America, 1750–1830* (1975); and Christine Stansell, *City of Women: Sex and Class in New York, 1789–1860* (1986). Gillian Lindt Gollin's *Moravians in Two Worlds* (1967), a social history of the period 1720–1820, includes material on handcrafts and artisans.

Not all early American artisans could hope eventually to become masters. Although the majority of enslaved African-Americans worked in agriculture, many northern slaves and more than a few in the South worked in skilled trades and industry. In such cases, free white workers often found themselves in competition with people who were bought or rented rather than employed. Edgar J. McManus has written on northern slavery, which was concentrated in urban centers, in *Black Bondage in the North* (1973) and *A History of Negro Slavery in New York* (1966). Arthur Zilversmit's state-by-state treatment includes details about the skills of slaves in *The First Emancipation: The Abolition of Slavery in the North* (1967). Robert S. Starobin surveys industries from iron forging to canal building, including manufactures, in *Industrial Slavery in the Old South* (1970). Claudia Dale Goldin's *Urban Slavery in the American South, 1820–1860: A Quantitative History* (1976), heavily econometric in its approach, highlights the objections of white artisans to their enslaved competitors.

Tools, Woodwork, Metalwork, and Machining

The importance of wood as an early American resource has only recently been appreciated. Metal's industrial connotations originate in wood-poor England; in forested North America, wood was both abundant and functional. Brooke Hindle began to remedy the neglect of wood-based technologies when he collected the essays in *America's Wooden Age: Aspects of Its Early Technology* (1975). On woodworking tools, both hand-held and mechanical, see *Tools and Technologies: America's Wooden Age* (1979), edited by Paul Kebabian and William Lipke; their use and their manufacture is treated in Kebabian's collaboration with Dudley Whitney, *American Woodworking Tools* (1979). Kenneth Roberts's catalogs *Planemakers and Other Edge Tool Enterprises in New York State in the Nineteenth Century* (1971), coauthored with Jane Roberts, and *Wooden Planes in Nineteenth Century America* (1975) provide illustrations and a wealth of detail which may be useful for reference. Furniture is one of the more studied wooden products; for example, Charles F. Montgomery includes material on such topics as tools, apprenticeship, and regional uses of wood in *American Furniture: The Federal Period, in the Henry Francis du Pont Winterthur Museum* (1966). One of the few studies of carpenters is Mark Erlich's *With Our Hands: The Story of Carpenters in Massachusetts* (1986). Some material on carpentry may also be found under "Building and Architecture" in Section 7, "Civil Engineering and Transportation."

The ax, wooden-handled and metal-headed, and of course used to cut down trees, bridges American woodwork and metalwork—and has been treated by Henry J. Kauffman in *American Axes: A Survey of Their Development and Their Makers* (1972). Another metal object critical to woodwork is the nail, which rapidly superseded alternatives such as wooden pegs in the construction of wooden structures; see *The Rise and Decline of the American Cut Nail Industry: A Study of the Interrelationships of Technology, Business Organization, and Management Techniques* (1983), by Amos J. Loveday, Jr.

More traditionally, the literature on metalwork treats machine tools, smithing, and production of metal mechanisms such as those in guns. On machine tools, the English *History of Machine Tools, 1700–1910* (1969), by W. Steeds, includes American developments. The construction of textile machinery receives the attention of Anthony F. C. Wallace in *Rockdale: The Growth of an American Village in the Early Industrial Revolution* (1978), and a broader mix of machine tool applications is described by John Lozier in *Taunton and Mason: Cotton Machinery and Locomotive Manufacture in Taunton, Massachusetts, 1811–1861* (1986). Louis C. Hunter includes a useful discussion of engine building in *Steam Power* (1985). Decorative metalwork has received at least one man's attention: Henry J. Kauffman provides us with *American Copper and Brass* (1979), *The Colonial Silversmith: His Techniques and His Products* (1974), and *The American Pewterer: His Techniques and His Products* (1970). Blacksmithing and tinsmithing have received minimal attention; the only sustained treatments I found are Jeannette Lasansky's informative, illustrated museum booklets *To Draw, Upset, and Weld: The Work of the Pennsylvania Rural Blacksmith, 1742–1935* (1980) and *To Cut, Piece, and Solder: The Work of the Pennsylvania Rural Tinsmith, 1778–1908* (1982).

Small arms production has been one of the classic examples used in descriptions of American factory mechanization, the original American System industry, and traditionally a success story. Traditional gunsmithing has not received similar attention, although one colonial case has been explored by Harold B. Gill, Jr., in *The Gunsmith in Colonial Virginia* (1974). The most important work on the nineteenth-century chapter of the story is Merritt Roe Smith's *Harpers Ferry Armory and the New Technology: The Challenge of Change* (1977). Smith explores the role of Army Ordnance commitment in the development of new technologies of factory organization and the role of worker response to the new methods, challenging notions both of the immediate profitability of the method and of American workers' enthusiasm for technological progress. Treatments of the subject may also be found in the works on the American System mentioned under the heading "General Manufacturing," above, and in Section 6, "Military Technology."

Clocks, Instruments, and the Technology of Science

Precision production of parts with close tolerances is also a feature of clock and instrument manufacture, although metal was by no means the only material used. Wood figured prominently in clockmaking until the mid-nineteenth century. For a general overview, see Chris H. Bailey's *Two Hundred Years of American Clocks and Watches* (1975). On specific clockmakers, see Kenneth D. Roberts's *Contributions of Joseph Ives to Connecticut Clock Technology, 1810–1862* (1971) and *Eli Terry and the Connecticut Shelf Clock* (1973) and Snowden Taylor's *The Developmental Era of Eli Terry and Seth Thomas Shelf Clocks* (1985). Donald R. Hoke treats wooden clock manufacture, among other cases, in *Ingenious Yankees: The Rise of the American System of Manufactures in the Private Sector* (1990). Silvio A. Bedini provides some material on scientific instruments in *Thinkers and Tinkers: Early American Men of Science* (1975).

Glass and Pottery

Ian M. G. Quimby has again edited a useful overview collection: *Ceramics in America* (1973) touches on topics from the range of American wares and taste to production technologies. More detail is available on the Pennsylvania ceramics industry: see Susan Myers, *Handcraft to Industry: Philadelphia Ceramics in the First Half of the Nineteenth Century* (1980), Jeannette Lasansky, *Central Pennsylvania Redware Pottery, 1780–1904* (1979), and Graham Hood, *Bonnin and Morris of Philadelphia: The First American Porcelain Factory, 1770–1772* (1972). Moving west, some information may be found in *Eighteenth-Century Ceramics from Fort Michilimackinac: A Study in Historical Archaeology* (1970) by J. Jefferson Miller II and Lyle M. Stone. On glass, see Arlene Palmer's *Wistars and Their Glass, 1739–1777* (1989) and Paul Gardner's *American Glass* (1977).

Paper, Printing, and Book Production

That paper is an important and illustrative case study in American industrialization is demonstrated by Judith A. McGaw in her account *Most Wonderful Machine: Mechanization and Social Change in Berkshire Paper Making, 1801–1885* (1987). In three sections entitled "Before the Machine," "The Machine," and "After the Machine," McGaw explores questions about which aspects of the process were mechanized, which workers performed which tasks, who made which decisions, and how the process changed in the transition from traditional to mechanized methods.

A preliminary chronology of the American development of the industry is provided by David C. Smith in his *History of Papermaking in the United States (1691–1969)* (1970).

Paper, of course, is just the first step in producing printed materials. Rollo G. Silver treats all aspects of the printing business in *The American Printer, 1787–1825* (1967). The career of one particular printer is recounted by Richard F. Hixson in *Isaac Collins: A Quaker Printer in Eighteenth Century America* (1968). Also useful is James Moran's overview *Printing Presses: History and Development from the Fifteenth Century to Modern Times* (1973). Other aspects of printing technology and products are treated in *Prints in and of America to 1850* (1973), edited by John D. Morse. John Tebbel's *History of Book Publishing in the United States*, I, *The Creation of an Industry, 1630–1865* (1972) provides a detailed summary of book manufacture. A more specific phase of production is treated by Hannah French in *Bookbinding in Early America: Seven Essays on Masters and Methods* (1986).

Leather

While apparently little work has been done on tanning, the New England shoe industry has received attention as another case in which to study industrialization. Change in the Massachusetts shoe industry came in the form of centralized workshops before any mechanization. Changes in shoemaking technology and shop and community organization are treated in Alan Dawley's *Class and Community: The Industrial Revolution in Lynn* (1976) and Paul G. Faler's *Mechanics and Manufacturers in the Early Industrial Revolution: Lynn, Massachusetts, 1780–1860* (1981). Mary H. Blewett focuses on gender roles in Lynn in *Men, Women, and Work: Class, Gender, and Protest in the New England Shoe Industry, 1780–1910* (1988). For more specific details on technological change, see Ross Thomson, *The Path to Mechanized Shoe Production in the United States* (1989).

Textiles

From a technological point of view, textile production, and particularly the factory-based New England approach, has received more attention from historians than any other American or British manufacture. Traditionally portrayed as the typical case of Industrial Revolution technology, New England textiles have only recently been put in perspective by studies not only of other industries but also of textile production in other regions.

The corporate model, financed by nonlocal investors, was just one mode of industrialization: many mills and shops, especially outside New England, were run on a smaller scale under proprietary ownership.

The more traditional New England–based interpretation is presented by Barbara Tucker in *Samuel M. Slater and the Origins of the American Textile Industry, 1790–1860* (1984). Thomas Dublin provides an alternative perspective on this traditional topic in his study *Women at Work: The Transformation of Work and Community in Lowell, Massachusetts, 1826–1860* (1979). Two collections of women mill workers' writings provide access to relevant sources: see Dublin's *Farm to Factory: Women's Letters, 1830–1860* (1981) and Philip S. Foner's *Factory Girls* (1977). In *The Coming of Industrial Order: Town and Factory Life in Rural Massachusetts, 1810–1860* (1983), Jonathan Prude describes the separation of new industrial communities from the older towns that spawned them. The implicit hegemony of the cotton textile industry is perfectly illustrated by the title of the otherwise useful *New England Mill Village, 1790–1860* (1982), which in fact treats cotton mills only; the collection, edited by Gary Kulik *et al.*, emphasizes a contextual approach to several aspects of the subject. The mid-nineteenth century cotton industry receives further treatment by David J. Jeremy in *Technology and Power in the Early American Cotton Industry: James Montgomery, the Second Edition of his "Cotton Manufacture" (1840), and the "Justitia" Controversy about Relative Power Costs* (1990). A peek at the New England woolen industry is provided by John Borden Armstrong in *Factory under the Elms: A History of Harrisville, New Hampshire, 1774–1969* (1970), and a broader view of American woolens is provided by Elizabeth Hitz in *A Technical and Business Revolution: American Woolens to 1832* (1986). For illustrations, see Steve Dunwell, *The Run of the Mill: A Pictorial Narrative of the Expansion, Dominion, Decline, and Enduring Impact of the New England Textile Industry* (1978). Also useful may be Grace Rogers Cooper, *The Copp Family Textiles* (1971), on Stonington, Connecticut, 1750 to 1850.

Studies of the mid-Atlantic region have been important in revising the historiography of the American Industrial Revolution. The interrelationships of community and indigenous industry, the roles of workers and proprietors, the balance between handwork and mechanization—such issues are all cast in a different light by works such as Philip Scranton's *Proprietary Capitalism: The Textile Manufacture at Philadelphia, 1800–1885* (1983), Cynthia Shelton's *Mills of Manayunk: Industrialization and Social*

Conflict in the Philadelphia Region, 1787–1837 (1986), and Anthony F. C. Wallace's *Rockdale: The Growth of an American Village in the Early Industrial Revolution* (1978).

Southern textile processing—as opposed to the widespread production of the raw materials—has received little attention, only partly because so much cotton was exported raw from that region. The exception is *The Textile Industry in Antebellum South Carolina* (1969), by Ernest McPherson Lander, Jr., although the works mentioned above on urban and industrial slavery provide some information.

Also, as David J. Jeremy has shown in his important study, *Transatlantic Industrial Revolution: The Diffusion of Textile Technologies between Britain and America, 1790–1830s* (1981), the diffusion of various technologies across the Atlantic, in both directions, was critical to the development of American textile production. The 1794 journal of one British traveler in America has been edited by Jeremy and republished in *Henry Wansey and His American Journal, 1794* (1970), and the letters of a British immigrant family of skilled workers are published in *The Hollingworth Letters: Technical Change in the Textile Industry, 1826–1837* (1969), edited by Thomas W. Leavitt. See also Section 13, "America and Technology Transfer."

Notable in all these citations is the complete neglect of colonial textiles (the earliest year treated in the studies mentioned is 1774). One exception is Florence M. Montgomery's useful dictionary *Textiles in America, 1650–1870* (1984), which defines and describes different types of fabrics and the uses to which they were commonly put. Limited materials on household-based spinning, weaving, and clothing manufacture may also be found in Section 4, "Household Technology."

Clothing production outside the home has received some attention, although mostly in the later period: see Claudia B. Kidwell's *Cutting a Fashionable Fit: Dressmakers' Drafting Systems in the United States* (1979) and Kidwell and Margaret Christman's *Suiting Everyone: The Democratization of Clothing in America* (1979). An unusual source has been reprinted in *The Weaver's Draft Book and Clothier's Assistant* (1792, 1979) by John Hargrove. On another item of apparel, the first chapter of David Bensman's study *The Practice of Solidarity: American Hat Finishers in the Nineteenth Century* (1985) treats the pre-1850 manufacture of hats from fur felt.

Chemicals

Early American chemical manufacture—including gunpowder, salt, bleach, dyes, and paint—has been another neglected area. Several of the studies mentioned in Section 3, "Agriculture," treat tar and potash, although not centrally. An exception is most (despite the dates in the title) of the study *The Bethlehem Oil Mill, 1745–1934: ... German Technology in Early Pennsylvania* (1984), by Carter Litchfield, Hans-Joachim Finke, Stephen Young, and Karen Zerbe Huetter. The early part of Reese V. Jenkins's *Images and Enterprise: Technology and the American Photographic Industry, 1839–1925* (1976) provides another small piece of a larger puzzle, to which Arthur H. Frazier's *Joseph Saxton and His Contributions to the Medal Ruling and Photographic Arts* (1975) may also contribute. Although it treats a later period, Norman B. Wilkinson's *Lammot du Pont and the American Explosives Industry, 1850–1884* (1984) may also be of use.

Other Manufacturing

Several studies of manufactures elude the above categories. William H. Armstrong presents a view of a relatively uncommon eighteenth-century craft in *Organs for America: The Life and Work of David Tannenberg* (1967). Stonework has received little attention, except as gravestones: see Diana Williams Combs, *Early Gravestone Art in Georgia and South Carolina* (1986), and Allan I. Ludwig, *Graven Images: New England Stonecarving and Its Symbols, 1650–1815* (1966).

II. EDUCATION, ORGANIZATION, AND SCIENCE

Technical education in early America took place in an array of settings, most of which were not formal schools. For children, common school education came early to include arithmetic, but for boys no further technical content was added until the later nineteenth century; boy's schools that did teach technical subjects were most often designed for the moral reform of juvenile delinquents. Girls who attended "dame schools" learned needlework along with common school subjects. Most children were expected to learn technical skills from their parents or, through apprenticeship, from another parentally designated adult chosen for the purpose. Formal education for adults tended to address highly specific audiences, in learned societies or, by the early nineteenth century, in organizations such as mechanics' institutes.

Several approaches illuminate early American ideas about education related to technology. Daniel Calhoun's innovative study *The Intelligence of a People* (1973) explores the breadth of early American thinking in sections on education, preaching, shipbuilding, and bridge building, covering the period from 1750 to 1870. In *"Let Virtue Be a Guide to Thee": Needlework in the Education of Rhode Island Women, 1730–1830* (1983), Betty Ring demonstrates that needlework samplers can tell us much about girls' education and literacy. Linda K. Kerber also provides a chapter on women's education in *Women of the Republic: Intellect and Ideology in Revolutionary America* (1980). Lawrence Cremin's *American Education: The Colonial Experience, 1607–1783* (1970) presents an overview of colonial education, including some technical education. Patricia Cline Cohen studies the rise of American mathematical thinking, a trait often observed by foreign visitors, in *A Calculating People: The Spread of Numeracy in Early America* (1983). Related topics are explored in two works by James H. Cassedy, *American Medicine and Statistical Thinking, 1800–1860* (1984) and *Demography in Early America: Beginnings of the Statistical Mind, 1600–1800* (1969). By the late eighteenth century, drawing was closely related to technical education; Peter C. Marzio provides some information in *The Art Crusade: An Analysis of American Drawing Manuals, 1820–1860* (1976). Studies of literacy (which may also be treated as technology) may add a useful comparison here; see, for example, Kenneth A. Lockridge, *Literacy in Colonial New England: An Enquiry into the Social Context of Literacy in the Early Modern West* (1974). Related topics may also be found in Section 4, "Household Technology," and Section 12, "Communication."

As mentioned above, much technological knowledge was relayed through apprenticeship, another understudied topic in the history of education, labor history, and history of technology. One exception is W. J. Rorabaugh's *Craft Apprentice: From Franklin to the Machine Age in America* (1986), which is heavily biased toward the relatively elite craft of printing, but Rorabaugh has culled qualitative evidence from an impressive list of archives. Ian M. G. Quimby's *Apprenticeship in Colonial Philadelphia* (1985), the publication in book form of his 1963 master's thesis, remains one of the most often cited works on the subject. Books such as Bernard Farber's *Guardians of Virtue: Salem Families in 1800* (1972) include material on apprenticeship patterns in individual communities. Many of the works included in Section 10, "Manufacturing and Labor," also provide some commentary on apprenticeship.

The better-known scientific and institutional side of the spread of technological knowledge is better documented and more often studied. Science generally included technology in this period, so books such as Alexandra Oleson and Sanborn C. Brown's collection *The Pursuit of Knowledge in the Early American Republic: American Scientific and Learned Societies from Colonial Times to the Civil War* (1976), Raymond Phineas Stearns's *Science in the British Colonies of America* (1970), or John C. Greene's *American Science in the Age of Jefferson* (1984) contain much useful material. More specifically technological is Bruce Sinclair's *Philadelphia's Philosopher Mechanics: A History of the Franklin Institute, 1824–1865* (1975), which provides a window into the interests of the technical elite and some of their attitudes toward more ordinary mechanics as it traces the institutional development of the institute. On individual scientists of technological interest, see Silvio A. Bedini's *Life of Benjamin Banneker* (1972) and *At the Sign of the Compass and Quadrant: The Life and Times of Anthony Lamb* (1984), Chandos Michael Brown's *Benjamin Silliman: A Life in the Young Republic* (1989), and Bern Dibner's *Benjamin Franklin, Electrician* (1976). On Thomas Jefferson's contributions, see Silvio A. Bedini, *Thomas Jefferson, Statesman of Science* (1990), and Samuel L[atham] Mitchill, *A Discourse on the Character and Services of Thomas Jefferson, More Especially as a Promoter of Natural and Physical Sciences* (1826, 1982). Edgar P. Richardson, Brooke Hindle, and Lillian B. Miller describe another versatile career, and explore relationships between fine arts, science, and technology, in *Charles Willson Peale and His World* (1983). *The Collected Papers of Charles Willson Peale and His Family, 1735–1885* (1980), edited by Lillian B. Miller, *The Collected Works of Count Rumford* (1968–1970), edited by Sanborn Brown, and *The Papers of Joseph Henry* (1972–), edited by Nathan Reingold *et al.*, are also useful. Further discussion of individual efforts may be found in Section 14, "Patents and Invention."

12. COMMUNICATION

Early American communication is important on several levels: first, communication involves technologies of various kinds, such as transportation, printing, or telegraphy, all of which had a profound impact on the people who used them; and, second, it promoted the spread, or "transfer" as it is usually called, of technologies. The present section comprises studies of mechanisms of communication; the communication of technological information is treated in the next section, "America and Technology Transfer."

Ian K. Steele explores transatlantic shipping routes and packet boats, refuting traditional assumptions about the ocean as a barrier to communication, in *The English Atlantic, 1675–1740: An Exploration of Communication and Community* (1986). David Cressy provides another approach to colonial connectedness to Europe, focusing more on kinship networks, in *Coming Over: Migration and Communication between England and New England in the Seventeenth Century* (1987).

On land, Allan R. Pred's *Urban Growth and the Circulation of Information: The United States System of Cities, 1790–1840* (1973) studies the influence of America's large cities on each other, for example, through newspapers, the postal service, commodity flow, and travel. Appropriate sections of Wayne E. Fuller's *American Mail: Enlarger of the Common Life* (1972) and Carl H. Scheele's more global *Short History of the Mail Service* (1970) provide details on the technology of the postal service. Studies of the American press are also useful: see Leonard W. Levy's *Emergence of a Free Press* (1985), Bernard Bailyn and John B. Hench's edited collection *The Press and the American Revolution* (1980), and Christopher L. Dolmetsch's *German Press of the Shenandoah Valley* (1984). Books provided another important means of communication, as explored by Edwin Wolf II in *The Book Culture of a Colonial American City: Philadelphia Books, Bookmen, and Booksellers* (1988), by the early part of Cathy N. Davidson's *Revolution and the Word: The Rise of the Novel in America* (1986), and by the essays in *Books in America's Past: Essays Honoring Rudolph H. Gjelsness* (1966), edited by David Kaser. William J. Gilmore's *Reading Becomes a Necessity of Life: Material and Cultural Life in Rural New England, 1780–1835* (1989) treats the role of reading in transmitting "mass culture." Note that the manufacture of newspapers and books is treated in Section 10, "Manufacturing and Labor."

13. AMERICA AND TECHNOLOGY TRANSFER

The technology of early America was as heterogeneous as were its people. Traditionally, much emphasis has been placed on the new inventions of the American people—usually meaning European settlers, and usually men. In fact, Americans of all backgrounds made ingenious adaptations of technology brought from elsewhere or borrowed from newfound neighbors. "Technology transfer," broadly defined to include both the geographical and intercultural migration of techniques and tools, is as integral a part of early American technology as is the "Yankee ingenuity" historians so often celebrate. Of course, in many cases the line between adaptation and invention is a fine one; related material may also be found in Section 14, "Patents and Invention."

Most neglected, not surprisingly, are the contributions of Native Americans and African-Americans to the technologies of the European Americans around them. In *The Skulking Way of War: Technology and Tactics among the Indians of New England* (1991), Patrick M. Malone explores not only the seventeenth-century transfer of firearms technology from the European Americans to the Native Americans but also the transfer of the techniques and strategies of forest warfare, a lasting legacy, from the Native Americans to the newcomers. Contributions to agricultural technologies by African and Native American groups are especially evident. For example, Peter H. Wood emphasizes the importance of African techniques such as rice cultivation in his discussion of slave-slaveowner relations in *Black Majority: Negroes in Colonial South Carolina from 1670 through the Stono Rebellion* (1974). Related materials may be found in Section 3, "Agriculture."

But even the more traditional terrain of industrial exchange between America and Europe awaits further exploration. In *The Transfer of Early Industrial Technologies to America* (1987), Darwin H. Stapleton provides an overview of the existing literature on the subject and presents a series of case studies. Several European-American case studies are also presented in the early part of *International Technology Transfer: Europe, Japan, and the USA* (1991), edited by David J. Jeremy. A model of the potential for larger studies, Jeremy's *Transatlantic Industrial Revolution: The Diffusion of Textile Technologies between Britain and America, 1790–1830s* (1981) focuses more narrowly on a single industry. Jeremy's work set ambitious new standards by exploring transfers in both directions and by emphasizing the role of skill, which migrated as people did. Carroll W. Pursell, Jr., in *Early Stationary Steam Engines in America: A Study in the Migration of a Technology* (1969), confronts the case of importation of a technology for which there was no initial perceived need and traces its accelerating development in the early nineteenth century. In *The John Bull: 150 Years a Locomotive* (1981), John H. White, Jr., explores the adaptations made to a British locomotive imported to the United States in 1831. Another case is illustrated by the early part of Geoffrey Tweedale's *Sheffield Steel and America: A Century of Commercial and Technological Interdependence, 1830–1930* (1987). Americans also went to Europe, especially in the nineteenth century, for scientific and technical training;

see, for example, Margaret W. Rossiter, *The Emergence of Agricultural Science: Justus Liebig and the Americans, 1840–1880* (1975). For information on some responses to this cultural exchange, see the early portion of Marvin Fisher's *Workshops in the Wilderness: The European Response to American Industrialization, 1830–1860* (1967), and J. F. C. Harrison's *Quest for the New Moral World: Robert Owen and the Owenites in Britain and America* (1969).

Material on immigrant groups often provides insight into the technologies people carried with them. Bernard Bailyn's *Peopling of British North America: An Introduction* (1986) and *Voyagers to the West: A Passage in the Peopling of America on the Eve of the Revolution* (1986) include information on characteristics such as craft skills and factors such as land policy in his account of who came where in North America. Charlotte Erikson's *Invisible Immigrants: The Adaptation of English and Scottish Immigrants in Nineteenth-Century America* (1972) is based on letters written home by immigrants, several of whom were artisans. One of the more lasting, and thus easier technologies to trace, is building and architecture, as illustrated by Charles Van Ravenswaay in *The Arts and Architecture of German Settlements in Missouri: A Survey of a Vanishing Culture* (1977). Log buildings provide another interesting case; see *The American Backwoods Frontier: An Ethnic and Ecological Interpretation* (1989), by Terry G. Jordan and Matti Kaups, and Jordan's *American Log Buildings: An Old World Heritage* (1985).

14. PATENTS AND INVENTION

Accounts of successful inventions, usually those which earned both a patent and fame, have long been a mainstay of the history of technology. More recently, scholars have begun to explore the complexities underlying these heroic tales. Attention has turned, for example, to the processes of invention: the definitions of problems to solve and the inventors' approaches to solving them. Notable among such studies, many of which treat the late nineteenth and early twentieth centuries, is Brooke Hindle's *Emulation and Invention* (1981). Hindle uses the early nineteenth-century cases of the steamboat and the telegraph to explore both the mental processes and the environment of invention, stressing the roles of visualization and manipulation of images in mechanical creativity. The patent system, too, was a complex institution; Carolyn C. Cooper's study of patent management in antebellum America, *Shaping Invention: Thomas Blanchard's Machinery and Patent Management in Nineteenth-Century*

America (1991), describes a social system within which personal interactions often determined not only the scope of a particular patent but also the methods of defining "invention" in general. For details on changing patent law, see *Supreme Court Justice Joseph Story: Statesman of the Old Republic* (1985), by R. Kent Newmyer, and *Genesis of American Patent and Copyright Law* (1967), by Bruce W. Bugbee. More traditional in its interpretive framework, Joseph and Frances Gies's *Ingenious Yankees: The Men, Ideas, and Machines That Transformed a Nation, 1776–1876* (1976) provides a useful overview of industrial change during the period it covers. *The Papers of William Thornton,* superintendent of the patent office from 1802 to 1828, when published will provide access to material on the early development of the patent system. Also useful may be the early parts of *Invention and Economic Growth* (1966), in which Jacob Schmookler uses aggregate patent activity to measure rates of inventiveness.

Fortunately for the student of early American technology, records of the patent system include the nonverbal documentation provided by patent models. A history of these inventions-in-miniature is provided by William and Marlys Ray in *The Art of Invention: Patent Models and Their Makers* (1974). More detailed examinations of patent models may be found in several Smithsonian Institution exhibit catalogs: see Robert C. Post, ed., *American Enterprise: Nineteenth-Century Patent Models* (1984), and Barbara Suit Janssen, ed., *Technology in Miniature: American Textile Patent Models* (1988) and *Icons of Invention: American Patent Models* (1990).

Works on specific inventors or inventions also illuminate the process of successful invention. Eugene S. Ferguson explores both the life of one notable figure and how inventive problems come to be defined in *Oliver Evans, Inventive Genius of the American Industrial Revolution* (1980). More scientifically oriented, but revealing of mid-nineteenth century patent office politics, is Robert C. Post's *Physics, Patents, and Politics: A Biography of Charles Grafton Page* (1976). Also telling a nineteenth-century story, Grace Rogers Cooper chronicles the beginnings of a machine rather than an inventor in *The Invention of the Sewing Machine* (1968, 1976). On the machinery of mechanized shoe production, including its invention, see Ross Thomson's study *The Path to Mechanized Shoe Production in the United States* (1989). Several of the articles in *Patents and Inventions* (1991), a special issue of *Technology and Culture* edited by Carolyn C. Cooper, treat early American topics. Esmond Wright's biography of Benjamin Franklin, *Franklin of Philadelphia* (1986), may be of use, although its focus is more political than technological. For a technological view of another political figure, see Silvio A. Bedini's *Thomas Jefferson and His Copying Machines* (1984).

15. TECHNOLOGY IN BUSINESS AND ECONOMIC HISTORY

Technology has been an integral part of American business throughout its history and as such has played a role in economic history as well. Economic and business historians, however, have often treated technology selectively—a minor character, it enters the scene when needed to further the story, but waits offstage at other times. Historians of technology, of course, have been equally likely to study production or transportation technologies without business and economic context. This section includes books that treat technology from an economic point of view, parts of which touch on early America, and books about business and economic history in early America, parts of which touch on technology. Many other sections of this bibliography contain related material: see especially Section 7, "Civil Engineering and Transportation"; Section 10, "Manufacturing and Labor"; and Section 13, "Communication."

The rapid industrialization of America, especially compared to Britain, has continued to interest economic historians: see, for example, Paul A. David's *Technical Choice, Innovation, and Economic Growth: Essays on American and British Experience in the Nineteenth Century* (1975). Several of the essays in Nathan Rosenberg's *Perspectives on Technology* (1976) and Joel Colton and Stuart Bruchey's edited *Technology, the Economy, and Society: The American Experience* (1987) treat the period before 1850, and parts of Rosenberg's *Technology and American Economic Growth* (1972) may also be useful.

Less specifically focused on technology, business history provides important background for studies of commercial technologies. Thomas C. Cochran presents a social history of American business, beginning in the colonial period, in *Business in American Life: A History* (1972). He deals more explicitly with technology in *Frontiers of Change: Early Industrialism in America* (1981). Joyce Appleby explores ideas about capitalism, commerce, and modernism in *Capitalism and a New Social Order: The Republican Vision of the 1790s* (1984). Philadelphia's merchant community and trade networks are explored by Thomas M. Doerflinger in *A Vigorous Spirit of Enterprise: Merchants and Economic Development in Revolutionary Philadelphia* (1986). The first part of *The Visible Hand: The Managerial Revolution in American Business* (1977), by Alfred D. Chandler, Jr., treats traditional business organization before 1840, although the focus of the book is on the transformation Chandler finds after that time. Also on the nineteenth century, Glenn Porter and Harold C. Livesay's *Merchants and Manufacturers: Studies in the Changing Structure of Nineteenth-Century Marketing* (1971) presents case studies of firms and industries beginning in 1815. James Hedges's study, *The Browns of Providence Plantations: The Nineteenth Century* (1968), focuses on the antebellum activities, in manufacturing and investment, of the firm of Brown and Ives.

More general economic histories, such as *Enterprise: The Dynamic Economy of a Free People* (1990) and *The Roots of American Economic Growth, 1607–1861: An Essay in Social Causation* (1968), by Stuart Bruchey; *A Prosperous People: The Growth of the American Economy* (1985), by Edwin J. Perkins and Gary M. Walton; *American Economic Growth: An Economist's History of the United States* (1972), by Lance E. Davis *et al.*; *The Course of American Economic Growth and Development* (1970), by Louis M. Hacker; or *The Economic Growth of the United States, 1790–1860* (1966), by Douglass North, also provide useful background. For the colonial era, John J. McCusker and Russell R. Menard offer a broad overview, providing access to many specialized studies, in *The Economy of British America, 1607–1789* (1991), as does Edwin Perkins in *The Economy of Colonial America* (1988). James Henretta synthesizes social and economic history, including treatment of the preindustrial economy, in *The Evolution of American Society, 1700–1815* (1973). Many of the sources in the documentary collection *America's Economic Heritage: From a Colonial to a Capitalist Economy, 1634–1900* (1983), edited by Meyer Weinberg, contain information about technology. In *Society and Economy in Colonial Connecticut* (1985), Jackson Turner Main discusses socioeconomic and occupational structure in one colony. Peter Temin explores the national economy in a later period in *The Jacksonian Economy* (1969). And Diane Lindstrom explores one region's transition from commerce to manufacturing in her study *Economic Development in the Philadelphia Region, 1810–1850* (1978).

Studies focusing on the economics of labor are of particular relevance to the history of technology. In *White Servitude in Colonial America: An Economic Analysis* (1981), David W. Galenson uses British records of indenture to explore servants' backgrounds, skills, and American destinations. Stephen Innes studies labor in one Massachusetts town in *Labor in a New Land: Economy and Society in Seventeenth-Century Spring-*

field (1983). And Mary M. Schweitzer focuses on the household as a unit of production and consumption in *Custom and Contract: Household, Government, and the Economy in Colonial Pennsylvania* (1987).

Books on Early American Technology

In the following listing, books are ordered alphabetically by first author or editor, except in the case of an individual's published papers, which are listed under the individual's name. The number in parentheses after each listing refers to the sections above in which the book is mentioned:

> 1. General Surveys and Bibliographies
> 2. American Surveys
> 3. Agriculture
> 4. Household Technology
> 5. Mining and Metals
> 6. Military Technology
> 7. Civil Engineering and Transportation
> 8. Settlement Patterns and Urban Growth
> 9. Power
> 10. Manufacturing and Labor
> 11. Education, Organization, and Science
> 12. Communication
> 13. America and Technology Transfer
> 14. Patents and Invention
> 15. Technology in Business and Economic History

Alvarez, Eugene. *Travel on Southern Antebellum Railroads, 1828–1860.* University, Ala., 1974. (7)

Ammerman, David L., and Philip D. Morgan, comps. *Books about Early America: 2001 Titles.* Williamsburg, Va., 1989. (1, 8)

Appleby, Joyce. *Capitalism and a New Social Order: The Republican Vision of the 1790s.* New York, 1984. (15)

Armstrong, John Borden. *Factory under the Elms: A History of Harrisville, New Hampshire, 1774–1969.* Cambridge, Mass., 1970. (10)

Armstrong, William H. *Organs for America: The Life and Work of David Tannenberg.* Philadelphia, 1967. (10)

Bailey, Chris H. *Two Hundred Years of American Clocks and Watches.* Englewood Cliffs, N.J., 1975. (10)

Baily, Francis. *Journal of a Tour in Unsettled Parts of North America in 1796 and 1797.* Ed. Jack D. L. Holmes. Carbondale, Ill., 1969. (2)

Bailyn, Bernard. *The Peopling of British North America: An Introduction.* New York, 1986. (13)

———. *Voyagers to the West: A Passage in the Peopling of America on the Eve of the Revolution.* New York, 1986. (13)

Bailyn, Bernard, and John B. Hench, eds. *The Press and the American Revolution.* Worcester, Mass., 1980. (12)

Baughman, James P. *Charles Morgan and the Development of Southern Transportation.* Nashville, Tenn., 1968. (7)

Baxandall, Rosalyn, Linda Gordon, and Susan Reverby, eds. and comps. *America's Working Women: A Documentary History.* New York, 1976. (2)

Bedini, Silvio A. *At the Sign of the Compass and Quadrant: The Life and Times of Anthony Lamb.* American Philosophical Society, *Transactions*, LXXIV, pt. 1. Philadelphia, 1984. (11)

———. *The Life of Benjamin Banneker.* New York, 1972. (11)

———. *Thinkers and Tinkers: Early American Men of Science.* New York, 1975. (10)

———. *Thomas Jefferson and His Copying Machines.* Charlottesville, Va., 1984. (14)

———. *Thomas Jefferson, Statesman of Science.* New York, 1990. (11)

Bender, Thomas. *Toward an Urban Vision: Ideas and Institutions in Nineteenth-Century America.* Lexington, Ky., 1975. (8)

Benes, Peter, ed. *Foodways in the Northeast.* Boston, 1984. (3)

Bensman, David. *The Practice of Solidarity: American Hat Finishers in the Nineteenth Century.* Urbana, Ill., 1985. (10)

Berkeley, Edmund, and Dorothy Smith Berkeley. *Dr. John Mitchell: The Man Who Made the Map of North America.* Chapel Hill, N.C., 1974. (7)

Berlin, Ira. *Slaves without Masters: The Free Negro in the Antebellum South.* New York, 1974. (8)

Binder, Frederick Moore. *Coal Age Empire: Pennsylvania Coal and Its Utilization to 1860.* Harrisburg, Pa., 1974. (5)

Blewett, Mary H. *Men, Women, and Work: Class, Gender, and Protest in the New England Shoe Industry, 1780–1910.* Urbana, Ill., 1988. (10)

Bonner, James C. *A History of Georgia Agriculture, 1732–1860.* Athens, Ga., 1964. (3)

Bouquet, Henry. *The Papers of Henry Bouquet.* 5 vols. Ed. S. K. Stevens *et al.* Harrisburg, Pa., 1972–1984. (6)

Bowler, R. Arthur. *Logistics and the Failure of the British Army in America, 1775–1783.* Princeton, N.J., 1975. (6)

Boydston, Jeanne. *Home and Work: Housework, Wages, and the Ideology of Labor in the Early Republic.* New York, 1990. (4)

Bradford, James C., ed. *Command under Sail: Makers of the American Naval Tradition, 1775–1850.* Annapolis, Md., 1985. (6)

Breen, T. H. *Tobacco Culture: The Mentality of the Great Tidewater Planters on the Eve of the Revolution.* Princeton, N.J., 1985. (3)

Brown, Alexander Crosby. *The Dismal Swamp Canal.* Chesapeake, Va., 1970. (7)

Brown, Chandos Michael. *Benjamin Silliman: A Life in the Young Republic.* Princeton, N.J., 1989. (11)

Brown, M. L. *Firearms in Colonial America: The Impact on History and Technology, 1492–1792*. Washington, D.C., 1980. (6)

Bruchey, Stuart. *Enterprise: The Dynamic Economy of a Free People*. Cambridge, Mass., 1990. (15)

——. *The Roots of American Economic Growth, 1607–1861: An Essay in Social Causation*. New York, 1968. (15)

Buel, Joy Day, and Richard Buel, Jr. *The Way of Duty: A Woman and Her Family in Revolutionary America*. New York, 1984. (4)

Bugbee, Bruce W. *Genesis of American Patent and Copyright Law*. Washington, D.C., 1967. (14)

Bushman, Claudia L. *"A Good Poor Man's Wife": Being a Chronicle of Harriet Robinson and Her Family in Nineteenth-Century New England*. Hanover, N.H., 1981. (4)

Calhoun, Daniel. *The Intelligence of a People*. Princeton, N.J., 1973. (11)

Calvert, Monte A. *The Mechanical Engineer in America, 1830–1910: Professional Cultures in Conflict*. Baltimore, 1967. (7)

Carp, E. Wayne. *To Starve the Army at Pleasure: Continental Army Administration and American Political Culture, 1775–1783*. Chapel Hill, N.C., 1984. (6)

Carr, Lois Green, Russell R. Menard, and Lorena S. Walsh. *Robert Cole's World: Agriculture and Society in Early Maryland*. Chapel Hill, N.C., 1991. (3)

Carr, Lois Green, Philip D. Morgan, and Jean B. Russo, eds. *Colonial Chesapeake Society*. Chapel Hill, N.C., 1988. (3)

Carroll, Charles F. *The Timber Economy of Puritan New England*. Providence, R.I., 1973. (3)

Carson, Jane. *Colonial Virginia Cookery: Procedures, Equipment, and Ingredients in Colonial Cooking*. 1966; Williamsburg, Va., 1985. (4)

Cassedy, James H. *American Medicine and Statistical Thinking, 1800–1860*. Cambridge, Mass., 1984. (11)

——. *Demography in Early America: Beginnings of the Statistical Mind, 1600–1800*. Cambridge, Mass., 1969. (11)

Chamberlain, Von Del. *When Stars Came Down to Earth: Cosmology of the Skidi Pawnee Indians of North America*. Los Altos, Calif., 1982. (3)

Chandler, Alfred D., Jr. *The Visible Hand: The Managerial Revolution in American Business*. Cambridge, Mass., 1977. (15)

Chaput, Donald. *The Cliff: America's First Great Copper Mine*. Kalamazoo, Mich., 1971. (5)

Christman, Margaret C. S. *Adventurous Pursuits: Americans and the China Trade, 1784–1844*. Washington, D.C., 1984. (7)

Clark, John G. *The Grain Trade in the Old Northwest*. Urbana, Ill., 1966. (3)

Clark, Mary Stetson. *The Old Middlesex Canal*. Melrose, Mass., 1974. (7)

Clark, William Bell, ed. *Naval Documents of the American Revolution*. Washington, D.C., 1964–. (6)

Clemens, Paul G. E. *The Atlantic Economy and Colonial Maryland's Eastern Shore: From Tobacco to Grain*. Ithaca, N.Y., 1980. (3)

Clinton, Catherine. *The Plantation Mistress: Woman's World in the Old South.* New York, 1982. (4)

Clowse, Converse D. *Economic Beginnings in Colonial South Carolina, 1670–1730.* Columbia, S.C., 1971. (3)

Cochran, Thomas C. *Business in American Life: A History.* New York, 1972. (15)

———. *Frontiers of Change: Early Industrialism in America.* New York, 1981. (2,15)

Coclanis, Peter A. *The Shadow of a Dream: Economic Life and Death in the South Carolina Low Country, 1670–1920.* New York, 1989. (3)

Cohen, Patricia Cline. *A Calculating People: The Spread of Numeracy in Early America.* Chicago, 1982. (11)

Colton, Joel, and Stuart Bruchey, eds. *Technology, the Economy, and Society: The American Experience.* New York, 1987. (15)

Combs, Diana Williams. *Early Gravestone Art in Georgia and South Carolina.* Athens, Ga., 1986. (10)

Condit, Carl W. *American Building: Materials and Techniques from the First Colonial Settlements to the Present.* Chicago, 1968. (7)

Cooke, Jacob E. *Tench Coxe and the Early Republic.* Chapel Hill, N.C., 1978. (10)

Cooper, Carolyn C. *Shaping Invention: Thomas Blanchard's Machinery and Patent Management in Nineteenth-Century America.* New York, 1991. (14)

———, ed. *Patents and Invention. Technology and Culture,* XXXII, no. 4 (October 1991). (14)

Cooper, Grace Rogers. *The Copp Family Textiles.* Washington, D.C., 1971. (10)

———. *The Invention of the Sewing Machine.* Smithsonian Institution Bulletin 254. Washington, D.C., 1968. Rev. and exp. as *The Sewing Machine: Its Invention and Development* (Washington, D.C., 1976). (14)

Cott, Nancy F. *The Bonds of Womanhood: "Woman's Sphere" in New England, 1780–1835.* New Haven, Conn., 1977. (4)

Coughtry, Jay. *The Notorious Triangle: Rhode Island and the African Slave Trade, 1700–1807.* Philadelphia, 1981. (7)

Cowan, Ruth Schwartz. *More Work for Mother: The Ironies of Household Technology, from the Open Hearth to the Microwave.* New York, 1983. (4)

Cowdrey, Albert E. *This Land, This South: An Environmental History.* Lexington, Ky., 1983. (8)

Cox, Thomas R., Robert S. Maxwell, Phillip Drennon Thomas, and Joseph J. Malone. *This Well-Wooded Land: Americans and Their Forests from Colonial Times to the Present.* Lincoln, Nebr., 1985. (3)

Coxe, Tench. *Papers of Tench Coxe in the Coxe Family Papers at the Historical Society of Pennsylvania* (microform); and Lucy Fisher West, *Guide to the Microfilm of the Papers of Tench Coxe in the Coxe Family Papers at the Historical Society of Pennsylvania.* Philadelphia, 1977. (10)

Crackel, Theodore J. *Mr. Jefferson's Army: Political and Social Reform of the Military Establishment, 1801–1809.* New York, 1987. (6)

Creighton, Margaret. *Dogwatch and Liberty Days: Seafaring Life in the Nineteenth Century.* Salem, Mass., 1983. (7)

Cremin, Lawrence A. *American Education: The Colonial Experience, 1607–1783.* New York, 1970. (11)

Cressy, David. *Coming Over: Migration and Communication between England and New England in the Seventeenth Century.* New York, 1987. (12)

Cromwell, Giles. *The Virginia Manufactory of Arms.* Charlottesville, Va., 1975. (6)

Cronon, William. *Changes in the Land: Indians, Colonists, and the Ecology of New England.* New York, 1983. (8)

———. *Nature's Metropolis: Chicago and the Great West.* New York, 1991. (8)

Cummings, Abbott Lowell. *The Framed Houses of Massachusetts Bay, 1625–1725.* Cambridge, Mass., 1979. (7)

Curtin, Philip D. *The Atlantic Slave Trade: A Census.* Madison, Wis., 1969. (7)

Cutcliffe, Stephen H., Judith A. Mistichelli, and Christine M. Roysdon. *Technology and Values in American Civilization: A Guide to Information Sources.* Detroit, Mich., 1980. (1)

Daly, George Anne, and John J. Robrecht. *An Illustrated Handbook of Fire Apparatus, with Emphasis on Nineteenth Century American Pieces.* Philadelphia, 1972. (7)

Danhof, Clarence H. *Change in Agriculture: The Northern United States, 1820–1870.* Cambridge, Mass., 1969. (3)

Dann, John C. *The Revolution Remembered: Eyewitness Accounts of the War for Independence.* Chicago, 1980. (6)

David, Paul A. *Technical Choice, Innovation, and Economic Growth: Essays on American and British Experience in the Nineteenth Century.* London, 1975. (15)

Davidson, Cathy N. *Revolution and the Word: The Rise of the Novel in America.* New York, 1986. (12)

Davis, Lance E., *et al. American Economic Growth: An Economist's History of the United States.* New York, 1972. (15)

Dawley, Alan. *Class and Community: The Industrial Revolution in Lynn.* Cambridge, Mass., 1976. (10)

Decker, Robert Owen. *Whaling Industry of New London.* York, Pa., 1973. (7)

Demos, John. *A Little Commonwealth: Family Life in Plymouth Colony.* New York, 1970. (4)

Dibner, Bern. *Benjamin Franklin, Electrician.* Norwalk, Conn., 1976. (11)

Ditz, Toby L. *Property and Kinship: Inheritance in Early Connecticut, 1750–1820.* Princeton, N.J., 1986. (4)

Doerflinger, Thomas M. *A Vigorous Spirit of Enterprise: Merchants and Economic Development in Revolutionary Philadelphia.* Chapel Hill, N.C., 1986. (15)

Dolmetsch, Christopher L. *The German Press of the Shenandoah Valley.* Columbia, S.C., 1984. (12)

Donnelly, Marian Card. *The New England Meeting Houses of the Seventeenth Century.* Middletown, Conn., 1968. (7)

Drinker, Elizabeth. *The Diary of Elizabeth Drinker.* Ed. Elaine Forman Crane. Boston, 1991. (4)

Dublin, Thomas. *Women at Work: The Transformation of Work and Community in Lowell, Massachusetts, 1826–1860.* New York, 1979. (10)

——, ed. *Farm to Factory: Women's Letters, 1830–1860*. New York, 1981. (10)

Dudden, Faye E. *Serving Women: Household Service in Nineteenth-Century America*. Middletown, Conn., 1983. (4)

Dudley, William S., ed. *The Naval War of 1812: A Documentary History*. Washington, D.C., 1985–. (6)

Duffy, John. *A History of Public Health in New York City, 1625–1866*. New York, 1968. (7)

Dunwell, Steve. *The Run of the Mill: A Pictorial Narrative of the Expansion, Dominion, Decline, and Enduring Impact of the New England Textile Industry*. Boston, 1978. (10)

Dwight, Timothy. *Travels in New England and New York (1821–1822)*. 4 vols. Ed. Barbara Miller Solomon. Cambridge, Mass., 1969. (2)

Earle, Carville V. *The Evolution of a Tidewater Settlement System: All Hallow's Parish, Maryland, 1650–1783*. University of Chicago, Department of Geography, Research Paper no. 170. Chicago, 1975. (3)

Ekirch, A. Roger. *Bound for America: The Transportation of British Convicts to the Colonies, 1718–1775*. Oxford, 1987. (3)

Erikson, Charlotte. *Invisible Immigrants: The Adaptation of English and Scottish Immigrants in Nineteenth-Century America*. Coral Gables, Fla., 1972. (13)

Erlich, Mark. *With Our Hands: The Story of Carpenters in Massachusetts*. Philadelphia, 1986. (10)

Fabel, Robin F. A. *The Economy of British West Florida, 1763–1783*. Tuscaloosa, Ala., 1988. (3)

Fairbanks, Jonathan, and Robert F. Trent. *New England Begins: The Seventeenth Century*. 3 vols. Boston, 1982. (2)

Faler, Paul G. *Mechanics and Manufacturers in the Early Industrial Revolution: Lynn, Massachusetts, 1780–1860*. Albany, N.Y., 1981. (10)

Faragher, John Mack. *Sugar Creek: Life on the Illinois Prairie*. New Haven, Conn., 1986. (8)

Farber, Bernard. *Guardians of Virtue: Salem Families in 1800*. New York, 1972. (11)

Ferguson, Eugene S. *Bibliography of the History of Technology*. Cambridge, Mass., 1968. (1)

——. *Oliver Evans, Inventive Genius of the American Industrial Revolution*. Greenville, Del., 1980. (14)

Fisher, Marvin. *Workshops in the Wilderness: The European Response to American Industrialization, 1830–1860*. New York, 1967. (13)

Fitch, John. *The Autobiography of John Fitch*. Ed. Frank D. Prager. Memoirs of the American Philosophical Society, CXIII. Philadelphia, 1976. (7)

Fitchen, John. *The New World Dutch Barn: A Study of Its Characteristics, Its Structural System, and Its Probable Erectional Procedures*. Syracuse, N.Y., 1968. (7)

Fleischmann, C. L. *Trade, Manufacture, and Commerce in the United States of America*. Trans. E. Vilim. Jerusalem, 1970. (2)

Folsom, Michael Brewster, and Steven D. Lubar, eds. *The Philosophy of*

Manufactures: Early Debates over Industrialization in the United States. Cambridge, Mass., 1982. (10)

Foner, Eric. *Tom Paine and Revolutionary America.* New York, 1976. (10)

Foner, Philip S., ed. *The Factory Girls.* Urbana, Ill., 1977. (10)

Foner, Philip S., and Ronald L. Lewis, eds. *The Black Worker: A Documentary History from Colonial Times to the Present,* I, *The Black Worker to 1869.* Philadelphia, 1979. (2)

Foreman, Henry Chandlee. *Early Nantucket and Its Whale Houses.* New York, 1966. (7)

————. *Old Buildings, Gardens, and Furniture in Tidewater Maryland.* Cambridge, Md., 1967. (7)

Fowler, William M., Jr. *Rebels under Sail: The American Navy during the Revolution.* New York, 1976. (6)

Fox-Genovese, Elizabeth. *Within the Plantation Household: Black and White Women of the Old South.* Chapel Hill, N.C., 1988. (4)

Franklin, Benjamin. *The Papers of Benjamin Franklin.* Ed. Leonard W. Labaree *et al.* New Haven, Conn., 1959–. (2)

Frazier, Arthur H. *Joseph Saxton and His Contributions to the Medal Ruling and Photographic Arts.* Washington, D.C., 1975. (10)

French, Hannah D. *Bookbinding in Early America: Seven Essays on Masters and Methods.* Worcester, Mass., 1986. (10)

Fries, Sylvia Doughty. *The Urban Idea in Colonial America.* Philadelphia, 1977. (8)

Fuller, Wayne E. *The American Mail: Enlarger of the Common Life.* Chicago, 1972. (12)

Galenson, David W. *White Servitude in Colonial America: An Economic Analysis.* New York, 1981. (3, 10, 15)

Gardner, Paul. *American Glass.* Washington, D.C., 1977. (10)

Garitee, Jerome R. *The Republic's Private Navy: The American Privateering Business as Practiced by Baltimore during the War of 1812.* Middletown, Conn., 1977. (6)

Garrison, J. Ritchie. *Landscape and Material Life in Franklin County, Massachusetts, 1770–1860.* Knoxville, Tenn., 1991. (8)

Garvan, Beatrice B., and Charles F. Hummel. *The Pennsylvania Germans: A Celebration of Their Arts, 1683–1850.* Philadelphia, 1982. (10)

Genovese, Eugene D. *Roll, Jordan, Roll: The World the Slaves Made.* New York, 1975. (4)

Gies, Joseph, and Frances Gies. *The Ingenious Yankees: The Men, Ideas, and Machines That Transformed a Nation, 1776–1876.* New York, 1976. (14)

Gilbert, David. *Where Industry Failed: Water-Powered Mills at Harper's Ferry, Virginia.* Charleston, W.Va., 1984. (9)

Gill, Harold B., Jr. *The Gunsmith in Colonial Virginia.* Williamsburg, Va., 1974. (10)

Gilman, Carolyn. *Where Two Worlds Meet: The Great Lakes Fur Trade.* St. Paul, Minn., 1982. (3)

Gilmore, William J. *Reading Becomes a Necessity of Life: Material and Cultural Life in Rural New England, 1780–1835.* Knoxville, Tenn., 1989. (12)

Glassie, Henry. *Folk Housing in Middle Virginia: A Structural Analysis of Historic Artifacts.* Knoxville, Tenn., 1975. (7)

———. *Pattern in the Material Folk Culture of the Eastern United States.* Philadelphia, 1968. (4)

Glick, Thomas F. *The Old World Background of the Irrigation System of San Antonio, Texas.* El Paso, Tex., 1972. (7)

Goldenberg, Joseph A. *Shipbuilding in Colonial America.* Charlottesville, Va., 1976. (7)

Goldin, Claudia Dale. *Urban Slavery in the American South, 1820–1860: A Quantitative History.* Chicago, 1976. (10)

Goldstein, Jonathan. *Philadelphia and the China Trade, 1682–1846: Commercial, Cultural, and Attitudinal Effects.* University Park, Pa., 1978. (7)

Gollin, Gillian Lindt. *Moravians in Two Worlds: A Study of Changing Communities.* New York, 1967. (10)

Goodwin, Gary C. *Cherokees in Transition: A Study of Changing Culture and Environment Prior to 1775.* University of Chicago, Department of Geography, Research Paper no. 181. Chicago, 1977. (8)

Gordon, Beverly. *Shaker Textile Arts.* Hanover, N.H., 1980. (4)

Gray, Ralph D. *The National Waterway: A History of the Chesapeake and Delaware Canal, 1769–1965.* Urbana, Ill., 1967. (7)

Graymont, Barbara. *The Iroquois in the American Revolution.* Syracuse, N.Y., 1972. (6)

Greenberg, David B., ed. *Land That Our Fathers Plowed: The Settlement of Our Country as Told by the Pioneers Themselves and Their Contemporaries.* Norman, Okla., 1969. (3)

Greene, John C. *American Science in the Age of Jefferson.* Ames, Iowa, 1984. (11)

Greene, Nathanael. *The Papers of General Nathanael Greene.* Ed. Richard K. Showman *et al.* Chapel Hill, N.C., 1976–. (6)

Grindle, Roger L. *Quarry and Kiln: The Story of Maine's Lime Industry.* Rockland, Maine, 1971. (5)

Hacker, Louis M. *The Course of American Economic Growth and Development.* New York, 1970. (15)

Haites, Erik F., James Mak, and Gary M. Walton. *Western River Transportation: The Era of Early Internal Development, 1810–1860.* Johns Hopkins University Studies in Historical and Political Science, 93d Ser., no. 2. Baltimore, 1975. (7)

Hamilton, Alexander. *The Papers of Alexander Hamilton.* 27 vols. Ed. Harold Syrett *et al.* New York, 1961–1987. (2)

Hampe, Edward C., Jr., and Merle Wittenberg. *The Lifeline of America: Development of the Food Industry.* New York, 1964. (3)

Hargrove, John. *The Weaver's Draft Book and Clothier's Assistant.* 1792; rpt., with introd. by Rita Adrosko. Charlottesville, Va., 1979. (10)

Harrison, J. F. C. *Quest for the New Moral World: Robert Owen and the Owenites in Britain and America.* New York, 1969. (13)

Harvey, Katherine A., ed. *The Lonaconing Journals: The Founding of a Coal and Iron Community, 1837–1840.* Philadelphia, 1977. (5)

Hayden, Dolores. *Seven American Utopias: The Architecture of Communitarian Socialism, 1790–1975.* Cambridge, Mass., 1976. (7)

Hazen, M. H., and R. M. Hazen. *Wealth Inexhaustible: A History of America's Mineral Industries to 1850.* New York, 1985. (5)

Hedges, James B. *The Browns of Providence Plantations: The Nineteenth Century.* Providence, R.I., 1968. (15)

Henretta, James. *The Evolution of American Society, 1700–1815: An Interdisciplinary Analysis.* Lexington, Mass., 1973. (15)

Henry, Joseph. *The Papers of Joseph Henry.* Ed. Nathan Reingold *et al.* Washington, D.C., 1972–. (11)

Herndon, G. Melvin. *William Tatham and the Culture of Tobacco: Including a Facsimile Reprint of an Historical and Practical Essay on the Culture and Commerce of Tobacco.* Coral Gables, Fla., 1969. (3)

Heusser, Albert H. *George Washington's Mapmaker: A Biography of Robert Erskine.* Ed. Hubert G. Schmidt. 1928; New Brunswick, N.J., 1966. (7)

Heyrman, Christine Leigh. *Commerce and Culture: The Maritime Communities of Colonial Massachusetts, 1690–1750.* New York, 1984. (7)

Hindle, Brooke. *Emulation and Invention.* New York, 1981. (7, 14)

———, ed. *America's Wooden Age: Aspects of Its Early Technology.* Tarrytown, N.Y., 1975. (10)

———, ed. *Material Culture of the Wooden Age.* Tarrytown, N.Y., 1981. (2)

Hindle, Brooke, and Steven Lubar. *Engines of Change: The American Industrial Revolution, 1790–1860.* Washington, D.C., 1986. (2)

Hines, Mary Anne, Gordon Marshall, and William Woys Weaver. *The Larder Invaded: Reflections on Three Centuries of Philadelphia Food and Drink.* Philadelphia, 1987. (4)

Hirsch, Susan E. *Roots of the American Working Class: The Industrialization of Crafts in Newark, 1800–1860.* Philadelphia, 1978. (10)

Hitz, Elizabeth. *A Technical and Business Revolution: American Woolens to 1832.* New York, 1986. (10)

Hixson, Richard F. *Isaac Collins: A Quaker Printer in Eighteenth Century America.* New Brunswick, N.J., 1968. (10)

Hodges, Graham Russell. *New York City Cartmen, 1667–1850.* New York, 1986. (7)

Hoffman, John N. *Anthracite in the Lehigh Regions of Pennsylvania, 1820–45.* Washington, D.C., 1968. (5)

Hoffman, Ronald, and Peter J. Albert, eds. *Arms and Independence: The Military Character of the American Revolution.* Charlottesville, Va., 1984. (6)

Hoke, Donald R. *Ingenious Yankees: The Rise of the American System of Manufactures in the Private Sector.* New York, 1990. (10)

Holmes, Oliver W., and Peter T. Rohrbach. *Stagecoach East: Stagecoach Days in the East from the Colonial Period to the Civil War.* Washington, D.C., 1983. (7)

Hood, Graham. *Bonnin and Morris of Philadelphia: The First American Porcelain Factory, 1770–1772.* Chapel Hill, N.C., 1972. (10)

Hounshell, David A. *From the American System to Mass Production, 1800–1932:*

The Development of Manufacturing Technology in the United States. Baltimore, 1984. (10)

Hubka, Thomas C. *Big House, Little House, Backhouse, Barn: The Connected Farm Buildings of New England.* Hanover, N.H., 1984. (7)

Hughes, J. Donald. *American Indian Ecology.* El Paso, Tex., 1983. (8)

Hughes, Sarah S. *Surveyors and Statesmen: Land Measuring in Colonial Virginia.* Richmond, Va., 1979. (7)

Hume, Ivor Noël. *A Guide to Artifacts of Colonial America.* New York, 1970. (2)

Hunter, Louis C. *A History of Industrial Power in the United States, 1780–1930,* I, *Waterpower in the Century of the Steam Engine.* Charlottesville, Va., 1979. (7, 9)

———. *A History of Industrial Power in the United States, 1780–1930,* II, *Steam Power.* Charlottesville, Va., 1985. (9, 10)

Hunter, Robert, and Edwin Dooley, Jr. *Claudius Crozet: French Engineer in America, 1790–1864.* Charlottesville, Va., 1989. (7)

Hurt, R. Douglas. *American Farm Tools: From Hand Power to Steam Power.* Manhattan, Kans., 1982. (3)

———. *Indian Agriculture in America: Prehistory to the Present.* Lawrence, Kans., 1987. (3)

Hutcheon, Wallace. *Robert Fulton: Pioneer of Undersea Warfare.* Annapolis, Md., 1981. (6)

Hutslar, Donald. *The Architecture of Migration: Log Construction in the Ohio Country, 1750–1850.* Athens, Ohio, 1986. (7)

Indiana Historical Society, ed. *Transportation and the Early Nation.* Indianapolis, Ind., 1982. (7)

Innes, Stephen. *Labor in a New Land: Economy and Society in Seventeenth-Century Springfield.* Princeton, N.J., 1983. (15)

———, ed. *Work and Labor in Early America.* Chapel Hill, N.C., 1988. (2)

Jackson, A. T. *Mills of Yesteryear.* El Paso, Tex., 1971. (9)

Janssen, Barbara Suit, ed. *Icons of Invention: American Patent Models.* Washington, D.C., 1990. (14)

———, ed. *Technology in Miniature: American Textile Patent Models.* Washington, D.C., 1988. (14)

Jefferson, Thomas. *The Papers of Thomas Jefferson.* Ed. Julian P. Boyd *et al.* Princeton, N.J., 1950–. (2)

———. *Thomas Jefferson's Farm Book, with Commentary and Relevant Extracts from Other Writings.* Ed. Edwin Morris Betts. 1953. Rpt. Charlottesville, Va., 1976. (3)

Jenkins, Reese V. *Images and Enterprise: Technology and the American Photographic Industry, 1839–1925.* Baltimore, 1975. (10)

Jensen, Joan M. *Loosening the Bonds: Mid-Atlantic Farm Women, 1750–1850.* New Haven, Conn., 1986. (4)

Jeremy, David J. *Technology and Power in the Early American Cotton Industry: James Montgomery, the Second Edition of his "Cotton Manufacture" (1840), and the "Justitia" Controversy about Relative Power Costs.* American Philosophical Society, Memoirs, CLXXXIX. Philadelphia, 1990. (9, 10)

———. *Transatlantic Industrial Revolution: The Diffusion of Textile Technologies between Britain and America, 1790–1830s*. Cambridge, Mass., 1981. (10,13)

———, ed. *Henry Wansey and His American Journal, 1794*. American Philosophical Society, Memoirs, LXXXII. Philadelphia, 1970. (10)

———, ed. *International Technology Transfer: Europe, Japan, and the USA*. Aldershot, Hants., 1991. (13)

Jervis, John B. *The Reminiscences of John B. Jervis, Engineer of the Old Croton*. Ed. with introd. by Neal Fitzsimons. Syracuse, N.Y., 1971. (7)

Johnstone, Paul. *Steam and the Sea*. Salem, Mass., 1983. (7)

Jonas, Manfred, and Robert Wells, eds. *New Opportunities in a New Nation: The Development of New York after the Revolution*. Schenectady, N.Y., 1982. (8)

Jones, Alice Hanson. *American Colonial Wealth: Documents and Methods*. 3 vols. New York, 1977. (4)

Jones, George Fenwick, *et al.*, eds. *Detailed Reports on the Salzburger Emigrants Who Settled in America . . . Edited by Samuel Urlsperger*. Trans. Herman J. Lacher *et al*. Athens, Ga., 1968–. (2)

Jones, Robert Leslie. *History of Agriculture in Ohio to 1880*. Kent, Ohio, 1983. (3)

Jordan, Terry G. *American Log Buildings: An Old World Heritage*. Chapel Hill, N.C., 1985. (13)

Jordan, Terry G., and Matti Kaups. *The American Backwoods Frontier: An Ethnic and Ecological Interpretation*. Baltimore, 1989. (13)

Kaser, David, ed. *Books in America's Past: Essays Honoring Rudolph H. Gjelsness*. Charlottesville, Va., 1966. (12)

Kasson, John F. *Civilizing the Machine: Technology and Republican Values in America, 1776–1900*. New York, 1976. (2)

Kauffman, Henry J. *American Axes: A Survey of Their Development and Their Makers*. Brattleboro, Vt., 1972. (10)

———. *American Copper and Brass*. New York, 1979. (10)

———. *The American Pewterer: His Techniques and His Products*. Camden, N.J., 1970. (10)

———. *The Colonial Silversmith: His Techniques and His Products*. New York, 1974. (10)

Kebabian, Paul, and William Lipke, eds. *Tools and Technologies: America's Wooden Age*. Burlington, Vt., 1979. (10)

Kebabian, Paul B., and Dudley Whitney. *American Woodworking Tools*. Boston, 1979. (10)

Kelsey, Darwin P., ed. *Farming in the New Nation: Interpreting American Agriculture, 1790–1840*. Washington, D.C., 1972. (3)

Kelso, William M. *Kingsmill Plantations, 1619–1800: Archaeology of Country Life in Colonial Virginia*. Orlando, Fla., 1984. (4)

Kennedy, Roger G. *Architecture, Men, Women, and Money in America, 1600–1860*. New York, 1985. (7)

Kerber, Linda K. *Women of the Republic: Intellect and Ideology in Revolutionary America*. Chapel Hill, N.C., 1980. (11)

Kidwell, Claudia B. *Cutting a Fashionable Fit: Dressmakers' Drafting Systems in the*

United States. Smithsonian Studies in the History of Technology no. 42. Washington, D.C., 1979. (10)

Kidwell, Claudia, and Margaret Christman. *Suiting Everyone: The Democratization of Clothing in America.* Washington, D.C., 1979. (4, 10)

Kim, Sung Bok. *Landlord and Tenant in Colonial New York: Manorial Society, 1664–1775.* Chapel Hill, N.C., 1978. (8)

Kirker, Harold. *The Architecture of Charles Bulfinch.* Cambridge, Mass., 1969. (7)

Kirker, Harold, and James Kirker. *Bulfinch's Boston, 1787–1817.* New York, 1964. (7)

Klinefelter, Walter. *Lewis Evans and His Maps.* American Philosophical Society, *Transactions,* N.S., LXI, pt. 7. Philadelphia, 1971. (7)

Kohn, Richard H. *Eagle and Sword: The Federalists and the Creation of the Military Establishment in America, 1783–1802.* New York, 1975. (6)

Kranzberg, Melvin, and Joseph Gies. *By the Sweat of Thy Brow: Work in the Western World.* New York, 1975. (1)

Kranzberg, Melvin, and Carroll W. Pursell, Jr. *Technology in Western Civilization.* 2 vols. New York, 1967. (1)

Krech, Shepard, III, ed. *Indians, Animals, and the Fur Trade: A Critique of "Keepers of the Game."* Athens, Ga., 1981. (3)

———, ed. *The Subarctic Fur Trade: Native Social and Economic Adaptations.* Vancouver, 1984. (3)

Kulik, Gary, *et al.,* eds. *The New England Mill Village, 1790–1860.* Cambridge, Mass., 1982. (10)

Kulikoff, Allan. *Tobacco and Slaves: the Development of Southern Cultures in the Chesapeake, 1680–1800.* Chapel Hill, N.C., 1986. (3)

Labaree, Benjamin. *Colonial Massachusetts: A History.* Millwood, N.Y., 1979. (3)

Laing, Alexander. *American Ships.* New York, 1971. (7)

Lander, Ernest McPherson, Jr. *The Textile Industry in Antebellum South Carolina.* Baton Rouge, La., 1969. (10)

Landing, James E. *American Essence: A History of the Peppermint and Spearmint Industry in the United States.* Kalamazoo, Mich., 1969. (3)

Lankton, Larry. *Cradle to Grave: Life, Work, and Death at the Lake Superior Copper Mines.* New York, 1991. (5)

Larkin, Daniel. *John B. Jervis, an American Engineering Pioneer.* Ames, Iowa, 1990. (7)

Lasansky, Jeannette. *Central Pennsylvania Redware Pottery, 1780–1904.* Lewisburg, Pa., 1979. (10)

———. *Pieced by Mother: Over One Hundred Years of Quiltmaking Traditions.* Lewisburg, Pa., 1987. (4)

———. *To Cut, Piece, and Solder: The Work of the Pennsylvania Rural Tinsmith, 1778–1908.* University Park, Pa., 1982. (10)

———. *To Draw, Upset, and Weld: The Work of the Rural Pennsylvania Blacksmith, 1778–1908.* Lewisburg, Pa., 1980. (10)

———, ed. *In the Heart of Pennsylvania: Symposium Papers.* Lewisburg, Pa., 1986. (4)

Latrobe, Benjamin Henry. *The Architectural Drawings of Benjamin Henry Latrobe.* 2 vols. Ed. Jeffrey A. Cohen and Charles E. Brownell. New Haven, Conn., 1994. (7)

———. *The Correspondence and Miscellaneous Papers of Benjamin Henry Latrobe.* 3 vols. Ed. John C. Van Horne *et al.* New Haven, Conn., 1984–1988. (7)

———. *The Engineering Drawings of Benjamin Henry Latrobe.* Ed. Darwin H. Stapleton. New Haven, Conn., 1980. (7)

———. *The Journals of Benjamin Henry Latrobe, 1799–1820: From Philadelphia to New Orleans.* Ed. Edward C. Carter II *et al.* New Haven, Conn., 1980. (7)

———. *Latrobe's View of America, 1795–1820: Selections from the Sketchbooks and Drawings of Benjamin Henry Latrobe.* Ed. Edward C. Carter II *et al.* New Haven, Conn., 1985. (7)

———. *Microfiche Edition to the Papers of Benjamin Henry Latrobe.* Ed. Thomas E. Jeffrey. Clifton, N.J., 1976. (7)

———. *The Virginia Journals of Benjamin Henry Latrobe.* 2 vols. Ed. Edward C. Carter II *et al.* New Haven, Conn., 1977. (7)

Laurens, Henry. *The Papers of Henry Laurens.* Ed. Philip Hamer *et al.* Columbia, S.C., 1968–. (2)

Laurie, Bruce. *Working People of Philadelphia, 1800–1850.* Philadelphia, 1980. (10)

Layton, Edwin T., Jr., ed. *Technology and Social Change in America.* New York, 1973. (2)

Leach, Douglas Edward. *The Northern Colonial Frontier, 1607–1763.* New York, 1966. (4)

Leavitt, Thomas W., ed. *The Hollingworth Letters: Technical Change in the Textile Industry, 1826–1837.* Cambridge, Mass., 1969. (10)

Lebsock, Suzanne. *The Free Women of Petersburg: Status and Culture in a Southern Town, 1784–1860.* New York, 1984. (4)

Lee, Jean Gordon. *Philadelphians and the China Trade, 1784–1844.* With essay by Philip Chadwick Foster Smith. Philadelphia, 1984. (7)

Lemon, James T. *The Best Poor Man's Country: A Geographical Study of Early Southeastern Pennsylvania.* Baltimore, 1972. (8)

Levitt, James H. *For Want of Trade: Shipping and the New Jersey Ports, 1680–1783.* Collections of the New Jersey Historical Society, XVII. Newark, 1981. (7)

Levy, Leonard W. *The Emergence of a Free Press.* New York, 1985. (12)

Lewis, Gene D. *Charles Ellet, Jr.: The Engineer as Individualist, 1810–1862.* Urbana, Ill., 1968. (7)

Lewis, Meriwether, and William Clark. *Atlas of the Lewis and Clark Expedition.* Ed. Gary E. Moulton. Vol. I of *The Journals of the Lewis and Clark Expedition.* Lincoln, Nebr., 1983. (7)

———. *The Journals of the Lewis and Clark Expedition.* Ed. Gary E. Moulton. Lincoln, Nebr., 1983–. (2)

Lewis, Ronald L. *Black Coal Miners in America: Race, Class, and Community Conflict, 1780–1980.* Lexington, Ky., 1987. (5)

———. *Coal, Iron, and Slaves: Industrial Slavery in Maryland and Virginia, 1715–1865.* Westport, Conn., 1979. (5)

Lewis, W. David. *Iron and Steel in America*. Greenville, Del., 1976. (5)

Lewis, W. David, and Walter Hugins. *Hopewell Furnace: A Guide to Hopewell Village National Historic Site, Pennsylvania*. National Park Handbook 124. Washington, D.C., 1983. (5)

Licht, Walter. *Working for the Railroad: The Organization of Work in the Nineteenth Century*. Princeton, N.J., 1983. (7)

Lindstrom, Diane. *Economic Development in the Philadelphia Region, 1810–1850*. New York, 1978. (15)

Litchfield, Carter, Hans-Joachim Finke, Stephen Young, and Karen Zerbe Huetter. *The Bethlehem Oil Mill 1745–1934: . . . German Technology in Early Pennsylvania*. Kemblesville, Pa., 1984. (10)

Littlefield, Daniel C. *Rice and Slaves: Ethnicity and the Slave Trade in Colonial South Carolina*. Baton Rouge, La., 1981. (3)

Lockridge, Kenneth A. *Literacy in Colonial New England: An Enquiry into the Social Context of Literacy in the Early Modern West*. New York, 1974. (11)

Lord, Philip L., Jr., with Martha A. Costello, illustrator. *Mills on the Tsatsawassa: A Guide for Local Historians*. Albany, N.Y., 1983. (9)

Loveday, Amos J., Jr. *The Rise and Decline of the American Cut Nail Industry: A Study of the Interrelationships of Technology, Business Organization, and Management Techniques*. Westport, Conn., 1983. (10)

Lozier, John. *Taunton and Mason: Cotton Machinery and Locomotive Manufacture in Taunton, Massachusetts, 1811–1861*. New York, 1986. (10)

Ludwig, Allan I. *Graven Images: New England Stonecarving and Its Symbols, 1650–1815*. Middletown, Conn., 1966. (10)

Lytle, Thomas G. *Harpoons and Other Whalecraft*. New Bedford, Mass., 1984. (7)

Macaulay, David. *Mill*. Boston, 1983. (9)

McCall, Edith. *Conquering the Rivers: Henry Miller Shreve and the Navigation of America's Inland Waterways*. Baton Rouge, La., 1984. (7)

Maccubbin, Robert P., and Peter Martin, eds., *British and American Gardens in the Eighteenth Century: Eighteen Illustrated Essays on Garden History*. Williamsburg, Va., 1984. (7)

McCullough, Robert, and Walter Leuba. *The Pennsylvania Main Line Canal*. York, Pa., 1973. (7)

McCusker, John J., and Russell R. Menard. *The Economy of British America, 1607–1789*. Rev. ed. Chapel Hill, N.C., 1991. (2, 15)

McGaw, Judith A. *Most Wonderful Machine: Mechanization and Social Change in Berkshire Paper Making, 1801–1885*. Princeton, N.J., 1987. (10)

[McLane, Louis]. *Documents Relative to the Manufactures in the United States of America*. U.S. Congress, House of Representatives, Executive Document no. 308, collected and transmitted by the Secretary of the Treasury. 3 vols. Washington, D.C., 1833. Rpt., New York, 1969. (10)

McLaughlin, Jack. *Jefferson and Monticello: The Biography of a Builder*. New York, 1988. (7)

McManis, Douglas R. *Colonial New England: A Historical Geography*. New York, 1975. (8)

McManus, Edgar J. *Black Bondage in the North*. Syracuse, N.Y., 1973. (10)

———. *A History of Negro Slavery in New York*. Syracuse, N.Y., 1966. (10)

Main, Gloria L. *Tobacco Colony: Life in Early Maryland, 1650–1720*. Princeton, N.J., 1982. (3)

Main, Jackson Turner. *Society and Economy in Colonial Connecticut*. Princeton, N.J., 1985. (15)

Malone, Patrick M. *Canals and Industry: Engineering in Lowell, 1821–1880*. Lowell, Mass., 1983. (9)

———. *The Skulking Way of War: Technology and Tactics among the Indians of New England*. Lanham, Md., 1991. (6, 13)

Martin, Calvin. *Keepers of the Game: Indian-Animal Relationships and the Fur Trade*. Berkeley, Calif., 1978. (3)

Marzio, Peter C. *The Art Crusade: An Analysis of American Drawing Manuals, 1820–1860*. Washington, D.C., 1976. (11)

Mayr, Otto, and Robert C. Post, eds. *Yankee Enterprise: The Rise of the American System of Manufactures*. Washington, D.C., 1981. (10)

Meinig, D. W. *The Shaping of America: A Geographical Perspective on Five Hundred Years of History*, I, *Atlantic America, 1492–1800*. New Haven, Conn., 1986. (8)

Miller, J. Jefferson, II, and Lyle M. Stone. *Eighteenth-Century Ceramics from Fort Michilimackinac: A Study in Historical Archaeology*. Washington, D.C., 1970. (10)

Miller, Nathan. *Sea of Glory: The Continental Navy Fights for Independence, 1775–1783*. New York, 1974. (6)

Minchinton, Walter, *et al.*, eds. *Virginia Slave-Trade Statistics, 1698–1775*. Richmond, Va., 1984. (7)

Mitchell, Robert D. *Commercialism and Frontier: Perspectives on the Early Shenandoah Valley*. Charlottesville, Va., 1977. (3)

Mitchill, Samuel L[atham]. *A Discourse on the Character and Services of Thomas Jefferson, More Especially as a Promoter of Natural and Physical Science* (1826). Rpt. Ed. Omer Allan Gianniny, Jr. Charlottesville, Va., 1982. (11)

Molloy, Peter M. *History of Metal Mining and Metallurgy: An Annotated Bibliography*. New York, 1986. (1, 5)

Montgomery, Charles F. *American Furniture: The Federal Period, in the Henry Francis du Pont Winterthur Museum*. New York, 1966. (10)

Montgomery, Florence M. *Textiles in America, 1650–1870*. New York, 1984. (10)

Moore, John Hebron. *Andrew Brown and Cypress Lumbering in the Old Southwest*. Baton Rouge, La., 1967. (3)

———. *The Emergence of the Cotton Kingdom in the Old Southwest: Mississippi, 1770–1860*. Baton Rouge, La., 1988. (3)

Moore, Warren. *Weapons of the American Revolution and Accoutrements*. New York, 1967. (6)

Moran, James. *Printing Presses: History and Development from the Fifteenth Century to Modern Times*. Berkeley, Calif., 1973. (10)

Morison, Elting E. *From Know-How to Nowhere: The Development of American Technology.* New York, 1974. (2)

Morse, John D., ed. *Prints in and of America to 1850.* Winterthur Conference Report, 1970. Charlottesville, Va., 1970. (10)

Mulholland, James A. *A History of Metals in Colonial America.* University, Ala., 1981. (5)

Mullin, Michael. *American Negro Slavery: A Documentary History.* Columbia, S.C., 1976. (3)

Murtagh, William J. *Moravian Architecture and Town Planning: Bethlehem Pennsylvania, and Other Eighteenth-Century Settlements.* Chapel Hill, N.C., 1967. (7)

Myers, Susan. *Handcraft to Industry: Philadelphia Ceramics in the First Half of the Nineteenth Century.* Smithsonian Studies in the History of Technology no. 43. Washington, D.C., 1980. (10)

Nabokov, Peter, and Robert Easton. *Native American Architecture.* New York, 1989. (7)

Nash, Gary B. *Forging Freedom: The Formation of Philadelphia's Black Community, 1720–1840.* Cambridge, Mass., 1988. (10)

Nelson, Lee. *The Colossus of 1812: An American Engineering Superlative.* New York, 1990. (7)

Neumann, George C. *The History of Weapons of the American Revolution.* New York, 1967. (6)

Newmyer, R. Kent. *Supreme Court Justice Joseph Story: Statesman of the Old Republic.* Chapel Hill, N.C., 1985. (14)

North, Douglass. *The Economic Growth of the United States, 1790–1860.* New York, 1966. (15)

Norton, Mary Beth. *Liberty's Daughters: The Revolutionary Experience of American Women, 1750–1800.* Boston, 1980. (4)

Oleson, Alexandra, and Sanborn C. Brown. *The Pursuit of Knowledge in the Early American Republic: American Scientific and Learned Societies from Colonial Times to the Civil War.* Baltimore, 1976. (11)

Olmsted, Frederick Law. *The Papers of Frederick Law Olmsted, I, The Formative Years, 1822 to 1852.* Ed. Charles Capen McLaughlin and Charles E. Beveridge. Baltimore, 1977. (2)

Olton, Charles S. *Artisans for Independence: Philadelphia Mechanics and the American Revolution.* Syracuse, N.Y., 1975. (10)

Palmer, Arlene. *The Wistars and Their Glass, 1739–1777.* Millville, N.J., 1989. (10)

Papenfuse, Edward C., and Joseph M. Coale III, comps. *The Hammond-Harwood House Atlas of Historical Maps of Maryland, 1608–1908.* Baltimore, 1982. (7)

Parks, Robert J. *Democracy's Railroads: Public Enterprise in Jacksonian Michigan.* Port Washington, N.Y., 1972. (7)

Paskoff, Paul F. *Industrial Evolution: Organization, Structure, and Growth of the Pennsylvania Iron Industry, 1750–1860.* Baltimore, 1983. (5)

Peale, Charles Willson. *The Collected Papers of Charles Willson Peale and His Family, 1735–1885.* Microform. Ed. Lillian B. Miller. Millwood, N.Y., 1980.

———. *The Selected Papers of Charles Willson Peale and His Family*. Ed. Lillian B. Miller *et al*. New Haven, Conn., 1983–.

Peckham, John M. *Fighting Fire with Fire*. Newfoundland, N.J., 1973. (7)

Perkins, Edwin J. *The Economy of Colonial America*. 2d ed. New York, 1988. (15)

Perkins, Edwin J., and Gary M. Walton. *A Prosperous People: The Growth of the American Economy*. Englewood Cliffs, N.J., 1985. (15)

Peterson, Charles E., ed. *Building Early America: Contributions toward the History of a Great Industry*. Radnor, Pa., 1976. (7)

———, ed. *The Rules of Work of the Carpenters' Company of the City and County of Philadelphia, 1786*. Princeton, N.J., 1971. (7)

Peterson, Harold L. *The Book of the Continental Soldier. . . .* Harrisburg, Pa., 1968. (6)

Peterson, Merrill D., ed. *Thomas Jefferson: A Reference Biography*. New York, 1986. (3)

Porter, Glenn, and Harold C. Livesay. *Merchants and Manufacturers: Studies in the Changing Structure of Nineteenth-Century Marketing*. Baltimore, 1971. (15)

Post, Robert C. *Physics, Patents, and Politics: A Biography of Charles Grafton Page*. New York, 1976. (14)

———, ed. *American Enterprise: Nineteenth-Century Patent Models*. New York, 1984. (14)

Powell, H. Benjamin. *Philadelphia's First Fuel Crisis: Jacob Cist and the Developing Market for Anthracite Coal*. University Park, Pa., 1978. (5)

Powell, Sumner Chilton. *Puritan Village: The Formation of a New England Town*. Middletown, Conn., 1965. (8)

Pred, Allan R. *Urban Growth and the Circulation of Information: The United States System of Cities, 1790–1840*. Cambridge, Mass., 1973. (12)

Prude, Jonathan. *The Coming of Industrial Order: Town and Factory Life in Rural Massachusetts, 1810–1860*. New York, 1983. (10)

Purdue, Robert E. *Black Laborers and Black Professionals in Early America, 1750–1830*. New York, 1975. (10)

Pursell, Carroll W., Jr. *Early Stationary Steam Engines in America: A Study in the Migration of a Technology*. Washington, D.C., 1969. (9, 13)

———, ed. *Readings in Technology and American Life*. New York, 1969. (2)

Pye, David. *The Nature and Art of Workmanship*. Cambridge, 1968. (10)

Pyne, Stephen J. *Fire in America: A Cultural History of Wildland and Rural Fire*. Princeton, N.J., 1982. (8)

Quimby, Ian M. G. *Apprenticeship in Colonial Philadelphia*. 1963; New York, 1985. (11)

———, ed. *Ceramics in America*. Winterthur Conference Report, 1972. Charlottesville, Va., 1973. (10)

———, ed. *The Craftsman in Early America*. New York, 1984. (10)

Ransom, James M. *Vanishing Ironworks of the Ramapos: The Story of the Forges, Furnaces, and Mines of the New Jersey–New York Border Area*. New Brunswick, N.J., 1966. (5)

Rasmussen, Wayne D., ed. *Agriculture in the United States: A Documentary History.* 4 vols. New York, 1975. (3)

Ray, William, and Marlys Ray. *The Art of Invention: Patent Models and Their Makers.* Princeton, N.J., 1974. (14)

Rediker, Marcus. *Between the Devil and the Deep Blue Sea: Merchant Seamen, Pirates, and the Anglo-American Maritime World, 1700–1750.* New York, 1987. (7)

Reinhart, Theodore R., ed. *The Archaeology of Shirley Plantation.* Charlottesville, Va., 1984. (4)

Reps, John W. *The Making of Urban America: A History of City Planning in the United States.* Princeton, N.J., 1965. (8)

———. *Tidewater Towns: City Planning in Colonial Virginia and Maryland.* Williamsburg, Va., 1972. (8)

Reynolds, Terry. *Stronger than a Hundred Men: A History of the Vertical Waterwheel.* Baltimore, 1983. (9)

Richardson, Edgar P., Brooke Hindle, and Lillian B. Miller. *Charles Willson Peale and His World.* New York, 1983. (11)

Rikoon, J. Sanford. *Threshing in the Midwest, 1820–1940: A Study of Traditional Culture and Technological Change.* Bloomington, Ind., 1988. (3)

Ring, Betty. *"Let Virtue Be a Guide to Thee": Needlework in the Education of Rhode Island Women, 1730–1830.* Providence, R.I., 1983. (4, 11)

Rink, Evald. *Technical Americana: A Checklist of Technical Publications Printed before 1831.* Millwood, N.Y., 1981. (1)

Risch, Erna. *Supplying Washington's Army.* Washington, D.C., 1981. (6)

Ristow, Walter W. *American Maps and Mapmakers: Commercial Cartography in the Nineteenth Century.* Detroit, 1985. (7)

Roberts, Kenneth D. *The Contributions of Joseph Ives to Connecticut Clock Technology, 1810–1862.* Bristol, Conn., 1971. (10)

———. *Eli Terry and the Connecticut Shelf Clock.* Fitzwilliam, N.H., 1972. (10)

———. *Wooden Planes in Nineteenth Century America.* Fitzwilliam, N.H., 1975. (10)

Roberts, Kenneth D., and Jane W. Roberts. *Planemakers and Other Edge Tool Enterprises in New York State in the Nineteenth Century.* Cooperstown, N.Y., 1971. (10)

Robinson, Willard B. *American Forts: Architectural Form and Function.* Urbana, Ill., 1977. (6)

Rock, Howard B. *Artisans of the New Republic: The Tradesmen of New York City in the Age of Jefferson.* New York, 1979. (10)

Rohrbough, Malcolm J. *The Land Office Business: The Settlement and Administration of American Public Lands, 1789–1837.* New York, 1986. (8)

Ronda, James P. *Lewis and Clark among the Indians.* Lincoln, Nebr., 1984. (4)

Rorabaugh, W. J. *The Craft Apprentice: From Franklin to the Machine Age in America.* New York, 1986. (11)

Rose, Willie Lee, ed. *A Documentary History of Slavery in North America.* New York, 1976. (3)

Rosenberg, Nathan. *Perspectives on Technology.* New York, 1976. (15)

——. *Technology and American Economic Growth.* New York, 1972. (15)

——, ed. *The American System of Manufactures: The Report of the Committee on the Machinery of the United States, 1855, and the Special Reports of George Wallis and Joseph Whitworth, 1854.* Chicago, 1969. (10)

Ross, Steven J. *Workers on the Edge: Work, Leisure, and Politics in Industrializing Cincinnati, 1788–1890.* New York, 1985. (10)

Rossiter, Margaret W. *The Emergence of Agricultural Science: Justus Liebig and the Americans, 1840–1880.* New Haven, Conn., 1975. (13)

Rosswurm, Steven. *Arms, Country, and Class: The Philadelphia Militia and "Lower Sort" during the American Revolution, 1775–1783.* New Brunswick, N.J., 1987. (6)

Rountree, Helen C. *The Powhatan Indians of Virginia: Their Traditional Culture.* Norman, Okla., 1989. (4)

Rouse, Parke, Jr. *The Great Wagon Road from Philadelphia to the South.* New York, 1973. (7)

Rumford, [Count]. *The Collected Works of Count Rumford.* 5 vols. Ed. Sanborn Brown. Cambridge, Mass., 1968–1970. (11)

Runyan, Timothy J., ed. *Ships, Seafaring, and Society: Essays in Maritime History.* Detroit, 1987. (7)

Russell, Howard S. *Indian New England before the Mayflower.* Hanover, N.H., 1980. (4)

——. *A Long, Deep Furrow: Three Centuries of Farming in New England.* Hanover, N.H., 1976. (3)

Rutman, Darrett B. *Husbandmen of Plymouth: Farms and Villages in the Old Colony, 1620–1692.* Boston, 1967. (3)

St. George, Robert Blair, ed. *Material Life in America, 1600–1860.* Boston, 1988. (2)

Salinger, Sharon V. *"To Serve Well and Faithfully": Labor and Indentured Servants in Pennsylvania, 1682–1800.* New York, 1987. (4)

Salsbury, Stephen. *The State, the Investor, and the Railroad: The Boston and Albany, 1825–1867.* Cambridge, Mass., 1967. (7)

Scheele, Carl H. *A Short History of the Mail Service.* Washington, D.C., 1970. (12)

Scheiber, Harry N. *Ohio Canal Era: A Case Study of Government and the Economy, 1820–1861.* Athens, Ohio, 1969. (7)

Schlebecker, John T. *Whereby We Thrive: A History of American Farming, 1607–1972.* Ames, Iowa, 1975. (3)

——, ed. *Eighteenth-Century Agriculture: A Symposium. Agricultural History,* XLIII, no. 1 (January 1969). (3)

Schmidt, Hubert G. *Agriculture in New Jersey: A Three-hundred-Year History.* New Brunswick, N.J., 1973. (3)

Schmookler, Jacob. *Invention and Economic Growth.* Cambridge, Mass., 1966. (14)

Schob, David E. *Hired Hands and Plowboys: Farm Labor in the Midwest, 1815–60.* Urbana, Ill., 1975. (3)

Schor, Joel, and Cecil Harvey. *A List of References for the History of Black Americans in Agriculture, 1619–1974.* Davis, Calif., 1975. (1,3)

Schultz, Stanley K. *Constructing Urban Culture: American Cities and City Planning, 1800–1920*. Philadelphia, 1989. (8)

Schweitzer, Mary M. *Custom and Contract: Household, Government, and the Economy in Colonial Pennsylvania*. New York, 1987. (15)

Scott, Elizabeth M. *French Subsistence at Fort Michilimackinac, 1715–1781: The Clergy and the Traders*. Mackinac Island, Mich., 1985. (3)

Scranton, Philip. *Proprietary Capitalism: The Textile Manufacture at Philadelphia, 1800–1885*. Cambridge, 1983. (10)

Sears, Joan Niles. *The First One Hundred Years of Town Planning in Georgia*. Atlanta, Ga., 1979. (8)

Shammas, Carole. *The Pre-Industrial Consumer in England and America*. Oxford, 1990. (4)

Shelton, Cynthia J. *The Mills of Manayunk: Industrialization and Social Conflict in the Philadelphia Region, 1787–1837*. Baltimore, 1986. (10)

Shepherd, James F., and Gary M. Walton. *Shipping, Maritime Trade, and the Economic Development of Colonial North America*. Cambridge, 1972. (7)

Silver, Rollo G. *The American Printer, 1787–1825*. Charlottesville, Va., 1967. (10)

Silver, Timothy. *A New Face on the Countryside: Indians, Colonists, and Slaves in South Atlantic Forests, 1500–1800*. New York, 1990. (8)

Sinclair, Bruce. *Early Research at the Franklin Institute: The Investigation into the Causes of Steam Boiler Explosions, 1830–1837*. Philadelphia, 1966. (9)

———. *Philadelphia's Philosopher Mechanics: A History of the Franklin Institute, 1824–1865*. Baltimore, 1974. (11)

Siracusa, Carl. *A Mechanical People: Perceptions of the Industrial Order in Massachusetts, 1815–1880*. Middletown, Conn., 1979. (10)

Sklar, Katherine Kish. *Catharine Beecher: A Study in American Domesticity*. New Haven, Conn., 1973. (4)

Slaski, Eugene R. *Poorly Marked and Worse Lighted: Being a History of the Port Wardens of Philadelphia, 1766–1907*. Allentown, Pa., [1980]. (7)

Smith, Barbara Clark. *After the Revolution: The Smithsonian History of Everyday Life in the Eighteenth Century*. New York, 1985. (4)

Smith, Billy G. *The "Lower Sort": Philadelphia's Laboring People, 1750–1800*. Ithaca, N.Y., 1990. (10)

Smith, David C. *History of Papermaking in the United States (1691–1969)*. New York, 1970. (10)

Smith, Duane. *Mining in America: The Industry and the Environment, 1800–1980*. Lawrence, Kans., 1987. (5)

Smith, Julia Floyd. *Slavery and Rice Culture in Low Country Georgia, 1750–1860*. Knoxville, Tenn., 1985. (3)

Smith, Merritt Roe. *Harpers Ferry Armory and the New Technology: The Challenge of Change*. Ithaca, N.Y., 1977. (10)

———, ed. *Military Enterprise and Technological Change: Perspectives on the American Experience*. Cambridge, Mass., 1985. (6)

Smith, Norman. *A History of Dams*. Secaucus, N.J., 1972. (7)

Smith, Paul H., ed. *Letters of Delegates to Congress, 1774–1789*. Washington, D.C., 1976–. (2)

Sobel, Mechal. *The World They Made Together: Black and White Values in Eighteenth-Century Virginia*. Princeton, N.J., 1987. (4)

Soltow, Martha Jane, and Mary K. Wery. *American Women and the Labor Movement, 1825–1974: An Annotated Bibliography*. 2d ed. Metuchen, N.J., 1976. (1)

Stackpole, Edouard A. *Whales and Destiny: The Rivalry between America, France, and Britain for Control of the Southern Whale Fishery, 1785–1825*. Amherst, Mass., 1972. (7)

Stansell, Christine. *City of Women: Sex and Class in New York, 1789–1860*. New York, 1986. (4, 10)

Stapleton, Darwin H. *The History of Civil Engineering since 1600: An Annotated Bibliography*. New York, 1986. (1, 7)

——— . *The Transfer of Early Industrial Technologies to America*. American Philosophical Society, Memoirs, CLXXVI. Philadelphia, 1987. (13)

Starobin, Robert S. *Industrial Slavery in the Old South*. New York, 1970. (5, 10)

Stearns, Raymond Phineas. *Science in the British Colonies of America*. Urbana, Ill., 1970. (11)

Steeds, W. *A History of Machine Tools, 1700–1910*. Oxford, 1969. (10)

Steele, Ian K. *The English Atlantic, 1675–1740: An Exploration of Communication and Community*. New York, 1986. (12)

Steffen, Charles G. *The Mechanics of Baltimore: Workers and Politics in the Age of Revolution, 1763–1812*. Urbana, Ill, 1984. (10)

Stein, Leon, and Philip Taft, eds. *Religion, Reform, and Revolution: Labor Panaceas in the Nineteenth Century*. New York, 1969. (10)

Stilgoe, John R. *Common Landscape of America, 1580 to 1845*. New Haven, Conn., 1982. (8)

Stiverson, Gregory A. *Poverty in a Land of Plenty: Tenancy in Eighteenth-Century Maryland*. Baltimore, 1977. (3)

Stover, John F. *History of the Baltimore and Ohio Railroad*. West Lafayette, Ind., 1987. (7)

Syrett, David. *Shipping and the American War, 1775–83: A Study of British Transport Organization*. London, 1970. (6)

Taylor, Snowden. *The Developmental Era of Eli Terry and Seth Thomas Shelf Clocks*. Fitzwilliam, N.H., 1985. (10)

Tebbel, John. *A History of Book Publishing in the United States, I, The Creation of an Industry, 1630–1865*. New York, 1972. (10)

Technology and Culture. "Current Bibliography in the History of Technology." Annually, 1965–. (1)

Temin, Peter. *Iron and Steel in Nineteenth-Century America: An Economic Inquiry*. Cambridge, 1964. (5)

——— . *The Jacksonian Economy*. New York, 1969. (15)

Thomas, Gertrude Z. *Richer than Spices: How a Royal Bride's Dowry Introduced*

Cane, Lacquer, Cottons, Tea, and Porcelain to England, and So Revolutionized Taste, Manners, Craftsmanship, and History in Both England and America. New York, 1965. (4)

Thomson, Ross. *The Path to Mechanized Shoe Production in the United States.* Chapel Hill, N.C., 1989. (10, 14)

Tucker, Barbara M. *Samuel Slater and the Origins of the American Textile Industry, 1790–1860.* Ithaca, N.Y., 1984. (10)

Tweedale, Geoffrey. *Sheffield Steel and America: A Century of Commercial and Technological Interdependence, 1830–1930.* New York, 1987. (13)

Tyler, John W. *Smugglers and Patriots: Boston Merchants and the Advent of the American Revolution.* Boston, 1986. (7)

Tyrell, Ian R. *Sobering Up: From Temperance to Prohibition in Antebellum America, 1800–1860.* Westport, Conn., 1979. (10)

Ulrich, Laurel Thatcher. *A Midwife's Tale: The Life of Martha Ballard, Based on Her Diary, 1785–1812.* New York, 1990. (4)

———. *Good Wives: Image and Reality in the Lives of Women in Northern New England, 1650–1750.* New York, 1982. (4)

Upton, Dell. *Holy Things and Profane: Anglican Parish Churches in Colonial Virginia.* Cambridge, Mass., 1986. (7)

Upton, Dell, and John Michael Vlach, eds. *Common Places: Readings in American Vernacular Architecture.* Athens, Ga., 1986. (4)

Valle, James E. *Rocks and Shoals: Order and Discipline in the Old Navy, 1800–1861.* Annapolis, Md., 1980. (6)

Van Deburg, William. *The Slave Drivers: Black Agricultural Labor Supervisors in the Antebellum South.* Westport, Conn., 1979. (3)

Van Ravenswaay, Charles. *The Arts and Architecture of German Settlements in Missouri: A Survey of a Vanishing Culture.* Columbia, Mo., 1977. (13)

Vignoles, K. H. *Charles Blacker Vignoles: Romantic Engineer.* New York, 1982. (7)

Vogel, Robert M. *Roebling's Delaware and Hudson Canal Aqueducts.* Washington, D.C., 1971. (7)

Wacker, Peter. *Land and People: A Cultural Geography of Preindustrial New Jersey: Origins and Settlement Patterns.* New Brunswick, N.J., 1975. (8)

Walker, Henry Pickering. *The Wagonmasters: High Plains Freighting from the Earliest Days of the Santa Fe Trail to 1880.* Norman, Okla., 1966. (7)

Walker, Joseph E. *Hopewell Village: A Social and Economic History of an Iron-making Community.* Philadelphia, 1966. (5)

Walker, Paul K. *Engineers of Independence: A Documentary History of the Army Engineers in the American Revolution, 1775–1783.* Washington, D.C., 1982. (6)

Wallace, Anthony F. C. *Rockdale: The Growth of an American Village in the Early Industrial Revolution.* New York, 1978. (10)

———. *St. Clair: A Nineteenth-Century Coal Town's Experience with a Disaster-prone Industry.* New York, 1987. (5)

Warner, Sam Bass, Jr. *The Private City: Philadelphia in Three Periods of Its Growth.* Philadelphia, 1968. (7, 8)

Weible, Robert, ed. *The World of the Industrial Revolution: Comparative and International Aspects of Industrialization.* North Andover, Mass., 1986. (9)

Weinberg, Meyer, ed. *America's Economic Heritage: From a Colonial to a Capitalist Economy, 1634–1900.* Westport, Conn., 1983. (15)

Weslager, C. A. *The Delaware Indians: A History.* New Brunswick, N.J., 1972. (4)

———. *The Log Cabin in America: From Pioneer Days to the Present.* New Brunswick, N.J., 1969. (7)

Whiffen, Marcus, and Frederick Koeper. *American Architecture, 1607–1976.* Cambridge, Mass., 1981. (7)

White, John H., Jr. *American Locomotives: An Engineering History, 1830–1880.* Baltimore, 1968. (7)

———. *The John Bull: 150 Years a Locomotive.* Washington, D.C., 1981. (7, 13)

White, Philip L. *Beekmantown, New York: Forest Frontier to Farm Community.* Austin, Tex., 1979. (3)

Whiteman, Maxwell. *Copper for America: The Hendricks Family and a National Industry, 1755–1939.* New Brunswick, N.J., 1971. (5)

Wilcox, Frank. *The Ohio Canals.* Ed. William A. McGill. Kent, Ohio, 1969. (7)

Wilentz, Sean. *Chants Democratic: New York City and the Rise of the American Working Class, 1788–1850.* New York, 1984. (10)

Wilkinson, Norman B. *Lammot du Pont and the American Explosives Industry, 1850–1884.* Charlottesville, Va., 1984. (10)

Winterthur Portfolio. Winterthur, Del.; Charlottesville, Va.; Chicago, 1979–. (10)

Wolf, Edwin, II. *The Book Culture of a Colonial American City: Philadelphia Books, Bookmen, and Booksellers.* Oxford, 1988. (12)

Wood, Peter H. *Black Majority: Negroes in Colonial South Carolina from 1670 through the Stono Rebellion.* New York, 1974. (13)

Woodman, Harold D. *King Cotton and His Retainers: Financing and Marketing the Cotton Crop of the South, 1800–1925.* Lexington, Ky., 1967. (3)

Wright, Esmond. *Franklin of Philadelphia.* Cambridge, Mass., 1986. (14)

Wright, Gavin. *The Political Economy of the Cotton South: Households, Markets, and Wealth in the Nineteenth Century.* New York, 1978. (3)

Wright, Gwendolyn. *Building the Dream: A Social History of Housing in America.* New York, 1981. (7)

Wright, J. Leitch, Jr. *Creeks and Seminoles: The Destruction and Regeneration of the Muscogulge People.* Lincoln, Nebr., 1986. (8)

Wright, Louis B., George B. Tatum, John W. McCoubrey, and Robert C. Smith. *The Arts in America: The Colonial Period.* New York, 1966. (10)

Wyckoff, William. *The Developer's Frontier: The Making of the Western New York Landscape.* New Haven, Conn., 1988. (8)

York, Neil Longley. *Mechanical Metamorphosis: Technological Change in Revolutionary America.* Westport, Conn., 1985. (2, 6, 10)

Zilversmit, Arthur. *The First Emancipation: The Abolition of Slavery in the North.* Chicago, 1967. (10)

Appendix: Brooke Hindle's pre-1966 Bibliography

To supplement Nina E. Lerman's preceding bibliography, it seemed appropriate to present the entries from Brooke Hindle's "Bibliography of Early American Technology." Abstracted from his narrative, they are presented here, without reediting, in the brief form that Hindle followed and in the original classifications. In a very few instances missing information has been supplied, and forthcoming works that did not come forth have been omitted.

Guides and Sources
General Surveys
American Surveys
Agriculture and Food Processing
Mining and Metals
Military
Civil Engineering and Transportation
   General Syntheses
   The Engineers
   Surveying
   Ships and Shipbuilding
   Steam and River Boats
   Canals
   Railroads
   Roads
   Bridges
   Building
   Other Engineering Works
Power
   Wind and Water
   Steam
Manufacturing
   Crafts and Craftsmen
   Tools, Woodwork, Metalwork, and Machining
   Clocks, Instruments, and the Technology of Science
   Glass and Pottery
   Paper and Printing
   Leather
   Textiles
   Chemicals
   Other Manufacturing
Heat, Light, and Electricity
Education, Organization, and Science
America and Europe

Patents and Invention
Technology in American History

GUIDES AND SOURCES

Allen, Zachariah, *The Practical Tourist, or Sketches of the State of the Useful
   Arts . . . in Great Britain, France, and Holland,* 2 vols. (Providence, 1832).
Baron Klinkowstrom's America, 1818–1820, trans. by Franklin D. Scott
   (Evanston, 1952).
Bell, Whitfield J., Jr., *Early American Science: Needs and Opportunities for Study*
   (Williamsburg, 1955).
Cappon, Lester J., and Stella M. Duff, eds., *Virginia Gazette Index, 1736–1780*
   (Williamsburg, 1950).
"Current Bibliography in the History of Technology," *Technology and Culture,* 5
   (1956), 138–48; (1965), 346–74.
Ferguson, Eugene S., "Contributions to Bibliography in the History of Technology,"
   published serially in *Technology and Culture,* 3 (1962), 73–84, 167–74, 298–306; 4
   (1963), 318–30; 5 (1964), 416–34, 578–94; 6 (1965), 99–107.
Garrett, Wendell D., and Jane N., annotated bibliography in Walter Muir Whitehill,
   The Arts in Early American History: Needs and Opportunities for Study (Chapel
   Hill, 1965).
Hamer, Philip M., *Guide to Archives and Manuscripts in the United States* (New
   Haven, 1961).
Holmes, Isaac, *An Account of the United States of America* (London, 1823).
La Rochefoucauld-Liancourt, Duc de, *Travels Through the United States of North
   America, 1795, 1796, 1797,* 4 vols. (2nd edn., London, 1800).
Larson, Henrietta M., *Materials for the Study of American Business History and
   Suggestions for Their Use* (Cambridge, Mass., 1948).
Reingold, Nathan, "Manuscripts Resources for the History of Science and
   Technology in the Library of Congress," Library of Congress, *Quarterly Journal of
   Current Acquisitions,* 17 (1960), 161–69.
——— , "The National Archives and the History of Science in America," *Isis,* 46
   (1955), 22–28.
——— , "U.S. Patent Office Records as Sources for the History of Invention and
   Technological Property," *Technology and Culture,* 1 (1969), 156–67.
Schoepf, Johann David, *Travels in the Confederation, 1783–1784,* trans. by Alfred J.
   Morrison, 2 vols. (Philadelphia, 1911).
Wansey, Henry, *The Journal of an Excursion to the United States of North America
   in the Summer of 1794* (Salisbury, 1796).

Albion (1822–50+).
American Journal of Science (1819–45).
American Magazine (1769).
American Mechanic's Magazine (1825–26).

American Museum (1787–92).
American Repertory of Arts, Science and Manufactures (1840–42).
Archives of Useful Knowledge (1810–13).
De Bow's Review (1846–50+).
Emporium of Arts and Sciences (1812–14).
Farmer and Mechanic (1844–50+).
Hunt's Merchants' Magazine and Commercial Review (1839–50+).
Journal of the Franklin Institute (1826–50+).
Massachusetts Magazine (1789–96).
Mechanics' Advocate (1846–48).
Mechanics' Magazine and Register of Inventions and Improvements (1833–37).
New York State Mechanic: A Journal of the Manual Arts, Trades, and Manufactures (1841–43).
Niles' Weekly Register (1811–49).
Pennsylvania Magazine (1775–76).
Port Folio (1810–27).
Quarterly Journal of Agriculture, Mechanics, and Manufacture (1834–35).
Rail Road Journal (1832–50+).
Railway Times (1849–50+).
Useful Cabinet (1808).

GENERAL SURVEYS

Daumas, Maurice, *Histoire Générale des Techniques* (Paris, 1962).
Derry, T. K., and Trevor I. Williams, *A Short History of Technology* (Oxford, 1961).
Encyclopaedia, 18 vols. (Philadelphia, 1798).
Finch, James Kip, *Engineering and Western Civilization* (New York, 1951).
———, *The Story of Engineering* (Garden City, 1960).
Forbes, Robert J., *Man the Maker* (N.Y., 1950, 1958).
Gillispie, Charles C., ed., *A Diderot Pictorial Encyclopedia of Trades and Industry,* 2 vols. (N.Y., 1959).
Kirby, Richard S., Sidney Withington, Arthur B. Darling, and Frederick G. Kilgour, *Engineering in History* (N.Y., 1956).
Rees, Abraham, *Cyclopaedia,* 39 vols. text + 6 vols. plates (Philadelphia, 1810–24).
Singer, Charles, E. J. Holmyard, A. R. Hall, and Trevor I. Williams, eds., *History of Technology* (London, 1954–58).
Ure, Andrew, *Dictionary of Arts, Manufactures, and Mines* (N.Y., 1842).
Usher, Abbot Payson, *A History of Mechanical Inventions* (Cambridge, Mass., 1929, 1954; Boston, 1959).
Wolf, A., *A History of Science, Technology and Philosophy in the 18th Century* (N.Y., 1939, 1961).
———, *A History of Science, Technology and Philosophy in the 16th and 17th Centuries* (London, 1935; N.Y., 1950, 1959).

AMERICAN SURVEYS

Antisell, T., *Hand-book of the Useful Arts* (N.Y., 1852).

Appleton's Dictionary of Machines, Mechanics, Engine-Work, and Engineering, 2 vols. (N.Y., 1852).

Baird, Spencer F., ed., *Iconographic Encyclopaedia of Science, Literature, and Art*, 6 vols. (N.Y., 1851–52).

Bishop, J. Leander, *A History of American Manufactures from 1608 to 1860*, 3 vols. (Phila., 1861–68).

Bolles, Albert S., *Industrial History of the United States* (Norwich, Conn., 1879).

Burlingame, Roger, *March of the Iron Men: A Social History of Union through Invention* (N.Y., 1938).

Calvert, Monte A., "American Technology at World Fairs, 1851–1876" (M.A. thesis, University of Delaware, 1962).

Clark, Victor S., *History of Manufactures in the United States, 1607–1860* (Washington, 1916).

Cole, Arthur Harrison, ed., *Industrial and Commercial Correspondence of Alexander Hamilton, Anticipating His Report on Manufactures* (Chicago, 1938).

Coxe, Tench, *Observations on the Agriculture, Manufactures and Commerce of the United States* (N.Y., 1789).

——, *A Statement of Arts and Manufactures . . . for the Year 1810* (Washington, 1814).

Curti, Merle, "America at the World Fairs, 1851–1893," *American Historical Review*, 55 (1950), 833–56.

Dupree, A. Hunter, *Science in the Federal Government* (Cambridge, Mass., 1957).

Freedley, Edwin T., *Philadelphia and Its Manufactures* (Phila., 1859).

Gallatin, Albert, *Report of the Secretary of the Treasury on the Subject of Public Roads and Canals* (Washington, 1808).

——, "Report on Manufactures, 1810," *American State Papers: Finance Vol. II* (Washington, 1832), 425–39.

Green, Constance McLaughlin, *History of Naugatuck, Connecticut* (New Haven, 1949).

Hazen, Edward, *The Panorama of Professions and Trades* (Phila., 1837).

Howe, Henry, *Memoirs of the Most Eminent American Mechanics* (N.Y., 1842).

Illes, George, *Leading American Inventors* (N.Y., 1912).

Johnson, Benjamin P., *Great Exhibition of the Industry of All Nations, 1851* (Albany, 1852).

Kaempffert, Waldemar, ed., *A Popular History of American Invention*, 2 vols. (N.Y., 1924).

Keir, Malcolm, ed., *Manufacturing: A Volume of Industries of America* (N.Y., 1928).

Kirby, Richard S., ed., *Inventors and Engineers of Old New Haven* (New Haven, 1939).

Lewis, Alonzo, and James R. Newhall, *History of Lynn* (Boston, 1865).

Lippincott, Isaac, *A History of Manufactures in the Ohio Valley to the Year 1860* (Chicago, 1914).

[McLane, Louis], U.S. Treasury Department, *Documents Relating to the Manufactures in the United States*, 2 vols. (Washington, 1833).

Oliver, John W., *History of American Technology* (N.Y., 1956).

Patter, C. E., *The History of Manchester* (Manchester, 1856).

Report of the Twenty-Second Exhibition of American Manufactures Held in the City of Philadelphia from the 19th to the 30th of October, Inclusive, 1852 by the Franklin Institute [Phila., 1853].

Rodgers, Charles T., *American Superiority at the World's Fair* (Phila., 1852).

Scharf, John Thomas, *History of Delaware, 1609–1888*, 2 vols. (Phila., 1888).

Scharf, John Thomas, and Thompson Westcott, *History of Philadelphia, 1609–1884*, 3 vols. (Phila., 1884).

Shlakman, Vera, *Economic History of a Factory Town: A Study of Chicopee, Massachusetts* (Northampton, Mass., 1935).

Smith, Thomas Russell, *The Cotton Textile Industry of Fall River, Massachusetts* (N.Y., 1944).

Stokes, I. N. Phelps, *Iconography of Manhattan Island*, 6 vols. (N.Y., 1895–1926).

Stone, Orra L., *History of Massachusetts Industries*, 4 vols. (Boston, 1930).

Struik, Dirk J., *Yankee Science in the Making* (Boston, 1948).

Stuart, Charles B., *Lives and Works of Civil and Military Engineers of America* (N.Y., 1871).

Swank, James M., *Progressive Pennsylvania* (Phila., 1908).

Trumbull, L. R., *A History of Industrial Paterson* (Paterson, 1882).

Van Slyck, J. D., *New England Manufacturers and Manufactories*, 2 vols. (Boston, 1879).

Weeden, William B., *Economic and Social History of New England, 1620–1789*, 2 vols. (Boston, 1890).

Weiss, Harry B. and Grace M., *Forgotten Mills of Early New Jersey: Oil, Plaster, Bark, Indigo, Fanning, Tilt, Rolling and Slitting Mills, Nail and Screw Making* (Trenton, 1969).

Whitworth, Sir Joseph, and George Wallis, *The Industry of the United States* (London, 1854).

Wright, Carroll D., *The Industrial Evolution of the United States* (N.Y., 1902).

AGRICULTURE AND FOOD PROCESSING

Albion, Robert G., *Forests and Sea Power* (Cambridge, Mass., 1926).

Ardrey, R. L., *American Agricultural Implements: A Review of Invention and Development in the Agricultural Implement Industry of the United States* (Chicago, 1894).

Bathe, Greville and Dorothy, *Oliver Evans: A Chronicle of Early American Engineering* (Phila., 1935).

Benoit, P.-M.-N., *Guide du Meunier et du Constructeur de Moulins* (Paris, 1830).

Bidwell, Percy W., and John I. Falconer, *History of Agriculture in the Northern United States, 1620–1860* (Washington, 1925).

Bordley, John Beale, *A Summary View of the Courses of Crops in the Husbandry of England and Maryland* (Phila., 1784).

Browne, C. A., *A Source Book of Agricultural Chemistry* (Waltham, 1944).

Carman, Harry J., ed., *American Husbandry* [London, 1775], (N.Y., 1939).

Clemen, R. A., *The American Livestock and Meat Industry* (N.Y., 1923).

Colman, Henry, *Agriculture and Rural Economy* (Boston, 1849).

Coyne, F. E., *The Development of the Cooperage Industry in the United States* (Chicago, 1940).

Craven, Avery O., *Soil Exhaustion as a Factor in the Agricultural History of Virginia and Maryland, 1606–1860* (Urbana, Ill., 1925).

Defebaugh, James Elliott, *History of the Lumber Industry of America*, 2 vols. (Chicago, 1906).

Edwards, Everett E., "American Agriculture—The First Hundred Years," U.S. Department of Agriculture, *Yearbook of Agriculture, 1940* (Washington, 1941).

Eliot, Jared, *Essays Upon Field Husbandry in New England and Other Papers, 1748–1762*, ed. by Harry J. Carman and Rexford G. Tugwell (N.Y., 1934).

Evans, Oliver, *The Young Mill-Wright and Miller's Guide* (Phila., 1795).

Gray, Lewis C., and E. K. Thompson, *History of Agriculture in the Southern United States to 1860*, 2 vols. (Washington, 1933).

Hughes, William Carter, *The American Miller, and Millwright's Assistant* (Detroit, 1850).

Hutchinson, William T., *Cyrus Hall McCormick*, 2 vols. (N.Y., 1930, 1936).

Jefferson, Thomas, "The Description of a Mouldboard of the Least Resistance," APS, *Transactions*, 4 (1799), 313–22.

Kendall, Edward C., *John Deere's Steel Plow* (U.S. National Museum, *Bulletin*, No. 218 [Washington, 1959], pp. 15–25).

Knittle, W. A., *Early Eighteenth Century Palatine Emigration: A British Redemptioner Project to Manufacture Naval Stores* (Phila., 1937).

Kuhlmann, Charles Byron, *The Development of the Flour-Milling Industry in the United States* (Boston and N.Y., 1929).

Lord, Eleanor L., *Industrial Experiments in the British Colonies of North America* (Baltimore, 1898).

McFarland, Raymond, *A History of the New England Fisheries* (N.Y., 1911).

Peters, Richard, *Agricultural Enquiries on Plaister of Paris* (Phila., 1797).

Quaintance, Hadley W., *The Influence of Farm Machinery on Production and Labor* (N.Y., 1904).

Richards, John, *A Treatise on the Construction and Operation of Wood-Working Machines: Including a History of the Origin and Progress of the Manufacture of Wood-Working Machinery* (London, 1872).

Rogin, Leo, *The Introduction of Farm Machinery in Its Relation to the Productivity of Labor in the United States during the Nineteenth Century* (Berkeley, 1931).

Ruffin, Edmund, *An Essay on Calcareous Manures* [1832], ed. by J. Carlyle Sitterson (Cambridge, Mass., 1961).

Starbuck, A., *History of the American Whale Fishery from Its Earliest Inception to the Year 1876* (Waltham, 1876).

Storck, John, and Walter D. Teague, *Flour for Man's Bread: A History of Milling* (Minneapolis, 1952).

Weiss, Harry B., and Robert J. Simm, *The Early Grist and Flouring Mills of New Jersey* (Trenton, 1956).

Weiss, Harry B. and Grace M., *The Early Snuff Mills of New Jersey* (Trenton, 1962).

Welsh, Peter C., "The Brandywine Mills: A Chronicle of Industry, 1763–1816," *Delaware History*, 7 (1956), 17–36.

Wik, Reynold M., *Steam Power on the American Farm* (Phila., 1953).

Woodward, Carl R., ed., *Ploughs and Politics: Charles Read of New Jersey and His Notes on Agriculture, 1715–1774* (New Brunswick, 1941).

MINING AND METALS

Aitchison, Leslie, *A History of Metals,* 2 vols. (London, 1960).

Bining, Arthur C., *Pennsylvania Iron Manufacture in the Eighteenth Century* (Harrisburg, 1938).

Boyer, Charles S., *Early Forges and Furnaces in New Jersey* (Phila., 1931).

Brown, James S., *Allaire's Lost Empire* (Freehold, N.J., 1958).

Bruce, Kathleen, *Virginia Iron Manufacture in the Slave Era* (N.Y., 1931).

Forges and Furnaces in the Province of Pennsylvania (Phila., 1914).

Harrison, William H., "The First Rolling Mill in America," American Society of Mechanical Engineers, *Transactions,* 2 (1881), 104–7.

Hartley, E. N., *Ironworks on the Saugus* (Norman, Okla., 1957).

Hasenclever, Adolf, *Peter Hasenclever, aus Remscheid* (Gotha, 1922).

Hasenclever, Peter, *The Remarkable Case of Peter Hasenclever* (London, 1773).

Hermelin, Samuel Gustaf, *Report about the Mines in the United States of America, 1783* (Phila., 1931).

Hewitt, Abram S., *A Century of Mining and Metallurgy in the United States* (Phila., 1876).

——, *On the Statistics and Geography of the Production of Iron* (N.Y., 1856).

Hunter, Louis C., "Influence of the Market upon Technique in the Iron Industry of Western Pennsylvania to 1860," *Journal of Economic and Business History,* 1 (1929), 241–81.

Ingalls, Walter Renton, *Lead and Zinc in the United States: Comprising an Economic History of the Mining and Smelting of the Metals* (N.Y., 1908).

Lathrop, William G., *The Brass Industry in Connecticut* (Shelton, Conn., 1909).

Latrobe, Benjamin Henry, *American Copper Mines* (n.p., n.d. [1800?]).

Lesley, J. P., *The Iron Manufacturer's Guide to the Furnaces, Forges and Rolling Mills of the United States* (N.Y., 1859).

Metallurgical Society of the American Institute of Mining, Metallurgical and
  Petroleum Engineers, *History of Iron and Steelmaking in the United States*
  (N.Y., 1961).
Norris, James D., *Frontier Iron: The Maramec Iron Works, 1826–1876* (Madison,
  Wis., 1964).
Overman, Frederick, *The Manufacture of Steel* (Phila., 1851).
Pearse, John B., *A Concise History of the Iron Manufacture of the American
  Colonies up to the Revolution, and of Pennsylvania until the Present Time*
  (Phila., 1876).
Pierce, Arthur D., *Family Empire in Jersey Iron* (New Brunswick, 1964).
———, *Iron in the Pines* (New Brunswick, 1957).
Rickard, Thomas A., *History of American Mining* (Phila., 1932).
Rowe, Frank H., *History of the Iron and Steel Industry in Scioto County*
  (Columbus, 1938).
Schoolcraft, Henry R., *A View of the Lead Mines of Missouri* (N.Y., 1819).
Schubert, H. R., *History of the British Iron and Steel Industry from c. 450 to A.D.
  1775* (London, 1957).
Sim, Robert J., and Harry B. Weiss, *Charcoal-Burning in New Jersey*
  (Trenton, 1955).
Smith, Cyril S., *A History of Metallography* (Chicago, 1960).
Swank, James M., *History of the Manufacture of Iron in all Ages and Particularly in
  the United States from Colonial Times to 1891* (2nd edn., Phila., 1892).
———, *Introduction to a History of Ironmaking and Coal Mining in Pennsylvania*
  (Phila., 1878).
Tunner, Peter, *A Treatise on Roll-Turning for the Manufacture of Iron* (N.Y., 1869).
Walker, Joseph E., *Hopewell Village: A Social and Economic History of an
  Iron-Making Community* (Phila., 1966).
Weiss, Harry B. and Grace M. *The Old Copper Mines of New Jersey*
  (Trenton, 1963).
Wertime, Theodore A., *The Coming of the Age of Steel* (Leiden, 1961).

MILITARY

Bushnell, David, "General Principles and Construction of a Submarine Vessel," APS,
  Transactions, 4 (1799), 303–12.
Carey, A. Merwyn, *American Firearms Makers* (N.Y., 1953).
Cohen, I. Bernard, "Science and the Revolution," *Technology Review*, 47 (1945),
  367–68, 374–88.
Couper, William, *Claudius Crozet* (Charlottesville, 1936).
Deyrup, Felicia J., *Arms Makers of the Connecticut Valley* (Northampton, 1948).
Edwards, William B., *The Story of Colt's Revolver: The Biography of Col. Samuel
  Colt* (Harrisburg, 1953).
Fitch, Charles H., "The Rise of a Mechanical Ideal," *Magazine of American History*,
  2 (1884), 516–27.

Forman, Sidney, "Early American Military Engineering Books," *Military Engineering*, 36 (1954), 93–95.

———, *West Point: A History of the United States Military Academy* (N.Y., 1950).

Fuller, Claud E., *Springfield Muzzle-Loading Shoulder Arms: A Description of the Flint Lock Muskets, Musketoons, Rifles, Carbines and Special Models from 1795 to 1865* (N.Y., 1930).

———, *The Whitney Firearms* (Huntington, W.Va., 1946).

Fulton, Robert, *Torpedo War and Submarine Explosions* (N.Y., 1810).

Gilbert, K. R., "The Ames Recessing Machine: A Survivor of the Original Enfield Rifle Machinery," *Technology and Culture*, 4 (1963), 207–11.

Goetzman, William H., *Army Exploration in the American West, 1803–63* (New Haven, 1959).

Green, Constance McLaughlin, *Eli Whitney and the Birth of American Technology* (Boston, 1956).

Haiman, Miecislaus, *Kosciuszko in the American Revolution* (N.Y., 1943).

Heusser, Albert H. *The Forgotten General: Robert Erskine* (Paterson, 1928).

Hicks, James E., *Nathan Starr (The First Official Sword Maker)* (Mt. Vernon, N.Y., 1940).

———, *U.S. Military Firearms, 1776–1956* (La Canada, Calif., 1962).

Hill, Forest G., *Roads, Rails, and Waterways: The Army Engineers and Early Transportation* (Norman, Okla., 1957).

Holt, W. Stull, *The Office of the Chief of Engineers of the Army: Its Non-Military History, Activities, and Organization* (Baltimore, 1923).

Jordan, Francis, Jr., *The Life of William Henry of Lancaster, Pa., 1729–1786* (Lancaster, 1910).

Kauffman, Henry, *The Pennsylvania-Kentucky Rifle* (Harrisburg, 1960).

Kite, Elizabeth S., *Brigadier General Louis Lebegue Duportail* (Baltimore, 1933).

Mahan, Dennis Hart, *An Elementary Course of Civil Engineering* (N.Y., 1837).

———, *A Treatise on Field Fortification* (N.Y., 1836).

Mirsky, Jeannette, and Allan Nevins, *The World of Eli Whitney* (N.Y., 1952).

Mordecai, Capt. Alfred, *Report on Experiments on Gunpowder made at the Washington Arsenal in 1843 and 1844* (Washington, 1845).

National Armories, The: A Review of the System of Superintendency, Civil and Military, Particularly with Reference to Economy, and General Management at the Springfield Armory (2nd edn., Springfield, Mass., 1852).

North, Ralph H. and S. N. D., *Simeon North, First Official Pistol Maker of the United States* (Concord, N.H., 1913).

Norton, Charles B., *American Inventions and Improvements in Breechloading Small Arms, Heavy Ordnance, Machine Guns, Magazine Arms* (Springfield, Mass., 1880).

Olmsted, Denison, *Memoir of Eli Whitney* (New Haven, 1846).

Peterson, Harold L., *The American Sword, 1775–1945* (New Hope, Pa., 1954).

———, *Arms and Armor in Colonial America, 1526–1783* (Harrisburg, 1956).

Roe, Joseph W., "Interchangeable Manufacture," Newcomen Society, *Transactions*, 17 (1937), 165–74.

Sawyer, John E., "The Social Basis of the American System of Manufacturing,"
    Journal of Economic History, 14 (1954), 361–79.
Tousard, Louis, *American Artillerist's Companion,* 3 vols. (N.Y., 1809).
Woodbury, Robert S., "The Legend of Eli Whitney," *Technology and Culture,* 1
    (1960), 235–51.

CIVIL ENGINEERING AND TRANSPORTATION

General Syntheses

Carey, Mathew, *A Connected View of the Whole Internal Navigation of the United
    States* (Phila., 1826).
Chevalier, Michel, *Histoire et Description des Voies de Communication aux
    États-Unis* (Paris, 1840).
Gerstner, Franz Anton von, *Die Innern Communicationen der Vereinigten Staaten
    von Nordamerica,* 2 vols. (Vienna, 1842–43).
Marestier, Jean Baptiste, *Mémoire sur les Bateaux à Vapeur des États-Unis* (Paris,
    1824); partially translated into English by Sidney Withington as *Memoir on
    Steamboats* (Mystic, Conn., 1959).
Nettels, Curtis P., *The Emergence of a National Economy: 1775–1815* (N.Y., 1962).
Poor, Henry V., *History of the Railroads and Canals of the United States*
    (N.Y., 1860).
Poussin, Guillaume Tell, *Atlas* (Paris, 1834).
———, *Chemins de Fer Américains* (Paris, 1836).
———, *Travaux d'Améliorations Intérieures* (Paris, 1834).
Ringwalt, J. L., *Development of Transportation Systems in the United States*
    (Phila., 1888).
Stevenson, David, *Sketch of the Civil Engineering of North America* (London, 1838).
Stucklé, Henri, *Voies de Communication aux États-Unis* (Paris, 1847).
Tanner, H. S., *A Description of the Canals and Rail Roads of the United States*
    (N.Y., 1840).
Taylor, George Rogers, *The Transportation Revolution: 1815–1860* (N.Y., 1951).

The Engineers

Abbott, Frederick K., "The Role of the Civil Engineer in Internal Improvements: The
    Contributions of the Two Loammi Baldwins, Father and Son, 1776–1838" (Ph.D.
    diss., Columbia, 1952, on Ann Arbor microfilm).
Calhoun, Daniel H., *The American Civil Engineer: Origins and Conflict*
    (Cambridge, Mass., 1960).
[Finch, James Kip], *Early Columbia Engineers* (N.Y., 1929).
Gallagher, H. M. Pierce, *Robert Mills: Architect of the Washington Monument,
    1781–1855* (N.Y., 1935).
Gilchrist, Agnes A., *William Strickland, Architect and Engineer, 1788–1854*
    (Phila., 1950).

Hamlin, Talbot, *Benjamin Henry Latrobe* (N.Y., 1955).

Jackson, Joseph, *Early Philadelphia Architects and Engineers* (Phila., 1922).

Kirby, Richard S., "Some Early American Civil Engineers and Surveyors," Connecticut Society of Civil Engineers, *Papers and Transactions,* 1930, pp. 26–47.

Latrobe, Benjamin Henry, *Journal . . . from 1796 to 1820* (N.Y., 1905).

Strickland, William, *Reports on Canals, Railways, Roads, and Other Subjects* (Phila., 1826).

Strickland, William, with Edward H. Gill and Henry R. Campbell, *Public Works of the United States of America* (London, 1841).

Stuart, Charles B., *Lives and Works of Civil and Military Engineers of America* (N.Y., 1871).

Surveying

Cajori, Florian, *The Chequered Career of Ferdinand Rudolph Hassler* (Boston, 1929).

Ellicott, Andrew, *Journal* (Phila., 1803).

Gibson, Robert, *A Treatise of Practical Surveying* (5th edn., Phila., 1789).

Gummere, John, *A Treatise on Surveying* (Phila., 1820, 1828, 1836).

Hassler, Ferdinand Rudolph, *Principal Documents Relating to the Survey of the Coast of the United States since 1816* (N.Y., 1834).

Hindle, Brooke, *David Rittenhouse* (Princeton, 1964).

Matthews, Catharine Van C., *Andrew Ellicott* (N.Y., 1908).

Moore, Samuel, *An Accurate System of Surveying* (Litchfield, 1796).

Odgers, Merle, *Alexander Dallas Bache* (Phila., 1947).

Pattison, William D., *Beginnings of the American Rectangular Land Survey System, 1784–1800* (Chicago, 1957).

Phillips, P. Lee, *Notes on the Life and Works of Bernard Romans* (Deland, Fla., 1924).

Ships and Shipbuilding

Albion, Robert G., *The Rise of New York Port (1815–60)* (N.Y., 1939).

——— , *Square Riggers on Schedule* (Princeton, 1938).

Chapelle, Howard I., *The Baltimore Clipper, Its Origin and Development* (Salem, 1930).

——— , *The History of American Sailing Ships* (N.Y., 1935).

——— , *The History of the American Sailing Navy* (N.Y., 1949).

Clark, Arthur H., *The Clipper Ship Era* (N.Y., 1911).

Griffiths, John W., *Treatise on Marine and Naval Architecture, or Theory and Practice Blended in Ship Building* (N.Y., 1851).

Hutchinson, William, *The Maritime History of Maine: Three Centuries of Shipbuilding and Seafaring* (N.Y., 1948).

Morison, Samuel E., *The Maritime History of Massachusetts, 1783–1860* (Boston, 1921).

Morrison, John H., *History of American Steam Navigation* (N.Y., 1903).

———, *History of New York Ship Yards* (N.Y., 1909).

———, *Iron and Steel Hull Steam Vessels of the United States, 1825–1905*.

Murray, Mungo, *A Treatise on Shipbuilding and Navigation* (2nd edn., London, 1764).

Report of the President and Trustees of the Ship Timber Bending Co. (N.Y., 1854).

Stuart, Charles B., *The Naval and Mail Steamers of the United States* (N.Y., 1853).

———, *The Naval Dry Docks of the United States* (N.Y., 1852).

Taylor, Isaac, *The Ship, or Sketches of the Vessels of Various Countries with the Manner of Building and Navigating Them* (Phila., 1854).

Tyler, David B., *The American Clyde: A History of Iron and Steel Shipbuilding on the Delaware from 1840 to World War I* (Newark, Del., 1958).

———, *Steam Conquers the Atlantic* (N.Y., 1939).

Steam and River Boats

Anderson, J. A., *Navigation of the Upper Delaware* (Trenton, 1913).

Baldwin, Leland D., *The Keelboat Age on Western Waters* (Pittsburgh, 1941).

Bathe, Greville, *An Engineer's Miscellany* (Phila., 1938).

———, *The Rise and Decline of the Paddle Wheel* (St. Augustine, Fla., 1962).

———, *Three Essays: A Dissertation on the Genesis of Mechanical Transport in America before 1800* (St. Augustine, Fla., 1960).

Bathe, Greville and Dorothy, *Oliver Evans: A Chronicle of Early American Engineering* (Phila., 1935).

Dickinson, Henry W., *Robert Fulton, Engineer and Artist* (London, 1913).

Evans, Oliver, "On the Origin of Steam Boats and Steam Waggons" (1812), in Arlan K. Gilbert, *Delaware History*, 7 (1957), 142–67.

Fitch, John, *The Original Steamboat Supported* (Phila., 1788).

Flexner, James Thomas, *Steamboats Come True: American Inventors in Action* (N.Y., 1944).

Hunter, Louis C., *Steamboats on the Western Rivers* (Cambridge, Mass., 1949).

Lane, Carl D., *American Paddle Steamboats* (N.Y., 1943).

MacFarlane, Robert, *History of Propellers and Steam Navigation* (N.Y., 1851).

Rumsey, James, *A Short Treatise on the Application of Steam* (Phila., 1788).

Turnbull, Archibald D., *John Stevens, an American Record* (N.Y., 1928).

Canals

Bache, A. D., *Report of Experiments on the Navigation of the Chesapeake and Delaware Canal by Steam* (Phila., 1834).

Colles, Christopher, *Proposal of a Design for the Promotion of the Interests of the United States of America* (N.Y., 1808).

Dunaway, Wayland F., *History of the James River and Kanawha Company* (N.Y., 1922).

Fulton, Robert, *A Treatise on the Improvement of Canal Navigation* (London, 1796).

Gilpin, Joshua, *A Memoir on the Rise, Progress, and Present State of the Chesapeake and Delaware Canal* (Wilmington, 1821).

Goodrich, Carter, Julius Rubin, H. Jerome Cranmer, and Harvey H. Segal, *Canals and American Economic Development* (N.Y., 1961).

Harrison, Joseph H., Jr., "Internal Improvements and the American Union from Washington to Van Buren" (Ph.D. diss., U. Va., 1954).

Latrobe, Benjamin Henry, *Letters to the Honourable Albert Gallatin . . . and Other Papers Relative to the Chesapeake and Delaware Canal* (Phila., 1808).

Laws of the State of New York, in Relation to the Erie and Champlain Canals, Together with the Annual Reports of the Canal Commissioners, and Other Documents, 2 vols. (Albany, 1825).

Putnam, James William, *The Illinois and Michigan Canal* (Chicago, 1918).

Reid, William J., "The Cape Cod Canal" (Ph.D. diss., Boston U., 1958).

Roberts, Christopher, *The Middlesex Canal* (Cambridge, Mass., 1938).

Rubin, Julius, *Canal or Railroad? Imitation and Innovation in the Response to the Erie Canal in Philadelphia, Baltimore, and Boston,* in APS, *Transactions,* New Ser., 51 (1961), Pt. VII.

Sanderlin, Walter S., *The Great National Project: A History of the Chesapeake and Ohio Canal* (Baltimore, 1946).

Scheiber, Harry N., "Internal Improvements and Economic Change in Ohio, 1820–60" (Ph.D. diss., Cornell U., 1962).

Shaw, Ronald E., *Erie Water West* (Lexington, Ky., 1966).

[Smith, William], *An Historical Account of the Rise, Progress, and Present State of Canal Navigation in Pennsylvania* (Phila., 1795).

[Strickland, William], *Communication from the Chesapeake and Delaware Canal Company* (Phila., 1823).

Watson, Elkanah, *History of the Rise, Progress, and Existing Conditions of the Western Canals in the State of New-York* (Albany, 1820).

Whitford, Noble E., *History of the Canal System of the State of New York,* 2 vols. (Albany, 1906).

Railroads

Borden, Simeon, *A System of Useful Formulae* (Boston, 1851).

Derrick, Samuel M., *Centennial History of the South Carolina Railroad* (Columbia, 1930).

Earle, Thomas, *A Treatise on Rail-roads and Internal Communications* (Phila., 1830).

Harlow, Alvin F., *Steelways of New England* (N.Y., 1946).

Herron, James, *A Practical Description of Herron's Patent Trellis Railway Structure* (Phila., 1841).

Hungerford, Edward, *A Century of Progress: History of the Delaware and Hudson Company* (Albany, 1925).

————, *The Story of the Baltimore and Ohio Railroad*, 2 vols. (N.Y., 1928).

Kistler, Thelma M., *The Rise of Railroads in the Connecticut River Valley* (Northampton, 1938).

Knight, Jonathan, and Benjamin H. Latrobe, [Jr.], *Report on the Plan of Construction of Several of the Principal Rail Roads in the Northern and Middle States* (Baltimore, 1838).

————, *Report upon the Locomotive Engines, and the Police Management of Several of the Principal Rail Roads in the Northern and Middle States* (Baltimore, 1838).

Lardner, Dionysius, *Railway Economy* (N.Y., 1850).

[Latrobe, Benjamin Henry, Jr.], *Description of a new form of Edge Rail, to be called the Z rail* (n.p., [1840]).

Long, S. H., *Rail Road Manual*, 2 vols. (Baltimore, 1828–29).

Loree, L. R., "Track," *Bulletin of the University of Wisconsin, Engineering Series*, 1 (1894), 1–24.

Oliver, Smith Hempstone, *The First Quarter-century of Steam Locomotives in North America* (Washington, 1956).

Roberts, S. W., *An Account of Portage Rail Road, over the Allegheny Mountain in Pennsylvania* (Phila., 1836).

Sellers, George Escol, *Improvements in Locomotive Engines, and Railways* (Cincinnati, 1849).

Sinclair, Angus, *Development of the Locomotive Engineer* (N.Y., 1907).

Starr, John W., Jr., *One Hundred Years of American Railroading* (N.Y., 1928).

Stevens, John, *Documents tending to Prove the Superior Advantages of Rail-Ways and Steam Carriages over Canal Navigation* (N.Y., 1812).

Stuart, Charles B., *Report on the Tonawanda Rail Road* (N.Y., 1852).

Thomson, Thomas R., *Check List of Publications on American Railroads before 1841* (N.Y., 1942).

Tredgold, Thomas, *A Practical Treatise* (London, 1835).

Vose, George L., *A Sketch of the Life and Works of George W. Whistler, Civil Engineer* (N.Y., 1887).

Watkins, Elfreth, "The Development of American Rail and Track," *U.S. National Museum Report, 1888–89* (Washington, 1889), 651–708.

Weale, John, *Ensamples [sic] of Railway Making; which although not of English Practice are submitted with Practical Illustrations to the Civil Engineer, and the British and Irish Public* (London, 1843).

White, John H., *Cincinnati Locomotive Builders, 1845–1868* (U.S. National Museum, *Bulletin*, No. 245 [Washington, 1965]).

Wilson, William Bender, *History of the Pennsylvania Railroad Company* (Phila., 1899).

Wooddy, William, *Experiments on Rail Roads, in England* (Baltimore, 1829).

Roads

Berkebile, Don H., *Conestoga Wagons in Braddock's Campaign, 1775* (U.S. National Museum, *Bulletin*, No. 218 [Washington, 1959], pp. 142–53).

Bloodgood, S. DeWitt, *A Treatise on Roads* (Albany, 1838).

Colles, Christopher, *A Survey of the Roads of the United States of America, 1789*, ed. by Walter W. Ristow (Cambridge, Mass., 1961).

Durrenberger, Joseph A., *Turnpikes* (Valdasta, Tex., 1931).

Gillespie, W. M., *A Manual of the Principles and Practice of Road-Making* (3rd edn., N.Y., 1849).

Hunter, Robert F., "Turnpike Construction in Antebellum Virginia," *Technology and Culture*, 4 (1963), 177–200.

Jackson, W. Turrentine, *Wagon Roads West: A Study of Federal Road Surveys and Construction in the Trans-Mississippi West, 1846–1869* (Berkeley, 1952).

Kingsford, W., *History, Structure, and Statistics of Plank Roads, in the United States and Canada* (Phila., 1852).

Shumway, George, *et al.*, *Conestoga Wagon, 1750–1850* (York, Pa., 1964).

Wood, Frederick J., *The Turnpikes of New England and Evolution of the Same* (Boston, 1919).

Bridges

Condit, Carl W., *American Building Art: The Nineteenth Century* (N.Y., 1960).

Culmann, K., "Der Bau der Eisernen Brücken in England und Amerika," *Allgemeine Bauzeitung mit Abbildung*, 17 (1852), 163–222.

Description of Ithiel Town's Improvement in the Construction of Wood and Iron Bridges, A (New Haven, 1821).

Edwards, Llewellyn N., *A Record of History and Evolution of Early American Bridges* (Orono, Me., 1959).

Haupt, Herman, *General Theory of Bridge Construction* (N.Y., 1851).

Peale, Charles W., *An Essay on Building Wooden Bridges* (Phila., 1797).

Pope, Thomas, *A Treatise on Bridge Architecture; in which the Superior Advantages of the Flying Pendent Lever Bridge are Fully Proved* (N.Y., 1811).

Post, Simeon S., "Treatise on Principles of Civil Engineering as Applied to the Construction of Wooden Bridges," *American Railroad Journal*, 32 (1859), 226.

Sloane, Eric, *American Barns and Covered Bridges* (N.Y., 1954).

Steinman, David B., *The Builders of the Bridge: The Story of John Roebling and His Son* (N.Y., 1945).

Steinman, David B., and Sara Ruth Watson, *Bridges and Their Builders* (N.Y., 1941).

Whipple, Squire, *An Elementary and Practical Treatise on Bridge Building* (orig. edn., 1847, N.Y., 1873).

Building

Asher, Benjamin, *The American Builder's Companion* (Boston, 1806).

——, *The Practical House Carpenter* (Philadelphia, 1812).

——, *The Practice of Architecture* (N.Y., 1917).

——, *A Reprint of the Country Builder's Assistant* (New York, 1917).

——, *The Rudiments of Architecture* (Boston, 1814).

Beirne, Rosamond Randall, and John Henry Scarff, *William Buckland, 1734–1774: Architect of Virginia and Maryland* (Baltimore, 1958).

Biddle, Owen, *The Young Carpenter's Assistant; or a System of Architecture Adapted to the Style of building in the United States* (Phila., 1805).

Bridenbaugh, Carl, *Peter Harrison: First American Architect* (Chapel Hill, 1949).

Catalog of the Measured Drawings and Photographs of the Survey in the Library of Congress (Washington, 1941), *Catalog Supplement* (Washington, 1959).

Condit, Carl W., *American Building Art: The Nineteenth Century.*

Downing, A. J., *Cottage Residences* (N.Y., 1844).

Fitch, James M., *American Building: The Forces that Shape It* (Boston, 1948).

Lesley, Robert W., *History of the Portland Cement Industry in the United States* (N.Y., 1924).

Meikleham, Robert, *Dictionary of Architecture*, 3 vols. (Phila., 1854).

Morrison, Hugh, *Early American Architecture* (N.Y., 1952).

Nicholson, Peter, *The Carpenter's New Guide* (10th edn., Phila., 1830).

Peterson, Charles E., "Early American House-Warming by Coal Fires," Society of Architectural Historians, *Journal*, 9 (1950), 21–23.

——, "Early American Prefabrication," *Gazette de Beaux-Arts*, 6th Ser., 33 (1948), 37–46.

——, "Notes on Copper Roofing in America to 1802," Society of Architectural Historians, *Journal*, 24 (1965), 313–18.

Swan, Abraham, *A Collection of Designs in Architecture* (Phila., 1775).

Wall, Alexander, *Books on Architecture Printed in America, 1775–1830* (N.Y., 1925).

Whiffen, Marcus, *The Eighteenth-Century Houses of Williamsburg: A Study of Architecture and Building in the Colonial Capital of Virginia* (Williamsburg, 1960).

——, *The Public Buildings of Williamsburg: Colonial Capital of Virginia* (Williamsburg, 1958).

Wightwick, George, *Hints to Young Architects* (1st Amer. edn., N.Y., 1847).

Wood, Charles B., III, "A Survey and Bibliography of Writings on English and American Architectural Books Published Before 1895," *Winterthur Portfolio*, 2 (1965), 127–37.

Other Engineering Works

Blake, Nelson M., *Water for the Cities: A History of the Urban Water-Supply Problem in the United States* (Syracuse, 1956).

Ewbank, Thomas, *A Descriptive and Historical Account of Hydraulic and Other Machines* (N.Y., 1850).

Hale, Nathan, Jr., *Proceedings before a Joint Special Committee of the Massachusetts Legislature* (Boston, 1845).

Latrobe, B. Henry, *A View of the Practicability and Means of Supplying the City of Philadelphia with Wholesome Water* (Phila., 1799).

Light Houses: Report of the Secretary of the Treasury on the Improvements in the Light House System (Washington, 1846).

Storrow, Charles S., *A Treatise on Water-Works for Conveying and Distributing Supplies of Water* (Boston, 1835).

Wegmann, Edward, *The Water Supply of the City of New York, 1658–1895* (N.Y., 1896).

Weston, William, *Report of William Weston, Esquire, on the Practicability of Introducing the Water of the River Bronx into the City of New York* ([N.Y.], 1799).

POWER

Wind and Water

Francis, James B., *The Lowell Hydraulic Experiments* (N.Y., 1855).

Frizell, Joseph R., "The Old-Time Water-Wheels of America," American Society of Civil Engineers, *Transactions,* 28 (1893), 237–49.

Hamilton, Edward Pierce, "Some Windmills of Cape Cod," Newcomen Society, *Transactions,* 5 (1925), 39–44.

Hunter, Louis C., "Origines des Turbines Francis et Pelton," *Revue d'Histoire des Sciences,* 17 (1964), 209–42.

Main, Charles T., "Evolution of the Transmission of Water Power," in E. Everton Foster, *Lamb's Textile Industries* (Boston, 1916), 1, 223–31.

Nancarrow, John, "Calculations Relating to Grist and Saw Mills," APS, *Transactions,* 4 (1799), 348–61.

Rawson, Marion Nicholl, *Little Old Mills* (N.Y., 1935).

Safford, Arthur T., and Edward Pierce Hamilton, "The American Mixed-Flow Turbine and Its Setting," American Society of Civil Engineers, *Transactions,* 85 (1922), 1237–1356.

Shelton, F. H., "Windmills, Picturesque and Historic: The Motors of the Past," *Journal of the Franklin Institute,* 187 (1919), 171–98.

Steam

Bartol, B. H., *A Treatise on the Marine Boilers of the United States* (Phila., 1851).

Bathe, Greville, *Citizen Genêt: Diplomat and Inventor* (Phila., 1946).

Bathe, Greville, with Dorothy Bathe, *Jacob Perkins, His Inventions, His Times, and His Contemporaries* (Phila., 1943).

Bathe, Greville and Dorothy, *Oliver Evans: A Chronicle of Early American Engineering* (Phila., 1935).

Cooper, Thomas, "An Account of the Steam Engine," *Emporium of Arts and Sciences,* New Ser., 2 (1814), 1–220, 333–85.

Dickinson, H. W., *A Short History of the Steam Engine,* with a new introduction by A. E. Musson (London, 1963).

Evans, Oliver, *Abortion of the Young Steam Engineer's Guide* (Phila., 1805), also published with a textual deletion as *The Young Steam Engineer's Guide* (Phila., 1805).

Graff, Frederick, illustrations in "The History of the Steam Engine in America," *Journal of the Franklin Institute,* 102 (1876), 253–68.

———, "Notice of the Earliest Steam Engines used in the United States," *Journal of the Franklin Institute,* 55 (1853), 269–71.

Hodge, P. R., *The Steam Engine, Its Origin and Gradual Improvement from the Time of Hero to the Present Day* (N.Y., 1840).

Latrobe, Benjamin Henry, "First Report in Answer to the Enquiry Whether Any and What Improvements Have Been Made in the Construction of Steam Engines in America," APS, *Transactions,* 6 (1809), 89–98.

Mitman, Carl W., "Stevens' Porcupine Boiler, 1804: A Recent Study," Newcomen Society, *Transactions,* 19 (1939), 165–71.

Nelson, William, *Josiah Hornblower; and the First Steam Engine in America* (Newark, 1883).

Pursell, Carroll W., Jr., "Stationary Steam Engines in America before the Civil War" (Ph.D. diss., California, 1963, on Ann Arbor microfilm).

Read, David, *Nathan Read: His Invention of the Multi-Tubular Boiler and Portable High-Pressure Engine, and Discovery of the True Mode of Applying Steam-Power to Navigation and Railways* (N.Y., 1870).

Thurston, Robert H., *A History of the Growth of the Steam Engine* (orig. edn., 1878; Ithaca, 1939).

MANUFACTURING

Crafts and Craftsmen

Belknap, Henry Wyckoff, *Trades and Tradesmen of Essex County, Massachusetts* (Salem, 1929).

Bridenbaugh, Carl, *The Colonial Craftsman* (N.Y., 1950).

Christensen, Erwin O., *The Index of American Design* (N.Y., 1950).

Clarke, Herbert, *The Apothecary in Eighteenth-Century Williamsburg* (Williamsburg, 1965).

De Matteo, William, *The Silversmith in Eighteenth-Century Williamsburg* (Williamsburg, 1956).

Dow, George F., *The Arts and Crafts in New England, 1704–1775* (Topsfield, Mass., 1927).

Gottesman, Rita S., *The Arts and Crafts in New York, 1726–1776; 1777–1799* (New-York Historical Society, *Collections,* 69, 81 [1938, 1954]).

Heuvel, Johannes, *The Cabinetmaker in Eighteenth-Century Williamsburg* (Williamsburg, 1958).

Klapper, August, *The Printer in Eighteenth-Century Williamsburg* (Williamsburg, 1964).

Payne, Lloyd, *The Miller in Eighteenth-Century Virginia* (Williamsburg, 1958).

Prime, Alfred C., *The Arts and Crafts in Philadelphia, Maryland, and South Carolina, 1721–1785; 1786–1800*, 2 vols. (Topsfield, Mass., 1929, 1932).

Rawson, Marion Nicholl, *Handwrought Ancestors* (N.Y., 1936).

Samford, C. Clement, *The Bookbinder in Eighteenth-Century Williamsburg* (Williamsburg, 1959).

Tattershall, Edward S., *The Wigmaker in Eighteenth-Century Williamsburg* (Williamsburg, 1959).

Tryon, Rolla Milton, *Household Manufactures in the United States, 1640–1860* (Chicago, 1917).

Tunis, Edwin R., *Colonial Craftsmen and the Beginnings of American Industry* (Cleveland, 1965).

Whitehill, Walter Muir, *The Arts in Early American History: Needs and Opportunities for Study* (Chapel Hill, 1965).

Williamson, Scott Graham, *The American Craftsman* (N.Y., 1940).

Tools, Woodwork, Metalwork, and Machining

Allison, Robert, "The Old and the New," American Society of Mechanical Engineers, *Transactions*, 16 (1895), 742–61.

Brewington, M. V., *Shipcarvers of North America* (Barre, Mass., 1962).

Burton, E. Milby, *Charleston Furniture, 1700–1825* (Charleston, 1955).

Deming, Edward, and Faith Andrews, *Shaker Furniture: The Craftsmanship of an American Communal Sect* (New Haven, 1937).

Gibb, George Sweet, *The Whitesmiths of Taunton: A History of Reed and Barton, 1824–1943* (Cambridge, Mass., 1943).

Goodman, W. L., *The History of Woodworking Tools* (London, 1964).

Gould, Mary Earle, *Early American Wooden Ware and Other Kitchen Utensils* (Rutland, 1962).

Goyne, Nancy A., "Britannia in America: The Introduction of a New Alloy and a New Industry," *Winterthur Portfolio*, 2 (1965), 160–96.

Holtzapffel, John Jacob, *Turning and Mechanical Manipulation*, 5 vols. (2nd edn., London, 1864–84).

Hubbard, Guy, "Development of Machine Tools in New England," *American Machinist*, 59–61 (1923–25), *passim*.

Hummel, Charles F., "English Tools in America: The Evidence of the Dominys," *Winterthur Portfolio*, 2 (1965), 27–46.

Kauffman, Henry J., *Early American Copper, Tin, and Brass* (N.Y., 1950).

Laughlin, Ledlie Irwin, *Pewter In America: Its Makers and Their Marks*, 2 vols. (Boston, 1940).

Mercer, Henry C., *Ancient Carpenters' Tools* (Doylestown, Pa., 1929).

Navin, Thomas R., *The Whitin Machine Works since 1831* (Cambridge, Mass., 1950).

Phillips, John Marshall, *Silver* (N.Y., 1949).

Richards, John, *Treatise on . . . Woodworking Machines* (London, 1872).

Roe, Joseph W., "Early American Mechanics," *American Machinist*, 41 (1914), 729–34, 903–8, 1077–82.

———, *English and American Tool Builders* (N.Y., 1916).

———, "Machine Tools in America," *Journal of the Franklin Institute*, 225 (1938), 499–511.

Rolt, L. T. C., *A Short History of Machine Tools* (London, 1965).

Sellers, George Escol, *Early Engineering Reminiscences (1815–1849) of George Escol Sellers* (U.S. National Museum, Bulletin, No. 238 [Washington, 1965]), ed. by Eugene S. Ferguson.

Sloane, Eric, *A Museum of Early American Tools* (N.Y., 1964).

Smith, Henry M., *Fifty Years of Wire Drawing* (Worcester, 1884).

Sonn, Albert H., *Early American Wrought Iron*, 3 vols. (N.Y., 1928).

Stauffer, David McNeely, *American Engravers on Copper and Steel* (N.Y., 1907).

Taber, Martha Van Hoosen, *A History of the Cutlery Industry in the Connecticut Valley* (Northampton, 1955).

Tooker, Elva, *Nathan Trotter: Philadelphia Merchant, 1787–1853* (Cambridge, Mass., 1955).

Wilbur, W. R., *History of the Bolt and Nut Industry of America* (Cleveland, 1905).

Wildung, Frank H., *Woodworking Tools at Shelburne Museum* (Shelburne, Vt., 1957).

Wilkinson, Israel, *Memoirs of the Wilkinson Family in America* (Jacksonville, Ill., 1869).

Woodbury, Robert S., *History of the Grinding Machine* (Cambridge, Mass., 1959).

———, *History of the Milling Machine* (Cambridge, Mass., 1960).

Wroth, Lawrence, C., *Abel Buell of Connecticut: Silversmith, Type Founder and Engraver* (New Haven, 1926).

Clocks, Instruments, and the Technology of Science

Bedini, Silvio A., *Early American Scientific Instruments and Their Makers* (U.S. National Museum, *Bulletin*, No. 231 [Washington, 1964]).

Bion, Nicholas, *The Construction and Principal Uses of Mathematical Instruments*, trans. by Edmund Stone (London, 1723).

Brewington, M. V., *The Peabody Museum Collection of Navigating Instruments with Notes on Their Makers* (Salem, 1963).

Burnap, Daniel, *Shop Records of Daniel Burnap, Clockmaker*, ed. by Penrose R. Hoopes (Hartford, 1958).

Cohen, I. Bernard, *Some Early Tools of Science* (Cambridge, Mass., 1950).

Cutbush, James, *Hydrostatics* (Phila., 1812).

Daumas, Maurice, *Les Instruments Scientifiques aux XVIIe et XVIIIe Siècles* (Paris, 1953).

[Dearborn, Benjamin], *The Patent Balance Compared with other Instruments for Weighing* (Phila., 1803).

Eckhardt, George H., *Pennsylvania Clocks and Clockmakers* (N.Y., 1955).

——, *United States Clock and Watch Patents, 1790–1890* (N.Y., 1960).

Griffinhagen, George B., *Tools of the Apothecary* (Washington, 1957).

Hare, Robert, *Strictures on a Publication Entitled Clark's Gas Blowpipe* (Phila., 1820).

Hindle, Brooke, *David Rittenhouse* (Princeton, 1964).

Hoopes, Penrose R., *Connecticut Clockmakers of the Eighteenth Century* (Hartford and N.Y., 1930).

Jerome, Chauncey, *History of the American Clock Business for the Past Sixty Years, and Life of Chauncey Jerome, Written by Himself* (N.Y., 1860).

Logan, James, *Philosophical Transactions*, 38 (1734), 441–50; 39 (1736), 404–5.

Multhauf, Robert P., comp., *A Catalogue of Instruments and Models in the Possession of the American Philosophical Society* (Phila., 1961).

Multhauf, Robert P., Amasa Holcomb, Julia Fitz Howell, F. W. Preston, and William J. McGrath, Jr., *Holcomb, Fitz, and Peate: Three 19th Century American Telescope Makers* (U.S. National Museum, *Bulletin,* No. 228 [Washington, 1962], pp. 156–84).

Palmer, Brooks, *The Book of American Clocks* (N.Y., 1950).

Reid, Thomas, *A Treatise on Clock and Watch Making* (Phila., 1832).

Rice, Howard C., Jr., *The Rittenhouse Orrery* (Princeton, 1954).

Smart, Charles E., *The Makers of Surveying Instruments in America since 1700* (Troy, 1962).

[Wall, George, Jr.], *Description with Instructions for the Use of Wall's Newly Invented Surveying Instrument, called the Trigonometer* (Phila., 1788).

Willard, John Ware, *A History of Simon Willard, Inventor and Clockmaker* (New Haven, 1860).

Workman, Benjamin, *Gauging Epitomized* (Phila., 1788).

Glass and Pottery

Burgess, Bangs, *History of Sandwich Glass* (Yarmouthport, Mass., 1925).

Harrington, J. C., *Glassmaking at Jamestown* (Richmond, 1952).

Heiges, George L., *Henry William Stiegel and His Associates: A Story of Early American Industry* ([Lancaster], 1948).

Hume, Ivor Noël, *Excavations at Rosewell in Gloucester County, Virginia, 1957–1959* (U.S. National Museum, *Bulletin,* No. 225 [Washington, 1962], 153–229).

Knittle, Rhea Mansfield, *Early American Glass* (N.Y., 1927).

Lee, Ruth Webb, *Sandwich Glass: The History of the Boston and Sandwich Glass Company* (Framingham Centre, Mass., 1939).

McKearin, George S. and Helen, *American Glass* (N.Y., 1941, 1948).

Quynn, Dorothy Mackay, "Johann Friedrich Amelung at New Bremen," *Maryland Historical Magazine,* 43 (1948), 155–79.

Revi, Albert Christian, *American Pressed Glass and Figure Bottles* (N.Y., 1964).

Scoville, Warren G., "Growth of the American Glass Industry to 1880," *Journal of Political Economy,* 52 (1944), 192–216.

Spargo, John, *The Potters and Potteries of Bennington* (Boston, 1926).

Watkins, Lura Woodside, *American Glass and Glassmaking* (N.Y., 1950).

———, *Early New England Potters and Their Wares* (Cambridge, Mass., 1950).

Wilson, Kenneth M., *Glass in New England* (Sturbridge, Mass., 1959).

Paper and Printing

Adams, Thomas F., *Typographia: A Brief Sketch of the Origin, Rise, and Progress of the Typographic Art; with Practical Directions for Conducting Every Department in an Office* (Phila., 1837).

Davis, Charles Thomas, *The Manufacture of Paper* (Phila. and London, 1886).

French, Hannah D., introduction to "Early American Bookbinding by Hand," in Helmut Lehmann-Haupt, ed., *Bookbinding in America* (Portland, Me., 1941).

Hancock, Harold B., and Norman B. Wilkinson, "The Gilpins and Their Endless Paper Machine," *Pennsylvania Magazine of History and Biography,* 81 (1957), 391–405.

Hoe, Robert, and Co., *Short History of the Printing Press* (n.p., 1902).

Hunter, Dard, *Papermaking in Pioneer America* (Phila., 1952).

———, *Papermaking: The History and Technique of an Ancient Craft* (N.Y., 1943).

Kainen, Jacob, *George Clymer and the Columbian Press* (San Francisco, 1950).

Maxson, Joseph W., Jr., "Nathan Sellers, America's First Large Scale Maker of Paper Moulds," *The Paper Maker,* 29 (1960), 1–16.

Moxon, Joseph, *Mechanick Exercises or the Doctrine of Handy-works Applied to the Art of Printing,* 2 vols. (N.Y., 1896).

Munsell, Joel, *A Chronology of Paper and Paper-Making* (Albany, 1857).

Ringwalt, J. Luther, *American Encyclopaedia of Printing* (Phila., 1871).

Samford, C. Clement, and John M. Hemphill II, *Bookbinding in Colonial Virginia* (Charlottesville, 1966).

Silver, Rollo G., *Typefounding in America, 1787–1825* (Charlottesville, 1965).

Thomas, Isaiah, *History of Printing in America,* 2 vols. (Worcester, 1810).

Weeks, Lyman Horace, *A History of Paper-Manufacturing in the United States, 1690–1916* (N.Y., 1916).

Wroth, Lawrence C., *The Colonial Printer* (2nd edn., Portland, Me., 1938).

Leather

Davis, Charles Thomas, *The Manufacture of Leather . . . To Which are Added Complete Lists of all American Patents for Materials, Processes, Tools, and Machines for Tanning, Currying, etc.* (Phila., 1885).

Hazard, Blanche E., *The Organization of the Boot and Shoe Industry in Massachusetts before 1875* (Cambridge, Mass., 1875).

Kennedy, David, *The Art of Tanning Leather* (N.Y., 1857).

McDermott, Charles H., ed., *A History of the Shoe and Leather Industries of the United States* (Boston, 1918).

Morfit, Campbell, *The Arts of Tanning, Currying and Leather Dressing . . . from the French of Julia de Fontenelle and F. Malpyre* (Phila., 1852).

Welsh, Peter C., "A Craft That Resisted Change: American Tanning Practices to 1850," *Technology and Culture,* 4 (1963), 299–317.

———, *Tanning in the United States to 1850: A Brief History* (U.S. National Museum, *Bulletin,* No. 242 [Washington, 1964]).

Textiles

Appleton, Nathan, *Introduction of the Power Loom, and Origin of Lowell* (Lowell, 1858).

Bagnall, William R., *The Textile Industries of the United States Including Sketches and Notices of Cotton, Woolen, Silk, and Linen Manufactures in the Colonial Period,* Vol. I (Cambridge, Mass., 1893).

Baird, Robert H., *The American Cotton Spinner, and Managers' and Carders' Guide* (Phila., 1887).

[Batchelder, Samuel], *Introduction and Early Progress of the Cotton Manufacture in the United States* (Boston, 1863).

Brockett, L. P., *The Silk Culture in America* (N.Y., 1876).

Chase, William H., *Five Generations of Loom Builders* (Hopedale, Mass., 1950).

Cole, Arthur Harrison, *The American Wool Manufacture,* 2 vols. (Cambridge, Mass., 1926).

Cole, Arthur H., and Harold F. Williamson, *The American Carpet Manufacture* (Cambridge, Mass., 1941).

Copeland, Melvin T., *The Cotton Manufacturing Industry of the U.S.* (Cambridge, Mass., 1912).

Correspondence between Nathan Appleton and John A. Lowell in Relation to the Early History of the City of Lowell (Boston, 1848).

Correspondence Relating to the Invention of the Jacquard Brussels Carpet Power Loom (Boston, 1868).

Fennelly, Catherine, *Textiles in New England, 1790–1840* (Sturbridge, Mass., 1961).

Foster, E. Everton, *Lamb's Textile Industries of the United States,* Vol. I (Boston, 1916).

Gibb, George Sweet, *The Saco-Lowell Shops: Textile Machinery Building in New England, 1813–1949* (Cambridge, Mass., 1950).

Hayes, John L., *American Textile Machinery* (Cambridge, 1879).

Homergue, John d', and Peter Stephen Duponceau, *Essays on American Silk* (Phila., 1830).

Justitia, *Strictures on Montgomery on the Cotton Manufactures of Great Britain and America* (Newburyport, 1841).

Leigh, Evan, *The Science of Modern Cotton Spinning,* 2 vols. (2nd edn., Manchester, 1873).

Montgomery, James, *A Practical Detail of the Cotton Manufacture of the United States of America and the State of the Cotton Manufacture of that Country Contrasted and Compared with that of Great Britain* (Glasgow, 1840).

——, *The Theory and Practice of Cotton Spinning; or the Carding and Spinning Master's Assistant* (2nd edn., Glasgow, 1833).

Navin, Thomas R., *The Whitin Machine Works since 1831* (Cambridge, Mass., 1950).

Nourse, Henry Stedman, "Some Notes on the Genesis of the Power Loom in Worcester County," American Antiquarian Society, *Proceedings,* 16 (1904), 22–46.

Parslow, Virginia D., *Weaving and Dyeing Processes in Early New York* (Cooperstown, 1949).

Reath, Nancy Andrews, *The Weaves of Hand-Loom Fabrics* (Phila., 1927).

Rogers, Grace L., *The Scholfield Wool-Carding Machines* (U.S. National Museum, *Bulletin,* No. 218 [Washington, 1959], pp. 1–14).

Walton, Perry, *The Story of Textiles* (Boston, 1925).

Ware, Caroline F., *The Early New England Cotton Manufacture* (Boston and N.Y., 1931).

Webber, Samuel, "Historical Sketch of the Commencement and Progress of the Cotton Manufacture in the United States up to 1876," *Manual of Power* (N.Y., 1879).

White, George S., *Memoir of Samuel Slater* (2nd edn., Phila., 1836).

Chemicals

Bottée, [J.-J.-A.], and [J.-R.-D.-A.] Riffaut, *Traité de l'Art de Fabriquer la Poudre à Canon* (Paris, 1811).

Clow, Nan L., *The Chemical Revolution: A Contribution to Social Technology* (London, 1952).

Cooper, Thomas, *A Practical Treatise on Dyeing, and Callicoe Printing; Exhibiting the Processes in the French, German, English, and American Practice of Fixing Colours on Woolen, Cotton, Silk and Linen* (Phila., 1815).

Dossie, [Robert], *Observations on the Pot-Ash brought from America* (London, 1767).

du Pont, E. I., "On the Manufacture of War and Sporting Powder in the United States," in B. G. du Pont, *E. I. du Pont de Nemours and Company, 1802–1902* (Boston and N.Y., 1920).

Dyer and Colour Maker's Companion, The (Phila., 1850).

Edelstein, Sidney M., "Origins of Chlorine Bleaching in America," *American Dyestuff Reporter,* 49 (1960), 254–63.

Fennell, James, *Description of the Principles and Plan of Proposed Establishments of Salt Works* (Phila., 1798).

Haber, L. F., *The Chemical Industry during the Nineteenth Century: A Study of the*

Economic Aspects of Applied Chemistry in Europe and North America (London, 1958).

Haigh, James, *The Dier's Assistant* (Poughkeepsie, 1813).

Hall, Harrison, *Hall's Distiller* (Phila., 1813).

Haynes, Williams, *American Chemical Industry,* Vol. I (N.Y., 1954).

Heusser, Albert H., *The History of the Silk Dyeing Industry in the United States* (Paterson, 1927).

Hussey, Miriam, *From Merchants to "Colour Men": Five Generations of Samuel Wetherill's White Lead Business* (Phila., 1956).

Krafft, Michael, *The American Distiller; or the Theory and Practice of Distilling* (Phila., 1804).

Lewis, W., *Experiments and Observations on American Potashes* (London, 1767).

Morfit, Campbell, *Chemistry Applied to the Manufacture of Soap and Candles* (Phila., 1847).

———, *Perfumery: Its Manufacture and Use* (Phila., 1853).

New York Essays upon the Making of Salt-petre and Gunpowder (N.Y., 1776).

Painter, Tilder, and Varnisher's Companion containing Rules and Regulations in Everything Relating to the Arts of Painting, Gilding, Varnishing, and Glass-Staining, The (Phila., 1850).

Pilkington, James, *The Artist's Guide and Mechanic's Own Book, Embracing the Portion of Chemistry Applicable to the Mechanic Arts* (N.Y., 1841).

Practical Treatise on Dyeing and Calico-Printing, A (N.Y., 1846).

Process for Extracting and Refining Salt-Petre, The (Phila., 1774).

Sonnedecker, Glenn, *Kremers and Urdangs History of Pharmacy* (3rd edn., Phila., 1963).

Stephens, Thomas, *The Method and Plain Process for making Pot-Ash* (London, [1755]).

[———], *The Rise and Fall of Pot-Ash in America* (London, 1758).

Van Gelder, Arthur P., and Hugo Schlatter, *History of the Explosives Industry in America* (N.Y., 1927).

Weiss, Harry B. and Grace M., *The Revolutionary Saltworks of the New Jersey Coast: With Notes on the Early Methods of Making Salt in New England, New York, Delaware, and Virginia* (Trenton, 1959).

Other Manufacturing

McKay, George L., *Early American Currency* (N.Y., 1944).

Noe, Sydney P., *The New England Willow Tree Coinages of Massachusetts* (N.Y., 1943).

———, *The Oak Tree Coinage of Massachusetts* (N.Y., 1947).

Perkins, Jacob, *The Permanent Stereotype Steel Plate* (Newburyport, Mass., 1806).

Stewart, Frank H., *History of the First United States Mint* (n.p., 1924).

HEAT, LIGHT, AND ELECTRICITY

Bull, Marcus, *Experiments to Determine the Comparative Value of the Principal Varieties of Fuel* (Phila., 1827).

Cooper, Thomas, *Some Information Concerning Gas Lights* (Phila., 1816).

Cummings, Richard O., *The American Ice Harvests: A Historical Study in Technology, 1800–1918* (Berkeley and Los Angeles, 1949).

Franklin, Benjamin, *An Account of the New Invented Pennsylvania Fire-Places* (Phila., 1744).

———, "On the Cause and Cure of Smoky Chimneys," APS, *Transactions*, 2 (1786), 1–36.

Hayward, Arthur H., *Colonial Lighting* (Boston, 1923).

Johnson, Walter R., *Report of the Navy Department . . . on American Coals, Applicable to Steam Navigation* (Washington, 1849).

Jones, Alexander, *Historical Sketch of the Electric Telegraph* (N.Y., 1852).

Mabee, Carleton, *The American Leonardo: A Life of Samuel F. B. Morse* (N.Y., 1943).

Sharlin, Harold I., *The Making of the Electrical Age: From the Telegraph to Automation* (N.Y., 1963).

Spence, Clark C., "Early Use of Electricity in American Agriculture," *Technology and Culture,* 3 (1962), 142–60.

Theory and Practice of Warming and Ventilating Public Buildings, Dwelling-houses, and Conservatories, The (London, 1825).

Thompson, Robert L., *Wiring a Continent (1832–1866)* (Princeton, 1947).

Walker, Charles V., *Electrotype Manipulation* (Phila., 1844).

EDUCATION, ORGANIZATION, AND SCIENCE

American Correspondence of the Royal Society of Arts, The, London, 1755–1840, microfilm (London, 1963).

Bailyn, Bernard, *Education in the Forming of American Society: Needs and Opportunities for Study* (Chapel Hill, 1960).

Bates, Ralph S., *Scientific Societies in the United States* (2nd edn., N.Y., 1958).

Bennett, Charles Alpheus, *History of Manual and Industrial Education up to 1870* (Peoria, 1926).

Bigelow, Jacob, *Elements of Technology* (Boston, 1829).

———, *Inaugural Address* (Boston, 1817).

———, *The Useful Arts,* 2 vols. (N.Y., 1847).

Bridenbaugh, Carl, *The Colonial Craftsman* (N.Y., 1950).

Calhoun, Daniel H., *The American Civil Engineer: Origins and Conflict* (Cambridge, Mass., 1950).

Cutbush, James, *The American Artist's Manual, or Dictionary of Practical Knowledge in the Application of Philosophy to the Arts and Manufactures,* 2 vols. (Phila., 1814).

Eaton, Amos, *Art without Science: or Mensuration, Surveying and Engineering,*

Divested of the Speculative Principles and Technical Language of Mathematics (Albany, 1830).

————, *Prodromus of a Practical Treatise on the Mathematical Arts* (Troy, 1838).

Eldredge, Leonard Charles, *The Mechanical Principia; Containing all the Various Calculations on Water and Steam Power and on the Different Kinds of Machinery Used in Manufacturing* (N.Y., 1848).

Ellis, William A., ed., *Norwich University, 1819–1911*, 3 vols. (Montpelier, Vt., 1911).

Fleming, Donald, "Latent Heat and the Invention of the Watt Engine," *Isis*, 43 (1952), 3–5.

Forman, Sidney, *West Point: A History of the United States Military Academy* (N.Y., 1950).

Gillispie, Charles C., "The Discovery of the Leblanc Process," *Isis*, 48 (1957), 152–67.

Goode, George Brown, *The Smithsonian Institution, 1846–1896: The History of Its First Half Century* (Washington, 1897).

Hindle, Brooke, *The Pursuit of Science in Revolutionary America* (Chapel Hill, 1956).

Hudson, Derek, and Kenneth W. Luckhurst, *The Royal Society of Arts, 1754–1954* (London, 1954).

Jernegan, Marcus W., *Laboring and Dependent Classes in Colonial America, 1607–1783* (Chicago, 1931).

Johnson, Walter R., *A Lecture on the Mechanical Industry and the Inventive Genius of America* (Baltimore, 1849).

————, *Notes on the Use of Anthracite in the Manufacture of Iron* (Boston, 1841).

————, "Observations on the Relations between Rolling and Dragging Friction," *Experiments on Rail-Roads* (Baltimore, 1829), 1–12.

————, *Report of the Navy Department . . . on American Coals Applicable to Steam Navigation* (Washington, 1849).

Kerker, Milton, "Science and the Steam Engine," *Technology and Culture*, 2 (1961), 381–90.

[MacKenzie, Colin], *MacKenzie's Five Thousand Receipts in all the Useful and Domestic Arts* (Phila., 1846).

Mann, Charles Riborg, *A Study of Engineering Education* (N.Y., 1918).

Multhauf, Robert P., "Sal Ammoniac: A Cast History in Industrialization," *Technology and Culture*, 6 (1955), 569–86.

Nason, Henry B., ed., *Biographical Record of the Officers and Graduates of the Rensselaer Polytechnic Institute, 1824–1886* (Troy, 1887).

Nicholson, John, *The Operative Mechanic and British Machinist; being a Practical Display of the Manufactories and Mechanical Arts of the United Kingdom*, 2 vols. (Phila., 1826).

Overman, Frederick, *Mechanics for the Millwright, Machinist, Engineer, Civil Engineer, Architect, and Student* (Phila., 1851).

Potter, Alonzo, *The Principles of Science Applied to the Domestic and Mechanic Arts* (Boston, 1840).

Price, Derek J. de Solla, "Is Technology Historically Independent of Science? A Study in Statistical Historiography," *Technology and Culture,* 6 (1965), 553–68.

Renwick, James, *Applications of the Science of Mechanics to Practical Purposes* (N.Y., 1844).

Rogers, William Barton, *An Elementary Treatise on the Strength of Materials* (Charlottesville, 1838).

Schofield, Robert E., "The Industrial Orientation of Science in the Lunar Society of Birmingham," *Isis,* 48 (1957), 408–15.

Weale's Catalogue of Works Recently Published on the Various Branches of Civil and Military Engineering, Architecture, Mechanics, Naval Architecture and Steam Navigation (London, 1848).

AMERICA AND EUROPE

Berthoff, Rowland Tappan, *British Immigrants in Industrial America, 1790–1950* (Cambridge, Mass., 1953).

Bond, Phineas, "Letters of Phineas Bond, British Consul at Phildelphia, to the Foreign Office of Great Britain, 1787, 1788, 1789," American Historical Association, *Annual Report for 1896,* Vol. I (Washington, 1897).

Carey, Mathew, *Reflections on the Subject of Emigration* (Phila., 1826).

Cooper, Thomas, *Some Information Respecting America* (London, 1794).

Ferguson, Eugene S., "On the Origin and Development of American Mechanical 'Know-How,' " *Mid-continent American Studies Journal,* 3 (1962), 3–16.

Hancock, Harold B., and Norman B. Wilkinson, "Joshua Gilpin: An American Manufacturer in England and Wales, 1795–1801," Newcomen Society, *Transactions,* 33 (1961), 15–28, 57–66.

Heaton, Herbert, "The Industrial Immigrant in the United States, 1783–1812," APS, *Proceedings,* 45 (1951), 519–27.

Henderson, W. O., *Britain and Industrial Europe, 1750–1870* (Liverpool, 1954).

Pursell, Carroll W., Jr., "Thomas Digges and William Pearce: An Example of the Transit of Technology," *William and Mary Quarterly,* 3d Ser., 21 (1964), 551–60.

Shaw, Ralph R., *Engineering Books Available in America Prior to 1830* (N.Y., 1933).

Timoshenko, Stephen P., *History of Strength of Materials* (N.Y., 1953).

Wilkinson, Norman B., "Brandywine Borrowings from European Technology," *Technology and Culture,* 4 (1963), 1–13.

PATENTS AND INVENTION

Berle, Alf K., and L. Sprague de Camp, *Inventions, Patents, and Their Management* (Princeton, 1959).

Bruchey, Stuart, *The Roots of American Economic Growth, 1607–1861: An Essay in Social Causation* (N.Y., 1965).

Byrn, Edward W., *Progress of Inventions in the Nineteenth Century* (N.Y., 1904).

Campbell, Levin H., *The Patent System of the United States* (Washington, 1891).

de Camp, L. Sprague, *The Heroic Age of American Invention* (Garden City, 1961).

Deller, Anthony William, "Social and Economic Impact of Patents," *Journal of the Patent Office Society,* 46 (1964), 424–57.

Evans, Oliver, *Exposition of Part of the Patent Law* (n.p., 1816).

———, *Oliver Evans to his Counsel* (n.p., 1817).

———, *Patent Right Oppression Exposed; or, Knavery Detected* (Philadelphia, 1813).

Federico, P. J., ed., "Outline of the History of the United States Patent Office," *Journal of the Patent Office Society,* 18 (1936), No. 7, pp. 1–251.

Fessenden, Thomas G., *Essay on the Law of Patents* (Boston, 1810).

Fiske, Bradley A., *Invention, the Master-Key to Progress* (N.Y., 1921).

Gilfillan, S. Collum, *Inventing the Ship* (Chicago, 1935).

———, *The Sociology of Invention* (Chicago, 1935).

Habakkuk, H. J., *American and British Technology in the Nineteenth Century: The Search for Labour-Saving Inventions* (Cambridge, Eng., 1962).

Hubert, Philip G., Jr., *Inventors* ("Men of Achievement Series") (N.Y., 1893).

Jewkes, John, David Sawers, and Richard Stillerman, *The Sources of Invention* (London, 1958).

Leggett, M. D., comp., *Subject Matter Index of Patents . . . 1790 to 1873,* 3 vols. (Washington, 1874).

Merton, Robert K., "Priorities in Scientific Discovery," *American Sociological Review,* 22 (1957), 635–59.

———, "Singletons and Multiples in Scientific Discovery," APS, *Proceedings,* 105 (1961), 470–92.

North, Douglass C., *The Economic Growth of the United States, 1790–1860* (Englewood Cliffs, N.J., 1961).

Phillips, Willard, *Law of Patents* (Boston, 1837).

Report of the Commissioner of Patents . . . during the Year 1843, 28th Cong., 1st Sess., Senate Doc. No. 150 (Washington, 1844).

Rossman, Joseph, *The Psychology of the Inventor* (Washington, 1931).

Rostow, W. W., *The Process of Economic Growth* (N.Y., 1952).

Strassman, W. Paul, *Risk and Technological Innovation: American Manufacturing Methods during the Nineteenth Century* (Ithaca, 1959).

[U.S. Patent Office], *List of Patents . . . 1790 to 1847* (Washington, 1847).

Vaughn, Floyd L., *Economics of our Patent System* (N.Y., 1925, 1956).

TECHNOLOGY IN AMERICAN HISTORY

Boorstin, Daniel J., *The Americans: The National Experience* (N.Y., 1965).

———, *The Image, or What Happened to the American Dream* (N.Y., 1962).

Etzler, J. A., *The Paradise within the Reach of all Men without Labor, by Powers of Nature and Machinery* (Pittsburgh, 1833).

Ewbank, Thomas, *The World a Workshop* (N.Y., 1855).

Giedion, Siegfried, *Mechanization Takes Command* (N.Y., 1954).

Kouwenhoven, John A., *Made in America: The Arts in Modern Civilization* (Garden City, 1948).

Levy, Leo B., "Hawthorne's 'Canal Boat': An Experiment in Landscape," *American Quarterly*, 16 (1964), 211–15.

Marx, Leo, *The Machine in the Garden: Technology and the Pastoral Ideal in America* (N.Y., 1964).

Meier, Hugo A., "Technology and Democracy, 1800–1860," *Mississippi Valley Historical Review*, 43 (1957), 618–40.

Mumford, Lewis, *Art and Technics* (N.Y., 1952).

———, *Technics and Civilization* (N.Y., 1934).

Nef, John U., *War and Human Progress: An Essay on the Rise of Industrial Civilization* (Cambridge, Mass., 1950).

Rosenbloom, Richard S., "Some 19th-Century Analyses of Mechanization," *Technology and Culture*, 5 (1964), 489–511.

Ure, Andrew, *The Philosophy of Manufactures* (London, 1835).

Webb, Walter P., *The Great Plains* (N.Y., 1931).

Index

Notes on the Contributors

Judith A. McGaw is Associate Professor of the History of Technology in the Department of History and Sociology of Science at the University of Pennsylvania. She is the author of *Most Wonderful Machine: Mechanization and Social Change in Berkshire Paper Making, 1801–1885*.

Carolyn C. Cooper is Research Affiliate at Yale University, where she has taught history of technology. She is the author of *Shaping Invention: Thomas Blanchard's Machinery and Patent Management in Nineteenth-Century America*.

Robert B. Gordon is Professor of Geophysics and Applied Mechanics at Yale University. He is the author (with Patrick M. Malone) of *The Texture of Industry, an Archaeological View of the Industrialization of North America* and (with T. C. Koopmans, W. D. Nordhaus, and B. J. Skinner) *Toward a New Iron Age? A Study of Patterns of Resource Exhaustion*.

Brooke Hindle was Director of and is now Historian Emeritus of the National Museum of American History of the Smithsonian Institution. He is the author of *The Pursuit of Science in Revolutionary America, 1735–1789* and editor of *America's Wooden Age: Aspects of Its Early Technology*.

Donald C. Jackson is Assistant Professor of History at Lafayette College. He is the author of *Great American Bridges and Dams*.

Susan E. Klepp is Professor of History at Rider University. She is the author of *Philadelphia in Transition: A Demographic History of the City and Its Occupational Groups, 1720–1830* and (with Billy G. Smith) *The Infortunate: The Voyage and Adventures of William Moraley, an Indentured Servant*.

Nina E. Lerman is Visiting Assistant Professor of History at Auburn University and is working on a book on skill acquisition and social boundaries in nineteenth-century urban technical education.

A. Michal McMahon is Associate Professor of History and Associate Director of the Institute for the History of Technology and Industrial Archaeology at West Virginia University. He is the author of *The Making of a Profession: A Century of Electrical Engineering in America*.

Sarah F. McMahon is Associate Professor of History and past Director of Women's Studies at Bowdoin College and the author of many essays on diet and the culture of food in rural New England and the pioneer Midwest.

Patrick W. O'Bannon is Associate/Principal Historian with Kise Franks & Straw, a cultural resources consulting firm in Philadelphia. He is the author of *Gateways to Commerce: The U.S. Army Corps of Engineers' Nine-Foot Channel on the Upper Mississippi River* (with William Patrick O'Brien and Mary Yeater Rathbun).

Robert C. Post is Curator at the National Museum of American History and Editor of *Technology and Culture*. He is the author of *Physics, Patents, and Politics: A Biography of Charles Grafton Page*.

Printed in the United States
200139BV00003B/25/A